©2006 Algrove Publishing Limited
ALL RIGHTS RESERVED.
No part of this book may be reproduced in any form, including photocopying, without permission in writing from the publishers, except by a reviewer who may quote brief passages in a magazine or newspaper or on radio or television.

Algrove Publishing Limited
36 Mill Street, P.O. Box 1238
Almonte, Ontario, Canada K0A 1A0

Telephone: (613) 256-0350
Fax: (613) 256-0360
Email: sales@algrove.com

Acknowledgements:

We wish to thank ©*Country Guide-Farm Business Communications* for giving us the permission to reprint this *1947, Eighth Edition, Farm Workshop Guide*. You can visit their website at www.agcanada.com.

Library and Archives Canada Cataloguing in Publication

      Farm workshop guide.

(Classic reprint series)
Reprint of the 8th new and rev. ed. published: Winnipeg : Extension
      Dept., Country Guide, 1947.
Includes index.
ISBN-10 1-897030-49-5
ISBN-13 978-897030-49-3

      1. Farm equipment. 2. Agricultural machinery. I. Series: Classic reprint series (Almonte, Ont.)

S676.F37 2006   631.3   C2006-903080-4

Printed in Canada
#2-2-07

Publisher's Note

Necessity is no longer the mother of invention, she is the grandmother. Most things invented today are designed to create markets, not to satisfy needs. It was not always so, particularly on our farms a half century ago.

Just coming out of WWII, there were still shortages everywhere and farmers, who had been "making do" since the crash of '29, had become expert at it. Their self-sufficiency was at its all-time peak and an impressive peak it was. With a hammer, a pair of pliers and a piece of wire, farmers could fix most things that broke and they were expert at transforming all the parts of expired equipment into things they could still use.

This workshop guide is a testament to the ingenuity of farmers. If the tongue on your belt buckle broke, you replaced it with a cotter pin. If you wanted to kill rats in the barn and minimize peripheral damage you replaced the lead slugs in your .22 cartridges with pieces of crayon. Farmers had achieved a level of independence they would never again see. Regrettably, that independence had been combined with poverty, and necessity had truly driven them to innovate. But they did it well and it is a pleasure to republish this small part of that innovative record.

Leonard G. Lee, Publisher
Almonte, Ontario
June 2006

WARNING

This is a reprint of a publication compiled in 1947. It describes what was done and what was recommended to be done in accordance with the knowledge of the day. It should be read in this light.

It is advisable to treat all corrosive, explosive or toxic materials with greater caution than is indicated here, particularly any materials that come in contact with the body. Treat all chemicals with respect and use them only in ways sanctioned by the sellers and the law.

Originally published in 1947 by:

Canada's Largest Monthly Rural Magazine

Reprinted with the kind permission of:
©Country Guide - Farm Business Communications

Algrove Publishing~Classic Reprint Series

Dependable Power for Your Farm
with a
Johnson Iron-Horse
GASOLINE ENGINE

Instant dependable power is at your command for operating all sorts of farm and home appliances. Lathes, feed mixers, grinders, washing machines, cream separators and countless other labour-saving appliances can be operated from the direct power take-off pulley of the Iron Horse engine.

The Iron Horse will operate a water pump, too, making running water available wherever and whenever you want it.

The Iron Horse is a gasoline fueled engine. It is engineer designed and precision built to give years of dependable trouble-free service. It's low in cost and economical in operation, soon paying for itself in time and labour saved, in added comfort and convenience. Write today for complete information and illustrated literature. Address enquiries to

JOHNSON MOTORS
PETERBORO - - CANADA

SECTION 1.
Work Bench and Machine Shop

Bench Saw Table

A small powered saw on the farm is very useful for making small articles and doing light repair jobs for either the house or the various needs of the farm in general. Such a saw need not be elaborate, to meet most of the jobs that are usually required. If lumber can be sawn to correct size, quickly and accurately, then much time and labor is saved and often the work is improved.

Saw arbors complete with bearings to suit saws of different sizes are now commonly seen in hardware and chain stores and illustrated in mail order house catalogues. Some of these are exceptionally well made, being equipped with ball-bearings.

However, suitable bearings, shafting and washers to make a saw mandrel can often be found on the farm. A discarded piston from an automobile or tractor engine will make a good bearing and support for a saw arbor. The wall or skirt of the piston may be cut away on one side to accommodate the "V" belt to the pulley if the pulley is placed in the centre between the two piston bushings when the piston is used for a grinding and buffing machine (see elsewhere in this publication). The advantage of placing the pulley between the two bushings is more even distribution of load and wear on the bearings. If the saw is placed on one side of the piston and the pulley on the opposite side, the chief advantages are easier and more simple assembly, and quick removal or replacement of the "V" belt. In order to provide clearance for the saw blade the piston should be mounted on a wood base made from a piece of 2x6-inch lumber about 8 inches long.

To simplify construction the saw-table is made immovable instead of tilting. The frame and legs are made from 1x1x⅛-inch angle steel. About nine feet of this will be required. The frame is easily made by cutting 90-degree notches and bending in a vise as shown in the sketch. Four pieces of angle steel, each nine inches long are cut off for the legs. Feet for the legs are formed by sawing and bending one-inch at one end of each piece. Before bending to form the feet, drill the five-sixteenths-inch holes as shown in the sketch, to be used for bolting the saw table on the bench.

A piece of hardwood six feet two inches long, planed to four inches wide and 1 to 1⅛ inches thick is required to make the top for the saw table. This is cut and finished to make four equal lengths of 18 inches. Two of the pieces are rabbetted ¾ x 3/16 inches to provide a groove for the adjustable guide to slide in. The four pieces are fastened to the frame with 1½x¼-inch flat headed stove bolts. If suitable drilling facilities are available, holes for the stove bolts can be drilled through the wood table and steel frame in one operation, otherwise the holes in the wood and the frame are drilled separately. This latter

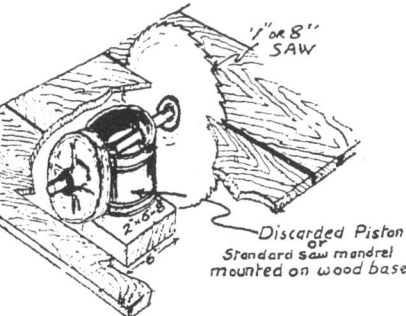

method requires more care and more time. Each of the four pieces are spaced to allow room for the adjustable guide to slide and for the saw blade. The saw requires one-quarter-inch space, while the two spaces for the guide are three-sixteenths inches each.

The adjustable guide is shown in detail in the sketch. A piece of ¾ x ¾ x ⅛-inch angle steel, 12 inches long, is used for the slide. Drill a one-quarter-inch hole one inch from one end of the slide and in the slide as shown in the sketch. Cut the head off a standard 1x¼-inch machine bolt. Insert the cut end in the one-quarter-inch hole in the slide and secure it by rivetting. A more secure method is to drill a three-sixteenths-inch hole instead of the one-quarter-inch hole in the slide end, and tap this to take a piece of a threaded one-quarter-inch rod or bolt and then rivetting it and filing smooth on the underside. The adjustable part of the guide is made from a piece of ⅛x2x2-inch angle steel six inches long. A curved slot having a one-inch radius is marked out. A centre punch mark is made exactly on the centre of the arc. Then centre punch marks are made five-sixteenths of an inch apart on the curved line. Drill one-quarter-inch holes exactly on each of the centre punch marks. The curved slot can then be finished by filing with a small rat-tail file and finally with a small half-round file. After the slot is made place the piece in

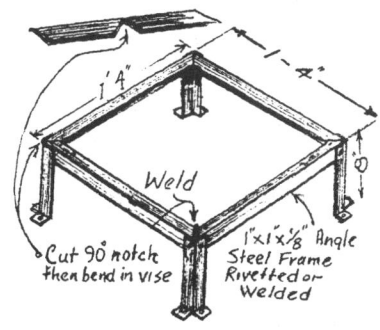

position on the slide so that the one-quarter-inch bolt protrudes through the slot. Clamp the 2x2-inch angle piece at exactly 90 degrees to the slide portion. Then drill a one-quarter-inch hole through both pieces at the centre point

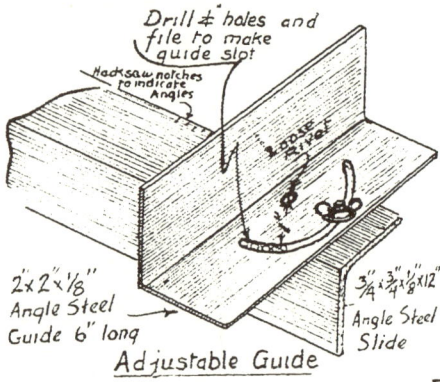

of the curved slot which is the centre punch mark previously made for marking the arc or curved line. The two pieces are then rivetted together just loose enough to allow the slide portion to pivot. A one-quarter-inch thumb screw and washer for the bolt through the slot completes this part. Angles can be indicated by sawing notches along the edge of the slide as shown in the sketch. A semi-circular protractor can be used to locate the position of the notches for the desired angles.

A ¼ h.p. or 1/3 h.p. electric motor can be used for a seven-inch or eight-inch saw for sawing wood up to one-inch thick. For thicknesses up to two inches ½ h.p. is required. When a line shaft driven by a gas engine is used, ½ h.p. to ⅝ h.p. is suitable for light work. Not less than ¾ h.p. is necessary for heavier work. The ordinary 1½ h.p. farm gas engine would be ideal power for driving most small powered tools from a line shaft. Whatever power is used, the speed of an eight-inch saw should be about 3,500 to 3,750 r.p.m. A four-inch pulley on an electric motor running at 1,750 r.p.m. will require a two-inch pulley on the saw mandrel to drive it at 3,500 r.p.m. A five-inch pulley on the same motor with a two and one-half inch pulley on the saw will operate the saw at the same speed but with less slippage due to the greater amount of contact of the belt with larger pulleys. Larger pulleys are therefore more desirable for most small power tools to make most efficient use of the limited amount of power provided by small fractional horse power motors or gas engines.—H.J.K.

Homemade Bandsaw

Except for a few small metal parts the bandsaw is constructed mainly of wood. The main frame is made up from rectangular pieces of wood which are commonly found on most farms or easily obtainable at local lumber yards. The entire machine can be constructed with ordinary hand tools. The two bandsaw wheels, however, are made accurate more easily if they can be turned and trued on a lathe. Similarly, the rectangular pieces for the frame can be cut quite accurately in very little time on a small homemade power saw such as is described elsewhere in this publication. The bandsaw table is fixed permanently to the main frame to maintain simplicity of construction. In actual practice, most of the bandsaw work is done on a horizontal table. If considerable bevel sawing is anticipated, the frame is suitable for inserting a trunnion and locking device for that purpose.

Good clear fir or pine is suitable for the main frame. If hardwood can be obtained, so much the better. The first job is to construct the main frame. Cut out all the frame pieces first. Plane and true each piece. Then glue and screw the pieces together as shown in the drawings. Four iron brackets are used to hold the saw frame rigid to the base. These are made from four pieces of 1½ x ¼-inch flat iron, nine inches long. Drill three three-sixteenths inch holes, three inches apart in each piece for round head screws one and one-half inches long. Bend each piece in a vise to form the bracket with one side six inches long and the other side three

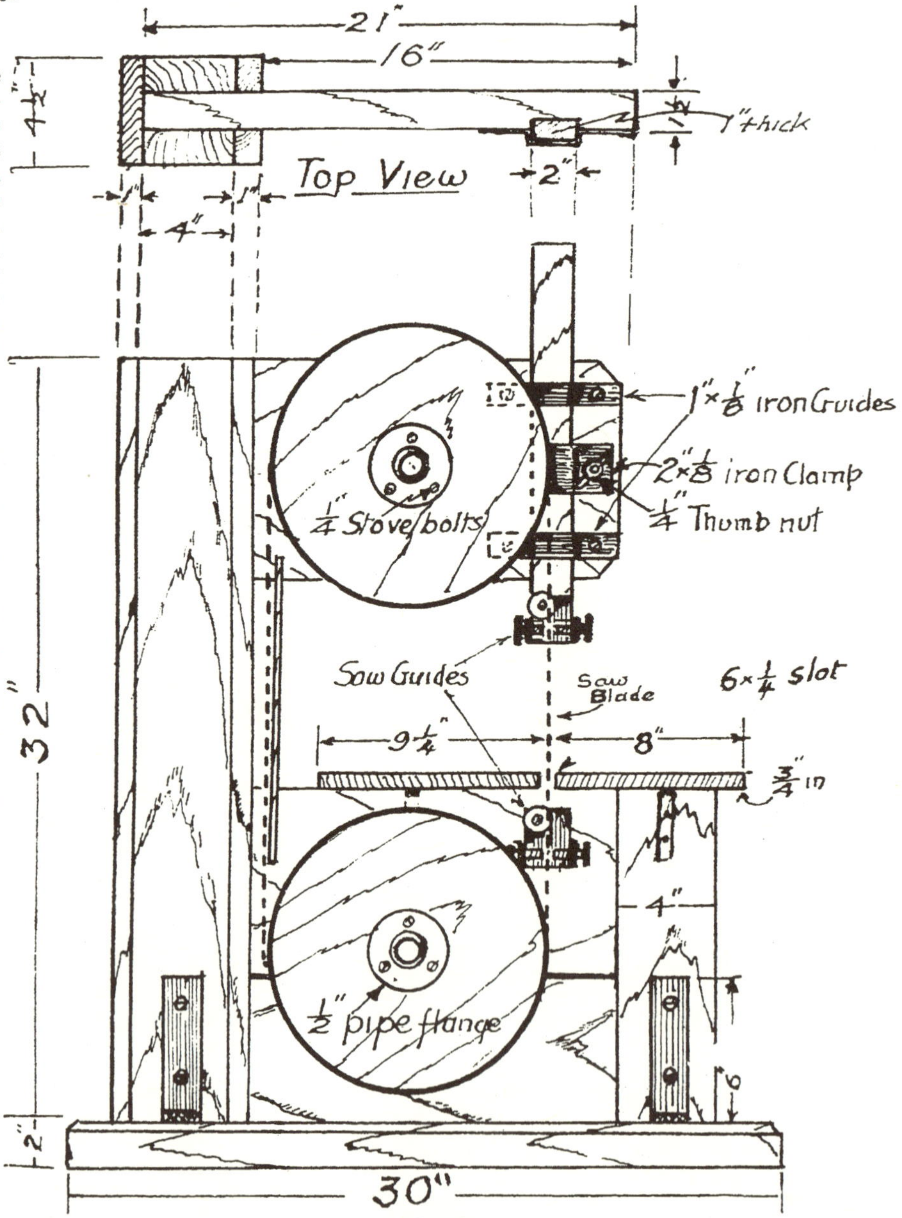

Left View.

inches long to serve as the foot. First glue and screw the base to the main frame. Drive the screws from the base to the main frame from the underside. Then attach the four iron brackets as indicated in the sketch.

The bandsaw wheels are of laminated construction and are twelve inches in diameter with a face of one and one-half inches if constructed with plain fir or pine. If plywood can be used the face or thickness can be reduced to one inch. If plain wood is used, cut out six discs 12¼ inches in diameter and one-half inch thick to make both bandsaw wheels. Glue and screw three discs together to make one wheel. The grain of the centre disc must be at right angles to the grain of the two outside discs to prevent warping and splitting. Bore a three-quarter inch hole through the centre of each wheel. A one-half inch pipe flange on each side of the wheel serves as a hub. The pipe flanges must be bolted exactly in the centre over the hole. Use one of the flanges to mark the position of the holes to receive one-quarter inch flat head stove bolts, two and one-half inches long. Soft iron rivets may be used instead of stove bolts. The pipe flanges are fitted with Ford Model "T" spindle bushings or similar bronze bushings. The bushings should fit tightly. After the bushings are pressed into the pipe flanges they should be reamed out to take a five-eighths inch shaft. The lower wheel bushing should fit the shaft snugly as this wheel is to be the drive wheel for the bandsaw. The bushing in the top driven wheel should be just loose enough to turn on the five-eighths inch steel bolt which is to be used with it. In order to true up the two wheels, the best job is done by fastening them on a five-eighths inch shaft or mandrel and turning them in a lathe. Each wheel is made exactly 12 inches in diameter with the face of each slightly crowned to make the bandsaw travel in the centre of the wheels. Standard one-inch rubber bands for 12-inch band-saw wheels can be purchased along with the proper cement for attaching them. If the face of the wheels are to be fitted with these rubber bands, a recess should be turned in the face of each wheel at the same time they are being trued in the lathe.

Two bearings are used for the shaft of the lower bandsaw wheel. A solid or a split pillar box bearing is used on the side next to the "V" belt drive pulley. This bearing is screwed on the 1½x6-inch member of the main frame as shown in the sketch. Another Ford Model "T" spindle bushing is used for the opposite side. Washers are used between the flange of the Ford bushing and the hub of the bandsaw wheel to align the wheel with the top wheel. Washers are used also between the pillar block bearing and the "V" belt drive pulley.

The top wheel may be raised or lowered to provide the proper tension for the bandsaw. The top wheel revolves freely on a 6x⅝-inch hexagon steel bolt. A 6x¼-inch tightener bolt is used to raise or lower the wheel. This is fitted

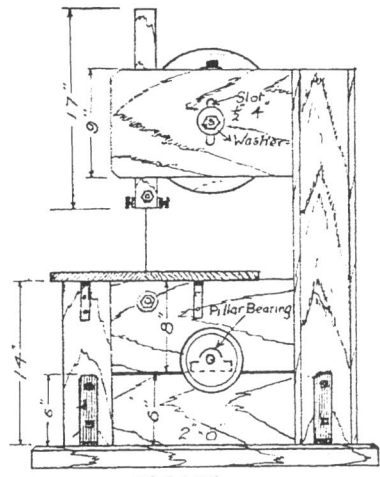

Right View.

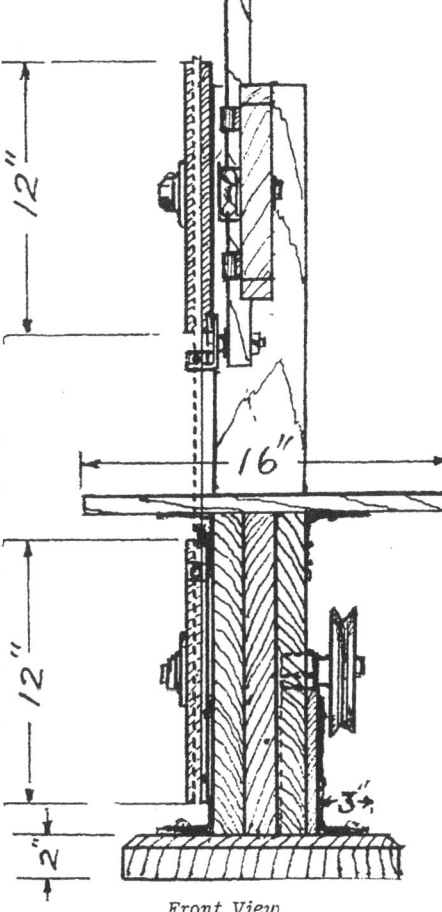

Front View.

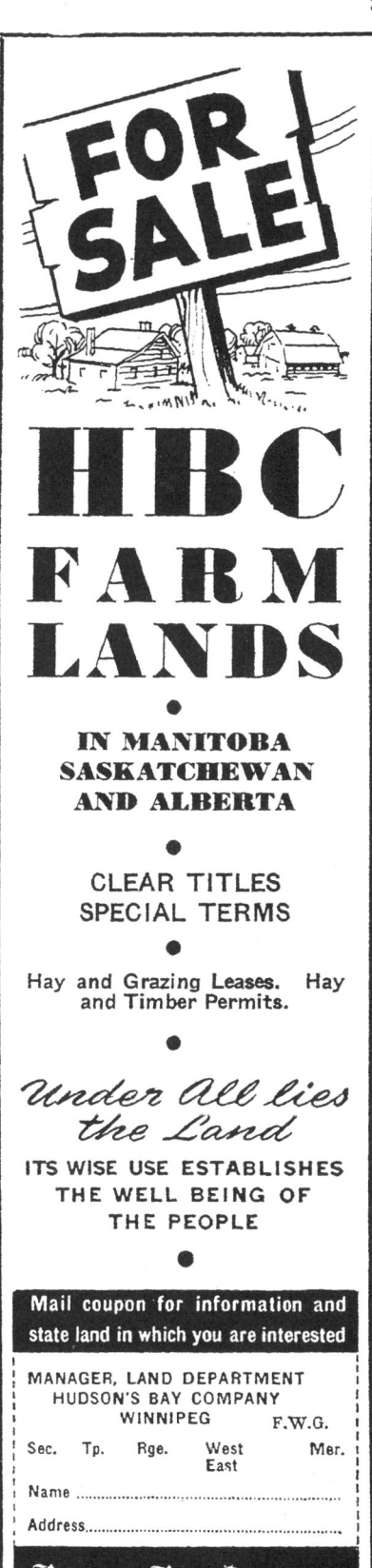

YOUR GUIDE for the FUTURE

FREE 186 PAGES

Read how to direct your ambition to achieve a fuller, more successful life. This FREE booklet contains information on many engineering subjects including examination courses listed below. Written by eminent authorities, these home study courses provide thorough training for ambitious men and women. Without obligation, write today for your free copy of "Engineering Opportunities" to the Canadian Institute of Science & Technology Limited, 262 Chester Building, 219 Bay Street, Toronto 1, Ontario.

Aeronautical Engineering
A.F.R.Ae.S. Exam.
Air Ministry Exams for Ground Engineers.
General Aeronautical.
Advanced Aeronautical and Aeroplane Design.
Aero Engines Course.
Aircraft Apprentice
Pilots' "B" License
Air Navigators' Certificates.

Civil, Mining, Structural and Automobile Engineering
A.M.I.C.E. Examination
Civil Engineering
Civil Engineering Specifications, Quantities
Road Construction
Surveying and Levelling
Hydraulics
A.M.I. Struct. E.
Structural Engineering
Structural Design
Reinforced Concrete
Heating and Air-Cond.
Royal Sanitary Institute Examination.
Sanitary Engineering
Institute of Builders Exam. (L.I.O.B.)
Building Construction and Drawing
Advanced Building Construction.
Railway Engineering
Special Combined Building Course
Architecture
Geology and Mineralogy.
Metallurgy.
Petroleum Technology
General Mining Practice
Mining Engineering
Automobile Engineering
High-speed Diesels
Electrical Equipment of Automobiles

Electrical Engineering
A.M.I.E.E. Examination
General Electrical
Alternating Current
Electrical Installations
Power House Design
Neon Lighting
Electric Traction
Electricity Supply
Electric Meters Measuring Instruments
Design and Manufacture of Electrical Machinery and Apparatus
Telegraphy Course
Telephony Course

Mechanical Engineering
A.M.I. Mech. E. Exam.
General Mechanical
Mechanical Drawing and Design
Works Manager
Die and Press Tool Work
Sheet Metal Work
Welding Course
Maintenance and Stationary Engineers
Refrigeration Course
Institution of Production Eng. (A.M.I.P.E.)
Commercial Eng. and Works Management

Plastics
Plastics Technology
Plant and Mould Design
Plastic Paints, Varnish, and Lacquer.
Natural and Synthetic Rubber.
Laminated Wood

General Educational
University of London Examinations
London Matriculation
Intermediate and Final B.Sc. (Pure Science)
Higher Mathematics
Practical Mathematics
General Education
Cost Accountant's Course
Complete Salesmanship
Advertising
Sales Manager's Course
Modern Languages

Radio and Allied Groups
A.M. Brit. I.R.E.
General Wireless
Advanced Wireless and High Frequency
Radio Servicing, Maintenance Repairs
Short Wave Radio
Practical Television
Sound Pictures

FREE 186 PAGES

MAIL THIS COUPON NOW

Canadian Institute of Science & Technology, Limited, 262 Chester Bldg., 219 Bay Street, Toronto 1, Ont.
Please forward free of cost or obligation of any kind your 186-page Handbook, "ENGINEERING OPPORTUNITIES."

Name..
Address...
....Age..............
Course Interested In............................

---

into a hole drilled in the steel bolt used as the axle as shown in the sketch. In order to secure the steel axle bolt a one-eighth-inch pin is fitted into the shaft between two five-eighths-inch washers. When the nut is tightened on one side the pin presses the washer against the slot on the opposite side, and leaves the wheel to turn freely.

The table is made of three-quarter-inch hardwood or plywood 18x16 inches.

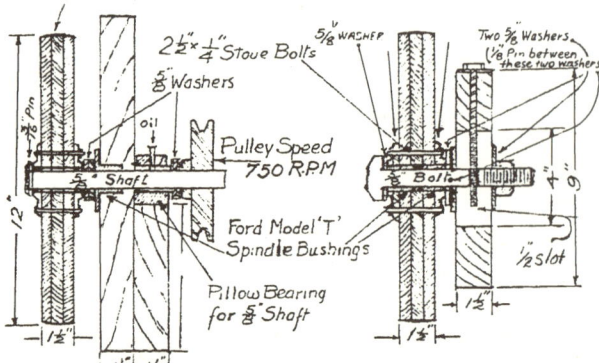

Lower Band Saw Pulley.   Upper Band Saw Pulley.

A slot six inches long and one-quarter-inch wide is cut for the saw blade. Note that the grain of the hardwood table is to be at right angles to the main frame to avoid splitting. Also the saw slot runs lengthwise with the grain. The table is glued and screwed on with one and one-half-inch No. 8 flat head screws in countersunk holes. It is a good plan to attach the saw-table with the bandsaw on the wheels so that the slot will be in the proper position for the saw blade. The table can be further supported and held to the main frame with four small iron brackets 3x3x½ inches. These are screwed on the underside of the table and against the frame. Two brackets are fastened on each side. The brackets can be purchased for a few cents.

Two adjustable bandsaw guides are required. These can also be purchased or homemade as shown in the sketch. One guide is fixed to the main frame below the table. The other is attached to a slide support above the table. This slide support is 17 inches long, two inches wide and one-inch thick. Hardwood is the best material for this slide support. The slide fits into a 2x½-inch groove in the upper arm of the main frame. The groove may be made before the main frame is assembled. However it is probably more accurately centred on the bandsaw after the bandsaw is in place. Two "U" shape guides are made from 1x⅛-inch flat iron. These should be attached so that the slide will move up and down with very little play. A "U" shaped clamp is made from 2x⅛-inch flat iron. This is placed between the two guides and held with two 2½x ¼-inch carriage bolts fitted with wing nuts.

The two home-made guides for the saw blades require careful construction but actually are not very difficult to make. A piece of flat iron 2x⅛ inches and 5¼ inches long is required for the "U" shaped bracket. Drill and tap the holes for the bolts before cutting and bending to shape. First mark the two portions to be cut out. Then make the centre punch marks for the three bolts and the other centre punch mark for the 1½x¼-inch wheel. Sizes of the bolts and cut-out portions are shown in the sketch. The wheel is made by cutting off a piece of 1½-inch shaft about five-sixteenths inches thick. This is then filed smooth on both sides down to one-quarter-inch in thickness. Drill a one-quarter-inch hole in the exact centre of the wheel. The wheel is fastened to the bracket with a one-quarter-inch machine screw just tight enough so that the wheel turns easily. A one-quarter-inch washer on each side of the wheel will provide clearance to enable the wheel to turn freely. To adjust the guides set them so that the back of the bandsaw barely touches the wheel when the saw is running idle. Next adjust the two one-quarter-inch side bolts so that they also just barely touch the sides of the saw blade and secure the bolts with locknuts as shown. If these two bolts can be brass or copper instead of iron or steel, so much the better.

If the machine is made exactly to size as shown in the drawings, an 85-inch bandsaw is required. However it is a good plan to measure the distance around the two bandsaw wheels with a

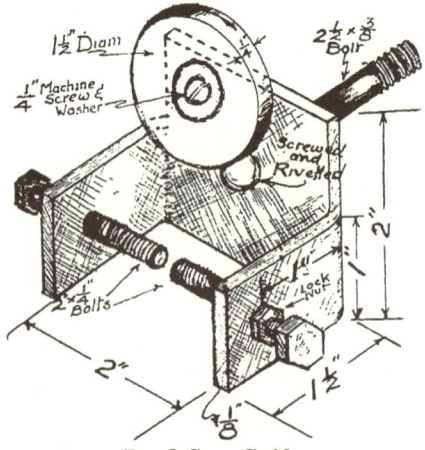

Band Saw Guide.

steel tape measure with the spindle of the top wheel in the centre of the slot. Bandsaws can be obtained in various widths from one-eighth to one-half-inch. The speed of the bandsaw should be about 2,250 feet per minute. The drive pulley speed for a 12-inch bandsaw wheel should be about 750 r.p.m. A one-quarter h.p. or one-third h.p. electric motor at 1,750 r.p.m. is suitable power. Or a belt from a line shaft with an "idler-tightener" pulley to serve as a clutch will be satisfactory. The "idler-tightener" pulley suggested and described for a homemade wood lathe shown also in this publication will serve the purpose.—H. J. Kemp.

A Small Power-Driven Hacksaw

Here is a small power-driven hacksaw which can be fastened on a bench or on a stand. If it is bolted on a bench it can be fastened directly behind a vise and may be swung up and back leaving the vise to be used for ordinary bench work.

Very little power is required. A 1/6 h.p. electric motor, such as a used washing machine motor would be satisfactory. Or it can be belted to a light line shaft driven by a small gas engine. A discarded piston or a block of hardwood of similar size fitted with bronze bushings is used as the main support for the drive shaft and all the parts to drive and control the motion of the hacksaw.

The size of the bushings in the piston or block will determine the diameter of the drive shaft to be used. A piece of discarded 1x1 inch square shaft, twenty inches long, from an old disc harrow will provide the slide bar for the hacksaw. This should be filed carefully to make all four sides as smooth and true as possible. The box-like slide, which

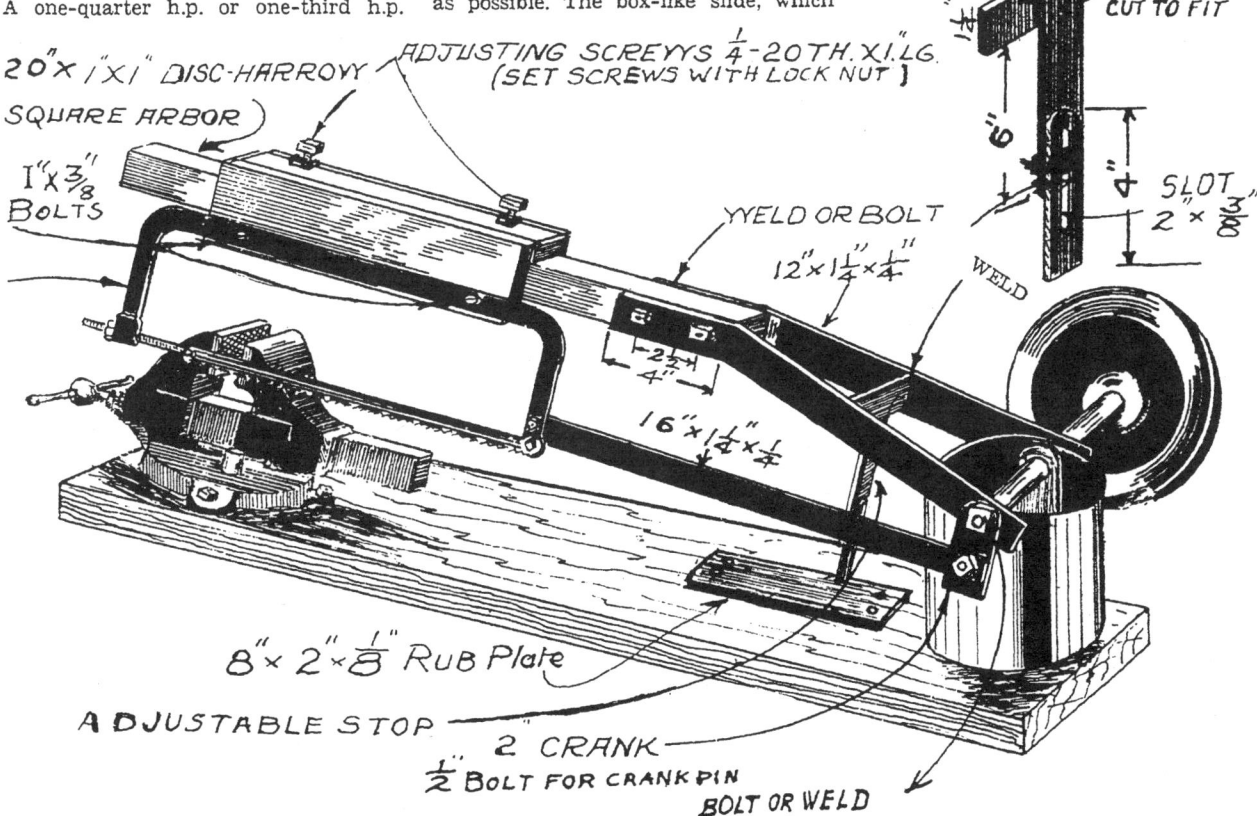

also serves as a holder for the hacksaw, is eight inches long. This is made up from ¼-inch flat iron or steel. Cut two pieces 8x1 inch, one piece 8x1½ inches and one piece 8x2 inches. One piece of 8x1 inch is drilled and tapped to receive two 1x¼-inch set screws. These set screws are used as adjusting screws to take up an excessive slack between the

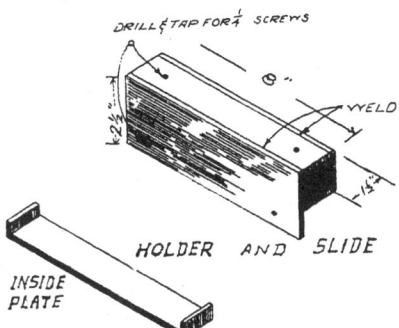

slide and slide bar and are held by lock-nuts. The cap centre on the end of the set screw must be ground off to leave a flat end to avoid scoring the top of the slide shaft. All the four pieces used for making the guide are welded together as shown in the sketch. Two pieces of 1¼x¼-inch flat iron, twelve inch long connect the square slide shaft to the drive shaft. Two 2x⅜ inch bolts, two and one-half inches apart are used to fasten these two pieces to the square slide shaft.

A hole is drilled in the opposite end of each piece to fit the drive shaft on which they are hinged. The crank is also made from a piece of flat iron or steel 1¼x¼x4 inches long. The crank may be held on the shaft between two nuts tightened firmly, or better still, by welding. The hole for the crank pin or wrist pin bolt is ½ inch in diameter and located on two inch centres from the drive shaft, to provide a four inch stroke. A one-half inch steel bolt is used to serve as the wrist pin for the crank. Here again the wrist pin bolt will be held more securely if the head is welded to the crank on the side next to the piston or block, so that the bolt protrudes through the hole to receive the connecting rod on one side with the welded head on the opposite side. The connecting rod is sixteen inches long and made from 1¼x¼-inch flat steel. A one-half inch hole is drilled at one end to fit the crank pin.

A ⅜-inch hole is drilled in the opposite end to receive a 1x⅜-inch bolt to fasten it to the hacksaw frame. The frame of a ten-inch hacksaw is secured to the guide with two 1x⅜-inch bolts. An adjustable stop is also made from 1¼x¼-inch flat iron or steel as detailed in the sketch. This is welded between the two hinged arms which support the square slide bar. The stop is made adjustable by means of a 2x⅜-inch slot in the lower piece and secured with a 1x⅜-inch carriage bolt, washer and nut. A piece of 2x⅛x8-inch flat iron screwed to the bench will serve as a rub plate for the adjustable stop.

The speed of the drive shaft pulley should be about 100 r.p.m. A fairly large drive pulley is recommended to obtain the relatively slow speed and also to provide sufficient contact of belt and pulley to prevent slippage. A twelve-inch "V" belt pulley is therefore recommended. Light power hacksaws of this type often cut best when the saw blades are placed in the saw frame with the teeth pointing towards the drive shaft end so that the sawing is done by pulling rather than by pushing, as when sawing by hand.—H. J. Kemp.

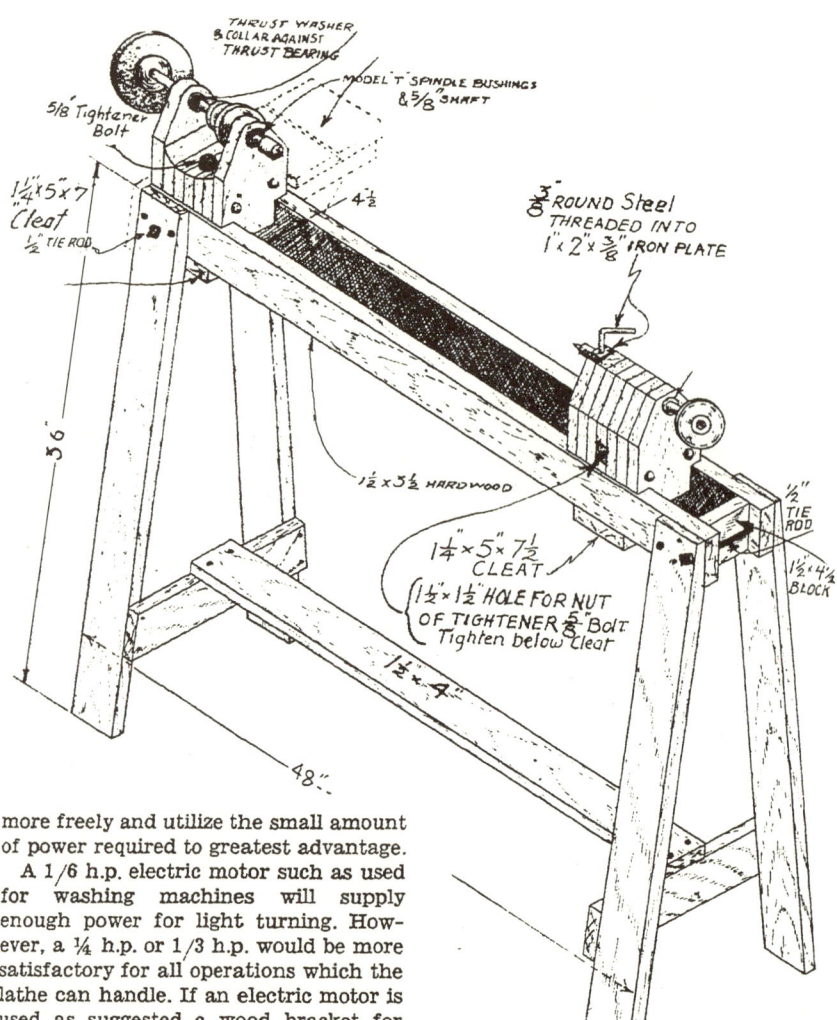

A Simple Wood Lathe

Here is a small lathe suitable for light wood-turning. It is of simple design and made mostly of wood so that it can be constructed with a few ordinary tools. The entire frame headstock and tailstock can be made with good clear pine or fir wood. However hardwood is preferable for the head, tailstock and the bed where accuracy is most essential. Ford Model "T" spindle bushings or similar bronze bushings will provide suitable bearings for the headstock and tailstock spindles. If ball bearings can be used instead of the plain bushings for the headstock, the machine will run more freely and utilize the small amount of power required to greatest advantage.

A 1/6 h.p. electric motor such as used for washing machines will supply enough power for light turning. However, a ¼ h.p. or 1/3 h.p. would be more satisfactory for all operations which the lathe can handle. If an electric motor is used as suggested a wood bracket for mounting the motor can be added as shown by the dotted lines in the sketch. A line shaft driven by a ½ h.p. gas engine can also be used to drive the lathe. Some sort of clutch is desirable to start and stop the lathe when driven by an engine powered line-shaft. An ordinary idler or tightener belt pulley can be made to serve as a clutch as shown in the sketch. This same simple device can be used for starting and stopping many other small powered machines.

While the legs and bed are of simple 2x4-inch (1½x3½ inches dressed) lumber construction, each piece should be carefully made and then glued and bolted together to make a good rigid job. The two 1½x3½-inch pieces which are used for the bed of the lathe should be of especially good clear lumber and planed perfectly straight on all sides. Oak or maple will make a good serviceable bed. In order to insure the two pieces for the bed being perfectly spaced and parallel to each other they should be first assembled and bolted together with 1½x4½-inch spacer blocks between the ways. The legs and braces may then be attached to the bed.

Both the headstock and tailstock are made with laminated 1½-inch lumber. Pine or fir can be used but maple or oak is preferable. The pieces may be cut to shape separately and then glued and bolted together. When cutting out the pieces and assembling them in this manner, they should all be cut a little oversize so that the outside surfaces can be made true to correct size and shape as soon as the glue has set. An alternative method is to cut all the pieces six inches wide and nine inches long and glue and bolt them together to form rectangular blocks. After the glue has set, the headstock and tailstock can be cut out of the laminated blocks and trued to correct size and shape. Ford Model "T" spindle bushings or other similar flanged bronze bushings are suggested for both the headstock and tailstock bearings as this type of bearing is very simple to install and replace when worn. The Model "T" bushings

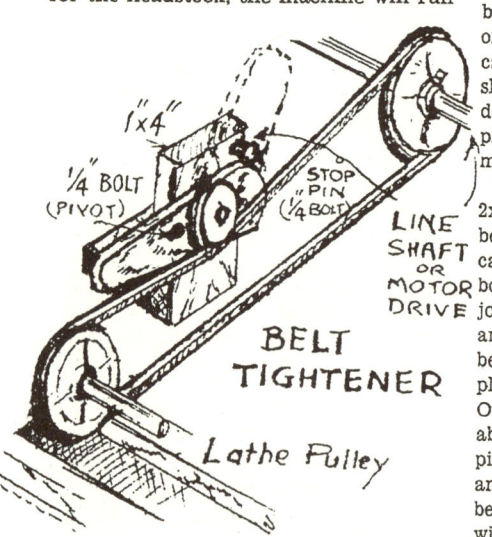

have a 9/16-inch hole. Three of these should be reamed out to fit a ⅝-inch shaft. Another similar bushing is threaded with a standard ⅝-inch tap (11 threads per inch). Two of the

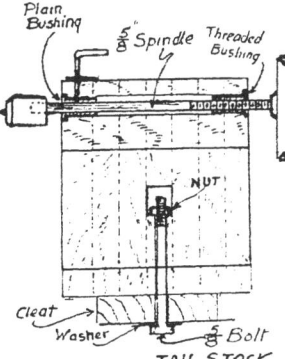
TAIL STOCK

reamed bushings are for the headstock. The other reamed bushing and the threaded bushing are used for the tailstock. These are all located as shown in the sketch. If ball bearings are desired for the headstock then a single race radial bearing such as a generator bearing can be used for the side closest to the tool rest. A double race radial ball bearing for this location would take a greater load and last longer. The opposite or outer bearing of the headstock may be a combination radial thrust ball bearing type or a taper roller bearing to take the thrust. A thrust washer and collar is set-screwed in place on the headstock spindle against the thrust side of the bushing or thrust bearing.

The headstock spindle is made from ⅝-inch cold rolled steel shafting. This size shafting, as well as collars and pulleys can be obtained at little cost from local hardware stores, mail order houses or chain stores. A sanding disc attachment to fit the outer end of the headstock spindle shaft can also be obtained from the same source, or a fine

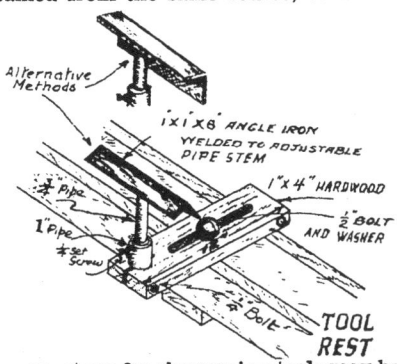
TOOL REST

emery stone for sharpening tools may be attached to the shaft instead of the sanding disc.

A piece of ⅝-inch shafting is also used for the tailstock spindle. This is threaded with a standard ⅝-inch disc, (11 threads per inch) to fit the threaded bushing in the tailstock. A four-inch "V" pulley with a ½-inch bore can be used for the handwheel on the end of the tailstock spindle. The ½-inch hole should be bored or reamed to 9/16-inch and then threaded with a ⅝-inch tap. The handwheel is then screwed on the threaded end of the tailstock spindle and screwed with the set screws already provided in the hub of the "V" pulley.

The groove in this "V" pulley can be filled in with a piece of an old "V" belt to provide a firm and comfortable grip for the hand. The cut ends of the belt can be secured with ⅛-inch round head wood screws if the die cast metal is not too hard. A hole should be drilled first to suit the size of screw used. If the metal is too hard, then ⅛-inch machine screws fitted into drilled and tapped holes are necessary. Cement, such as Duco, liquid solder or other plastic type of adhesives can also be used for securing the belt in the groove in addition to the screws used for the two ends.

The drive and tailstock centres can be made from one-inch round cold rolled steel shafting or they can be purchased.

Both the headstock and tailstock are clamped to the bed by means of a ⅝-inch bolt and cleat. The headstock is usually clamped permanently into position. Then metal shims are used to align the headstock with the tailstock and bed. The tailstock is easily moved by loosening the bolt below the cleat. The nut for this bolt is located in a recess in the tailstock as shown in the sketch.

The tool rest is shown separately in the sketch. It is also readily adjusted to the desired working position by means of a ½-inch carriage bolt and cleat below the bed. A piece of hardwood 1x4x12 inches serves for the base of the tool rest. The ends are reinforced with ¼-inch bolts to prevent splitting at one end and to clamp the piece of one-inch pipe used for the tool rest post. A slot ½x6 inches in the base is required for the ½-inch tightener bolt and washer. Note the tool rest post is made up of two pieces of pipe. A piece of ¾-inch pipe is made to slide into a piece of one-inch pipe. Some filling will be necessary to make the pieces telescope with a close sliding fit. A ¼-inch nut is welded on the side of the inch pipe over a ¼-inch hole to take a ¼-inch set screw to hold the tool rest at the desired height. A piece of angle iron or steel 1x1x⅛-inch and 8 inches long is welded or brazed on the top of the piece of ¾-inch pipe for the tool rest. This can be welded in position with sides flat or with angle vertical as shown. If a longer tool rest is required a larger sliding base with two adjustable posts to support a piece of 1x1-inch angle iron 12 inches long or longer may be constructed.—H.J.K.

FILES for the FARM

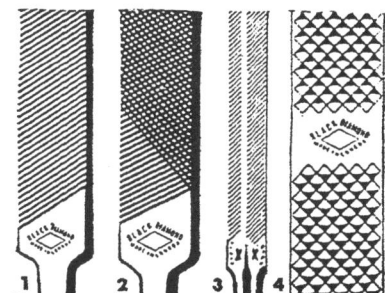

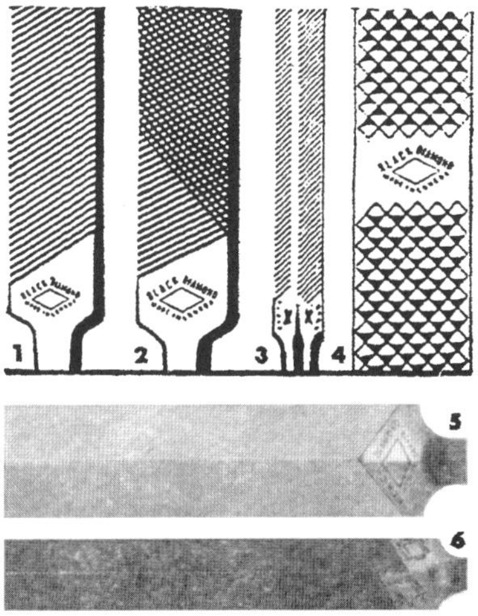

Keep an assortment of Black Diamond files around and they'll repay their cost many times over in time saved and work accomplished. Here are some popular types for keeping farm implements in repair, and saws and edged tools sharp:

1. MILL file for sharpening edged tools. 2. FLAT file for sharpening earth-working implement parts, and for general repairs. 3. SLIM TAPER files (various sizes) for filing handsaws. 4. HORSE RASP for dressing hoofs. 5. WEBSAW file for filing pulpwood saws. 6. CANTSAW file for "M"-tooth crosscuts.

Quality Black Diamond files are not expensive. Get them at your hardware or farm-equipment merchant's.

Free literature on saw filing. Address:
NICHOLSON FILE CO., PORT HOPE, ONT.

Please help to protect our forests

Bench Lathe for Wood Turning — Metal Construction

This sturdy wood turning lathe is made up from 2x4-inch channel steel, four engine connecting rods fitted with bushings, a small amount of ⅝ or ½-inch shafting, plain collars, a four step "V" belt pulley and a few simple fittings. It can be bolted down on a bench or on

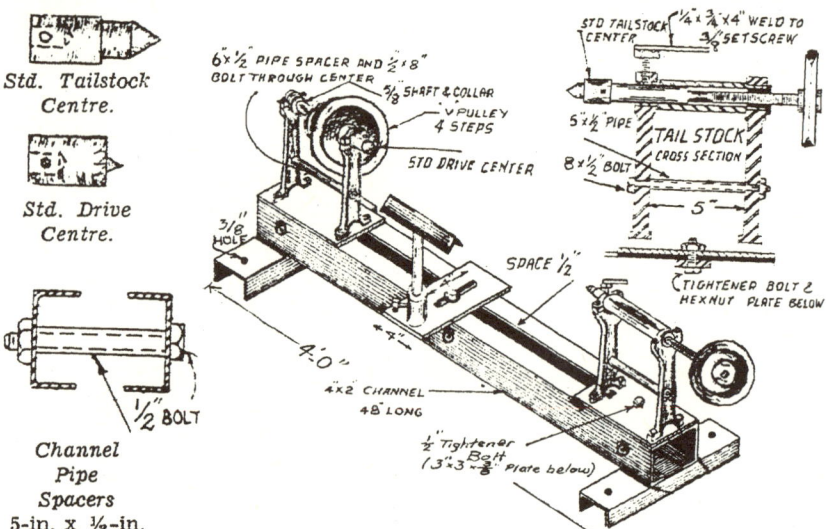

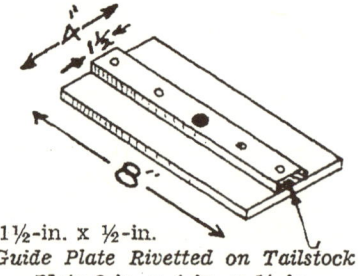

1½-in. x ½-in.
Guide Plate Rivetted on Tailstock Plate 8-in. x 4-in. x ¼-in.

a simple wood, or 2x2x¼-inch angle iron stand. The four foot bed and the two feet are made from a piece of 2x4x-3/16 or ¼-inch channel iron or steel, 10 feet six inches long. These are cut with a hacksaw. The two four-foot pieces which form the bed are spaced with three pieces of ½-inch pipe five inches long and held together with 6x-½-inch bolts. The two pieces of 2x4-inch channel iron used for the feet are made 15 inches long. Four ⅜-inch holes are drilled in each piece. The two centre holes are used for bolting the feet to the bed with 1x⅜-inch bolts. The holes at the ends are used for bolting the machine on a bench or stand. In order to make the bed more rigid the members may be spot welded with an electric welding outfit. (Electric welding is preferred since it is not so likely to warp the metal and make the bed surface uneven).

Four old automobile connecting rods are used for making both the headstock and tailstock. These are bolted on a piece of ¼-inch iron plate as shown in the drawing. After the headstock and tailstock are completely assembled and checked for alignment, the connecting rods can also be spot welded to the plates with an electric welder. The connecting rods are further made more rigid by bolting a piece of ½-inch pipe between them as shown.

The bearings for the headstock consist of two ⅝-inch flanged bronze bushings. Model "T" Ford spindle bushings will serve the purpose. These should be reamed out, if necessary, to fit a ⅝-inch shaft after the bushings have been pressed in place. A piece of ⅝-inch steel shafting 12 inches long is used for the spindle. A four-step "V" belt pulley and two ⅝-inch collars to fit the shaft can be purchased. A "wear" washer is placed between the collar and the flange of each bronze bearing. The "V" belt pulley is centred and the collars then secured with the safety set screws which are usually supplied with the collars. To make the headstock spindle turn easier under a load, a thrust bearing may be added. This should be placed between the collars and the washer. The headstock spindle will require a hollow type spur centre. Both the spur centre and the tailstock cup centre can be purchased. These are fitted with safety set screws so that they can easily be attached to the shafts.

The tailstock requires the most work and considerable care. The underside of the tailstock plate is fitted with a strip of iron or steel eight inches long by 1½x½ inches to serve as a guide between the "ways" of the bed. This guide should be filed to a close sliding fit and then rivetted to the bottom of the plate. In order to be sure of alignment of the headstock and tailstock the fastening of the slide in its permanent position may be left until the remainder of the tailstock is completed. Then the tailstock can be set on the ways and aligned with the headstock, with the guide plate held by the tightening down bolt. The rivet holes to be made in the tailstock plate can then be marked from the bottom through the rivet holes already drilled in the 8x1½x½-inch guide plate.

The housing in which the tailstock spindle is held may be made with a piece of ½-inch pipe, eight inches long reamed out to fit the ⅝-inch shaft. One end of the pipe is fitted with a bronze or steel bushing one-inch long bored and tapped to take a standard ½-inch thread to fit the thread part of the spindle. If a piece of 1⅛-inch cold rolled steel shafting can be machined to make the spindle housing, so much the better. The ends are machined down in a lathe to fit the connecting rod bearings with the shoulders snug against the side of the bearings. The shafting is then bored 7/16-inch and then counterbored to ⅝-inch, leaving about 1½ inches of the 7/16 hole to be threaded to fit the spindle. The spindle is made from ⅝-inch shafting 12 inches long. About five inches at one end is machined down to ½-inch and threaded. If machining cannot be done, then a six-inch length of ½-inch steel rod which has been threaded can be screwed into the end of a piece of ⅝-inch shaft and secured with a ⅛-inch pin. One end of the spindle is fitted with a standard cup centre with a hollow end and set screw. A ⅜-inch steel cap screw with a piece of flat iron 4x¾x¼-inches

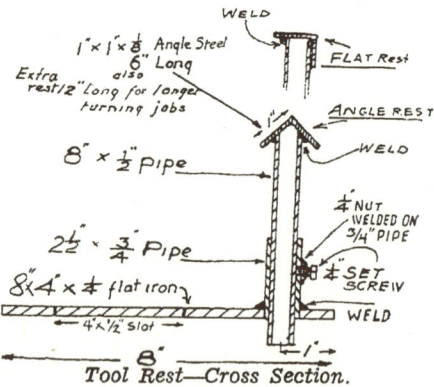

Tool Rest—Cross Section.

is welded on the cap to provide a lever to serve for locking the spindle. A ⅜-inch nut is welded over the front connecting rod for the tightener bolt. A four-inch "V" belt pulley with the groove filled in with a piece of "V" belt will make a good hand wheel. The pulley is fastened on the threaded end by either screwing it on or by securing it between two lock-nuts.

The slide rest construction is relatively simple as shown in the cross-section drawing. Two types of rest plates are shown made with 1x1x⅛-inch angle steel. The flat rest permits closer distance between the edge of the turning tool and the work and is generally preferred. The rest plates may be made in different lengths up to 10 inches to suit different kinds of work. For longer rests two supports are required. The supports are made with ½-inch pipe which slide into a piece of ¾-inch pipe or bushing fitted and welded on a piece of 8x4x¼-inch plate. A 4x½-inch slot is made in the plate for adjustment to working position. Thus the slide rest is provided with both vertical and horizontal adjustment.

A ¼ to 1/3 h.p. electric motor will supply sufficient power. If the motor shaft is fitted with the same size four-step pulley as on the lathe, the belt can be easily shifted with the motor fastened on a hinged bracket. If a gas engine driven line shaft is used a step-cone pulley exactly of the same size as on the lathe can be fixed on the line shaft. If the line shaft is too large for the bore of the pulley, then a counter shaft fitted with the step-cone pulley can be used between the line-shaft and the lathe. In order to facilitate changing the belt, a belt tightener arranged to serve as a clutch also may be used. This belt tightener is shown in use with other small power wood working tools and is described elsewhere in this publication.

A Bench Drill

This is a drill which I found very easy to construct. A casting from a plow coulter (A) is bolted to a 4x4 and held rigid by means of two steel braces. A shaft the size of the hole in the casting is drilled and keyed to a wheel. A clamp is fitted on below the casting, a small hole having been drilled in the shafting to afford a better grip for the set screw. A hole is drilled in the end of the shaft and another one into it from the side. Another clamp is then fitted to the shaft and a set screw fitted into this second hole holds the drill or bit in place. The 4x4 slides up and down, being held to the wall by the bench (B) and a steel strap iron (C) bolted to the wall. Pressure is obtained by putting weights on the steel bracket (D).—W. J. Loreburn. Sask.

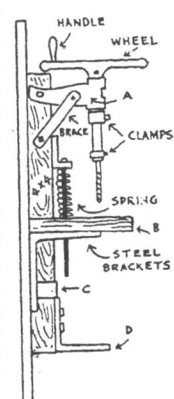

ALOX PRODUCTS

"ALOX" Linseed Oil Cake Meal for Livestock Feeding

Contains high content of "fats" valuable to the health and growth of farm animals.

"ALOX" Pure Raw Linseed Oil for Livestock Medicinal Purposes

Always keep a supply of pure high quality "ALOX" Linseed Oil on hand.

"ALOX" Pure Linseed Oils for Lasting Paint Jobs

Use "ALOX" Linseed Oils for a better surfaced and more weather resistant paint job.

The Alberta Linseed Oil Co. Ltd.
MEDICINE HAT, ALBERTA

SOME SIMPLE DEVICES AND MOVEMENTS USED FOR DESIGNING OR CONSTRUCTING MACHINES

INTRODUCTORY

The interest in the "Farm Workshop Guide" indicates a need for some information or suggestions to those who are especially interested in devising equipment for use on their own farms. There are many jobs around the farm which could be done quicker, better or more easily if suitable equipment was available. A large number of mechanically minded farmers have already contributed a great deal to the development of modern agricultural equipment. In most cases, the inspiration arose from an urgent need on their own farms. The modern demand for more efficient farm machines urges the need of even greater inventive thought.

The object here is to supply but a few illustrations of simple mechanical movements and devices along with some suggestions for designing equipment with the hope that some will be stimulated to make a greater study of this kind of work. Mechanical details are best conveyed by drawings and illustrations. A close study of sketches and blue prints helps to develop a sense of vision of something that may not yet be made, but is to be constructed. This sense of "preview" is invaluable to the inventor.

There are six simple machines which have provided man with an advantage in getting work done. Most readers will have already been acquainted with them in their early years at school. These are: 1, Lever; 2, Wheel and Axle; 3, Pulley; 4, Inclined Plane; 5, Wedge; 6, Screw. The wheel and axle, and the pulley are really continuous levers. The wedge and the screw may be classed as inclined planes. Thus the six orders of simple machines may be reduced to only two for practical purposes. These are the levers and inclined planes.

Machines are made up of a collection of levers and inclined planes which have been shaped and adapted to do specific work. These then contribute to the complete purpose for which the machine is constructed. Pulleys, gears and ratchets are forms of levers. Screws, cams and eccentrics are considered as inclined planes.

To be a successful designer one must spend considerable time in thinking and devising methods and processes. It is a good plan to develop a habit of making notes and sketches of any mechanical movement that comes to ones notice. Better still, if one can acquire a facility for making mental notes and then drawing them up later. These can then be filed away for future reference. When doing this, it will be found that while machines seem to be improved or replaced by newer ones, the basic principles of the lever and inclined plane are still there.

How should one design a machine? Since we are concerned with designers or inventors, it is most likely that each one will develop his own system. However, here are some considerations which all designers keep in mind:

1. The kind of work which is to be done.

2. Capacity-amount of work per day or hour which is to be accomplished.

3. Suitable speeds for various parts of the machine.

4. Size and weight.

5. Method of driving.

When the above are determined then consideration is to be given to materials and parts to be used. The parts may include bearings, gearing, shafting, collars, pulleys, clutches, cams, ratchets, methods of adjustments, etc. Provision for lubrication must not be overlooked.

At this stage is then the time to consider various combinations of mechanical movements and devices. When the machine is to be made of parts and materials from other machines, then the plans must be modified to include them.

To be a good mechanical draftsman will help a great deal in formulating plans of machines. However, it is not necessary to be a professional draftsman in order to design a homemade machine for farm use. Clear sketches of ample size with clear figures to indicate correct measurements may suffice to round out the ideas in the mind of the designer, so that he will have little trouble in constructing the machine especially if he builds it himself. Precise blueprinted plans are essential in shops and factories where the actual work is to be carried out by machine specialist rather than the designer.

The few mechanical devices and movements shown are grouped for comparison. Very little explanation is offered as the reader will then get most benefit by studying the outline more closely.

Belt and Friction Drives

Belts and flat face pulleys are one of the oldest means for transmitting power between revolving shafts. It is the most suitable means to transmit power between shafts which are a considerable distance apart. We think of belts as being either of the flat or "V" shape type. Chains and ropes may also be classed as forms of belts. Belts as they are known today are made from various materials such as leather, rubber, canvas, and woven cotton. Flat belts are usually run on crown-faced pulleys, the crown being provided to cause the belt to run on the centre of the pulley face. The "V" belts have come into greater use in recent years. There is less slippage with "V" belts as compared with the flat belts. Where considerable power is to be transmitted, two or more "V" belts are used. There are various forms of friction drives. They have the disadvantage of requiring an excessive amount of pressure on the bearings in order to maintain the minimum of slippage. Where they can be operated with a small amount of power and are to be subject to light loads the simplicity of the friction drive has a definite advantage.

1. A flat face three-step cone pulley changes speed drive. "V" belts and pulleys are also used in this manner.

2. A cone pulley drive which provides for gradual increase or decrease of speed by shifting the belt right or left.

3. A method of driving two shafts at an angle. Two idler pulleys are required to give directions to the belts. The centre of the pulleys on the angle shafts must be in direct line with the outer rim of the idler pulleys.

4. A method of belting two pulleys when they are either too close, or not in line to permit a normal direct drive. As in Fig 3 two idler pulleys are required.

5. A method of driving two shafts at right angles to each other.

6. Bevelled friction drive. Two bevelled pulleys with wear-resisting lagging used to drive shafts at an angle depending on the angle of the bevelled pulley faces.

7. Variable speed friction drive. A large flat disc is faced with leather, rubber or other composition. The face of a smaller pulley is held in contact with the disc to transmit power. As the smaller pulley is moved from the outer edge to the centre, the speed is decreased. At the centre it is neutral. When it moves across the centre to the opposite side, the direction is reversed and the speed increased as it moves to the outer edge.

8. A grooved friction drive. A rubber disc with a "V" shaped face runs in the groove of a "V" pulley. It is suitable only for very light work as on sewing machines, counting devices and some feed mechanisms.—H. J. Kemp.

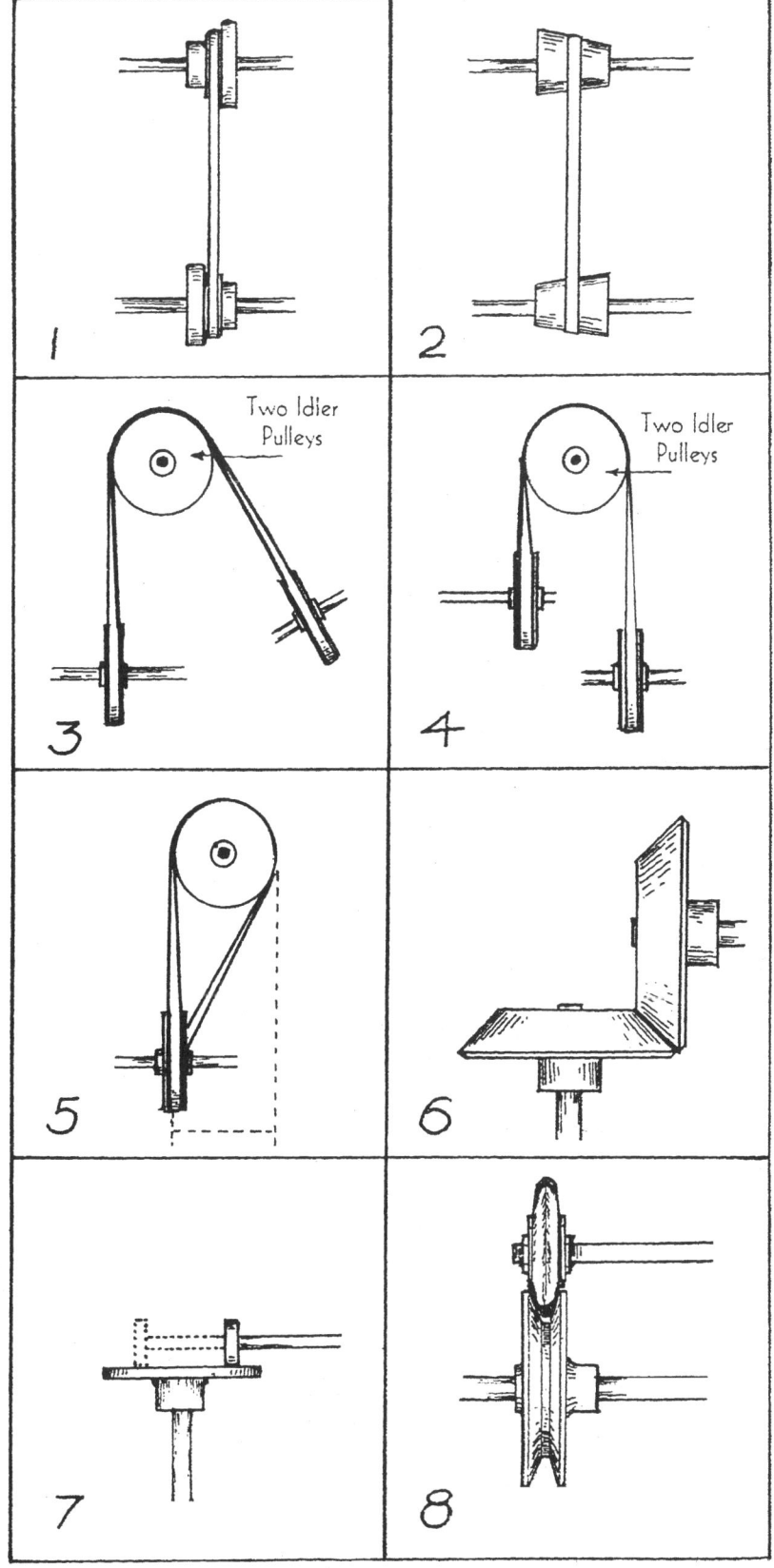

Simple Types of Clutches

Friction clutches are not used extensively on field implements since they would be subject to rapid wear due to sand and dust from the soil. They are more suitable for stationary machines such as threshers, tractor or engine drives. Some friction clutches are made to automatically become engaged when a prescribed speed has been attained. The positive types of clutches used for farm implements are generally of the most simple construction. These are usually of the square or spiral jaw type. Their chief disadvantage is that they must be engaged before power is applied, or they must be constructed to stand considerable strain if they are engaged when the power source is already in motion.

9—The Common Jaw Clutch
One half of the clutch is usually cast or fastened to a gear pulley or other driven member. The opposite side slides on a key or pin. A forked lever engages a groove or flange of the slidable member so that the clutch jaws may be meshed or released.

10—The Ratchet Clutch
Ratchet teeth are provided so that the clutch will slip when the motion is reversed. The spiral jaw type of clutch operates in a similar manner. The chief difference is that the spiral jaw clutch has larger and fewer teeth.

11—The Slip Clutch
Shallow "V" grooves are used for the contact surfaces. Strong spring pressure is applied to enable the clutch to transmit sufficient power for the work to be done, but to allow the grooved surfaces to slip when the load becomes excessive due to obstruction, etc. The slip clutch therefore is also a safety device.

12 and 13—The Pin Clutch
The sliding member is a disc or bar shaped casting which has one or two pins and a groove projection for engaging with a forked lever. The pins are brought into contact with holes or slots in the opposite disc shaped member which is fastened to the shaft. Occasionally the pins are made to contact two projections on the perimeter of the fixed disc instead of holes or slots. This arrangement allows the slidable member to be moved into contact position more readily. The sketch (No. 12) shows the simplest form of pin clutch in that only one pin is used and contacts a fixed bar casting. The dotted lines indicate the arrangement for a two pin bar clutch. Fig. 13 is of the two pin disc type with either holes or projections to effect contact.

14—The Friction Cone Clutch
This clutch is made up of two cones. One is made to fit into the other. Cone clutches have a tendency to stick. The angle of the cone should be made large enough to avoid sticking.

15—The Self Acting Friction Clutch
Sometimes called a cam clutch. The cam-like members operate inside of a special flanged pulley or cup shaped casting. Springs (not shown) are often added so that the cams operate as a speed governing device. This type of clutch is also used as a ratchet device so that power is transmitted only when motion is in one direction, and slippage occurs when the motion is reversed.

16—Roller Clutch or Ratchet
The rollers run free when the drive shaft runs one way and contacts almost instantly when the motion is reversed. This is an excellent device for controlling feed mechanism in minute gradations.—H. J. Kemp.

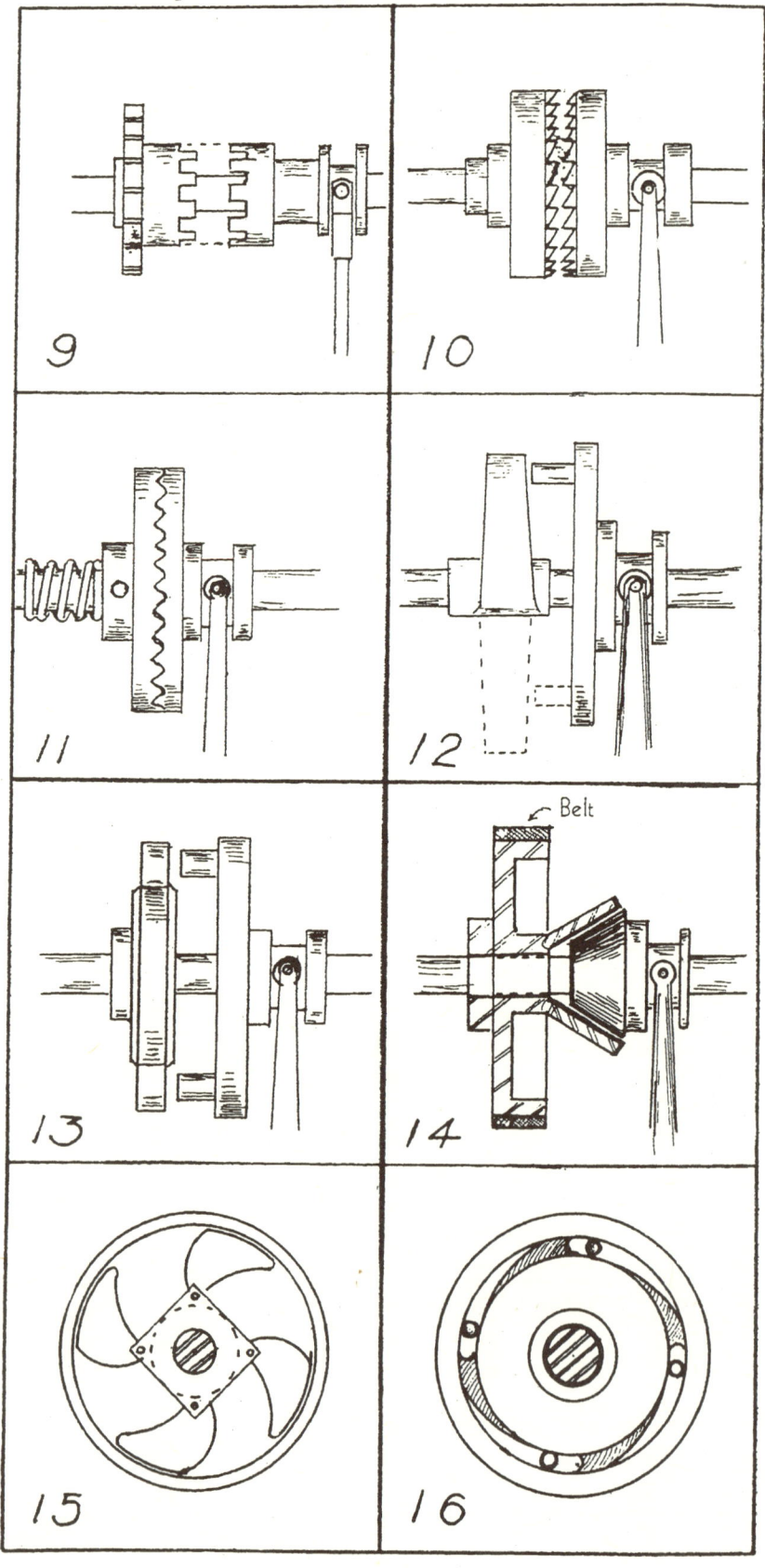

Various Cranks and Eccentrics

The simplest form of crank is the single crank as represented by an ordinary hand crank and is sometimes called a bell crank. When used on power driven machines they may be either of the bar or the disc type. To reduce vibration counter balance weights are added to the end which is opposite to the crank pin end. There are actually many forms of cranks. The eccentric may be considered as essentially a crank. The eccentric is really an enlarged crank pin mounted on or included with the drive shaft. The bearing around the eccentric is then called a strap. Like ordinary crank pin bearings, the eccentric bearings may be either solid or split. Rollers are sometimes added to reduce the friction of large bearing surfaces of the eccentric

17—The Common Bell Crank

When extended to provide weight as indicated by the dotted lines, it is called a balanced bell crank, provided the weight equals that of the crank arm and pin.

18—A Single Crank

Sketch "A" illustrates a bent crank. It is used mainly on farm implements and machines. Such a crank is formed by a die press. The bearing surfaces are machined. Bent cranks have the advantage of retaining continuous grain and strength. Sketch "B" is a square drop forged crank such as used on gas engines. The entire crank arm is generally forged as a solid unit. The recess for the connecting rod bearing is machined out. Similar cranks are sometimes built up.

19—An Eccentric Crank

Sketch "A" shows a small crank mounted on the crank pin of the larger crank. The small crank is set at an angle as shown in Sketch "B" to provide a timed movement, such as for opening and closing steam valves, or other purposes where special timing is required. It is also useful for providing a short reciprocating motion where timing is not a factor.

20—The Disc Crank

The disc portion is usually cast so as to include sufficient weight (Shown at A) to counter balance the weight and load at the crank pin. The load may be a connecting rod or a sickle type mower knife as used on mowers and other harvesting machines. The disc crank is frequently used to help to counterbalance straw decks in threshing machines.

21—The Common Eccentrics

Eccentrics are made in different ways. The sketch shows a solid sheave type with adjustable bearing straps. They are used frequently where there is no room for a crank, or where special timing of the crank or eccentric movement is required. Because of its large bearing surface it is very suitable for imparting rapid vibration movements or for applying great pressure as required for large die presses.

22—The Yoke Strap Eccentric

This eccentric is similar to Fig. 21 in that both employ a circular eccentric cam. The strap shown in Fig. 22 is called a yoke. The yoke permits the eccentric to drive directly against the sliding bar held by guides.

23—The Accelerated Eccentric

The extra rocker lever "A" is added to increase the motion and travel at "B".

24—Cam Lever Lock

The eccentric lever cam is used for gripping cable rope, belts and for sack holders.—H. J. Kemp.

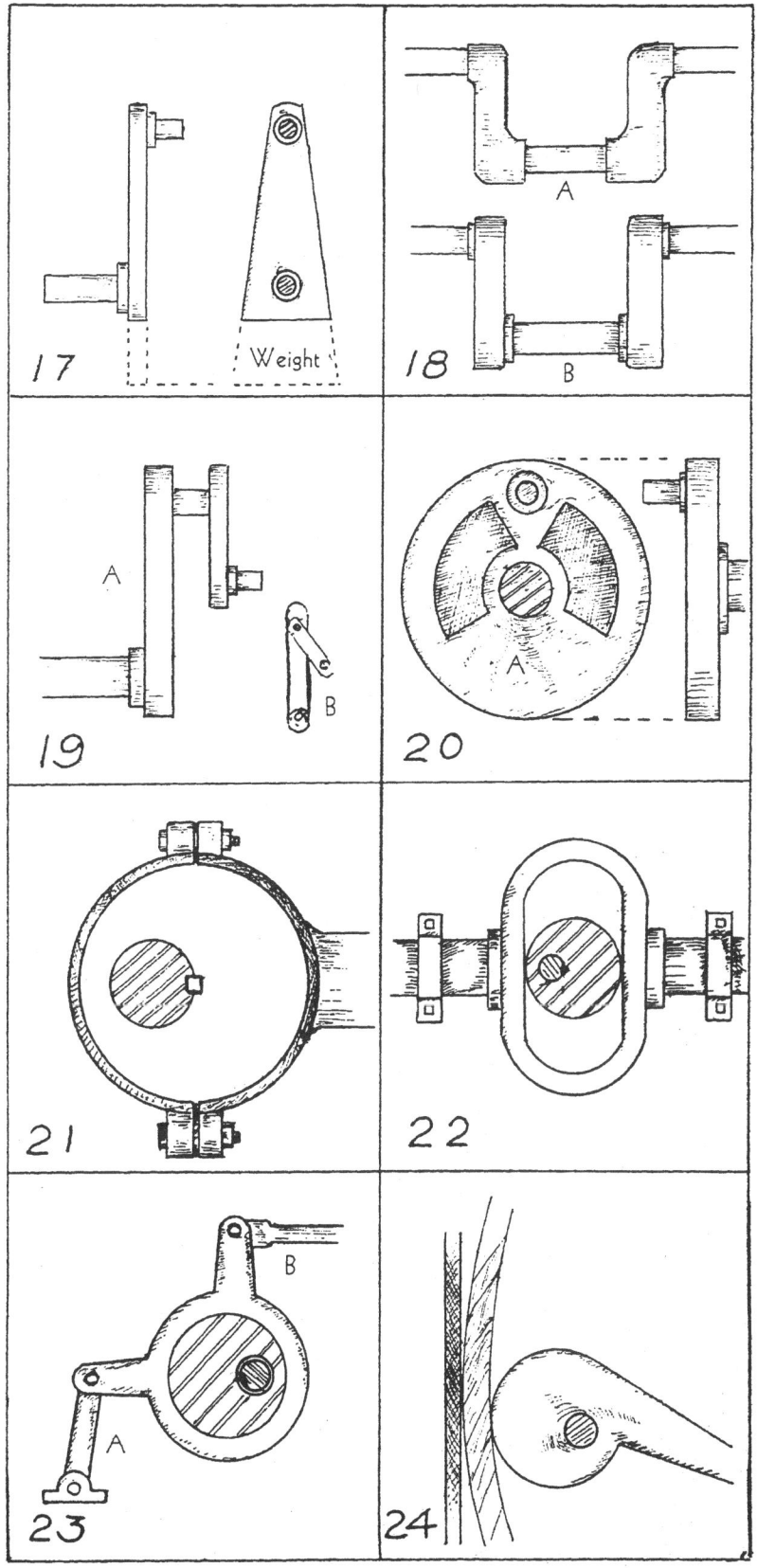

Cams and Followers

Cams differ from eccentrics mainly in that they give motion in one direction only or they provide irregular or intermittant motion depending largely on their shapes. While cams do not transmit much power, they can be made in a great variety so that a plain circular motion of a shaft or a reciprocating motion can be almost immediately changed to variable or intermittant speeds in any direction. The shapes of the cams however must be carefully plotted and made to give satisfactory results where extreme accuracy in timing and travel is required.

25—The Heart Cam
The heart-shaped cam may be made either open or closed. The open type is cut out to form a solid heart-shaped member. The closed type consists of a heart-shaped opening cut into the side of a circular disc.

A uniform rectilinear motion to a sliding rod between guides is provided by either the open or closed heart-shaped cam. The method for laying out a heart-shaped cam is indicated by the dotted lines in the sketch. To lay out the cam, a circle is drawn, the diameter of which is equal to the length of travel required for the rod. Any number of circles spaced equal distance apart are drawn inside the first circle. Radial lines are then drawn thirty degrees apart to divide the circles. The intersections of the circles and radii are marked so that the heart-shape may be drawn as shown in the sketch.

26—Elliptical Cam
Two kinds of followers are shown. Follower "A" is designed to provide an oscillating motion. Follower "B" provides direct motion.

27—Curved Slotted Cam for Vibrating Motion
A variable vibrating motion is given to the straight lever by this form of slotted cam.

28—Curved Slotted Cam for Rectilinear Motion
The same type of cam as in Fig. 27 but applied so as to convert a vibrating motion of the cam shaft into a variable rectilinear motion to a sliding bar held by guides.

29—Jump Cam
This type of cam can be used in several ways. It can serve as a release when a prescribed amount of accumulated pressure or weight is applied at "W". It may be used to impart uniformly timed motions to a ratchet gear. It can also provide rapid or slow vibrating motions.

30—Wiper Cam
The cam at "A" transmits a variable vibrating motion to the straight lever arm from a uniform vibrating shaft. The cam is shaped to impart the desired motion. At "B" the wiper engages a flat disc and imparts a uniform or variable rectilinear motion to the disc shaft "C" and at the same time causes the shaft to rotate.

31—Cylinder or Drum Cam
A groove in the face of the cylinder or drum imparts a reciprocating rectilinear motion to a sliding bar from a circular motion of the cam shaft. A spiral projection on the face of the drum is sometimes used instead of the groove.

32—Straight Slotted Cam
The wrist pin of a disc crank has a uniform circular motion. The travel of the pin up and down the slot provides a slow motion in one direction and a quick return motion.—H. J. Kemp.

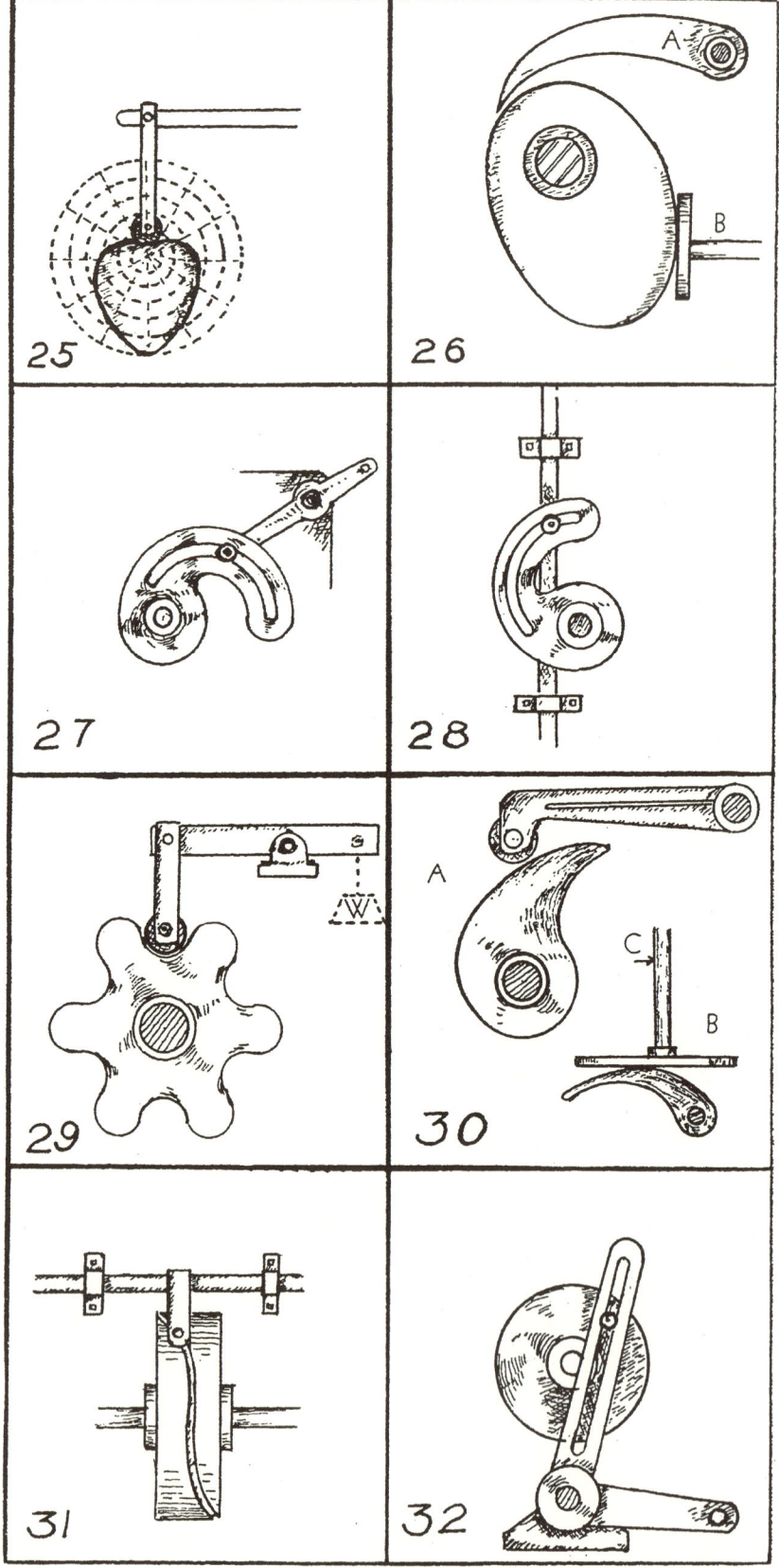

Devices for Adjusting

Few machines if any can be made without suitable means for adjustments for various purposes. Practically all machines require some means of adjustment in order to make them perform their functions satisfactorily. Certain parts are made adjustable to meet variable conditions of the work to be done. Adjustments are required to correct alignment. In the case of experimental or semi-experimental machines and homemade devices, provision for experimental adjustment is a good precaution. Few new or homemade machines operate perfectly when they are first put to work. After adjustments have been made and the machine works satisfactorily, the builder is then able to build a much better machine or improve his initial machine if he desires to do so. The more simple the means of adjustment are, the less will be their original cost and what is more important is the less will be the cost of time and effort in affecting adjustment.

33—Clamp for Adjusting Rod
A piece of strap iron and a thumb screw clamps the control rod for holding an air outlet valve.

34—A Ratchet Rod
With teeth which engage a knife edge is more readily shifted and is more suitable when adjustments are to be made frequently. This same decise is suitable for various feed systems required to deliver materials at a prescribed rate or volume in a given time.

35—Clamped Lever Arm
Lever arms often require adjustment on a shaft so that their motion may be increased or decreased. The split hub allows for minute adjustment and grips the shaft around its entire circumference. Set screws in solid hubs are a more popular form mainly due to their lower cost. Two set screws set at a forty-five degree angle to each other ensure the minimum of slippage.

36—Screw Adjustment for a Lever
This screw adjustment may be added to the adjustable lever shown in Fig. 35 to provide further means of adjustment. In effect, the combined adjustments are equal to micrometer adjustment.

37—Screw Rest
Here, a forged lever is clamped to a shaft in its approximate desired position. Final adjustment is made by means of the adjustable screw.

38—Spur Gear and Rack
This form of adjustment has many uses such as controlling slide openings, raising or lowering working parts of machines as saw arbors, power drill spindles, horizontal movements such as in lathe carriages. It is also used for raising or lowering farm implements to adjust their working position according ot the type of field work to be done.
Modified forms will be found on threshing machines for adjusting cylinder concaves, levelling combine harvester threshers for side hill work.

39—Slotted Brace Adjustment
This is another common means of adjustments to provide effective leverage. It is also used to hold parts of machines in required positions to do effective work, such as wind control adjustments of fans or blowers or holding seed separation screens at the most effective angles of work.

40—Protractor Adjustment with Indicator
A lever is set at any desired angle, the positions being indexed in degrees around the semi-circular slot.—H.J.K.

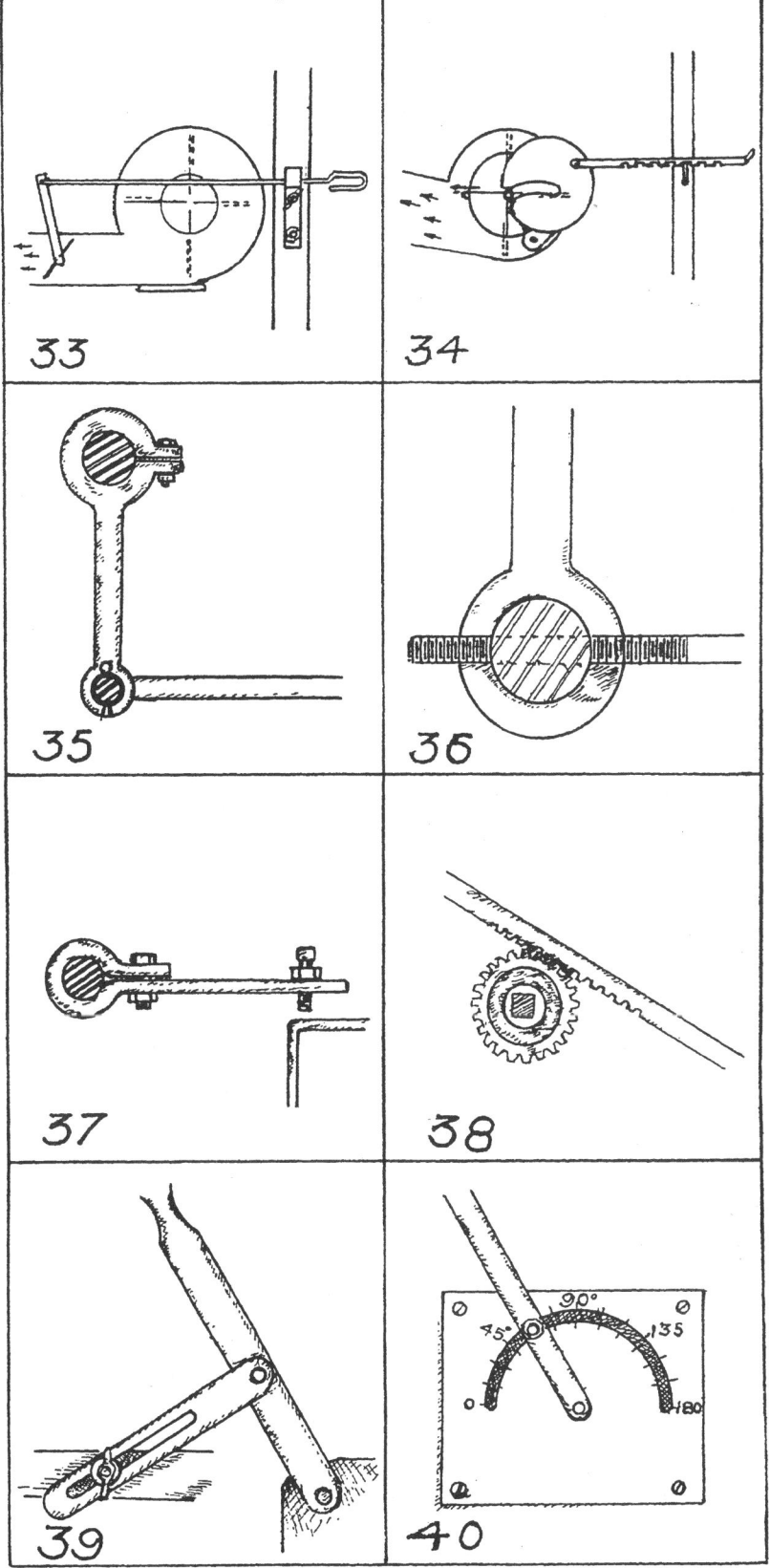

Devices for Adjusting

A GOOD HOME is Your Best INVESTMENT

Head Office - Winnipeg, Man.
Retail Yards in Manitoba and Saskatchewan

Soldering

A good soldering outfit properly used will repay its cost several times over each year in repairing milk cans, tinware, roofing, gutters, spouts, electric wiring, etc. Contrary to the general belief, the work is quite easy if a few simple precautions are observed; and the more skillful one becomes, the more

practical uses he finds for it. Anyone can do successful soldering if he will study each job carefully and see that the following essentials are secured:

1. Keep soldering copper clean, at the right heat, and properly tinned.
2. Scrape and clean thoroughly the surfaces to be soldered.
3. Use the proper flux for the given surfaces.
4. Heat both surfaces above the melting point of the solder, by applying the solder copper or by preheating with the blow torch.
5. Fasten or hold the work so solder will run into the joint to be united.

Heating the Soldering Copper

First smooth and clean the point of the soldering copper by pounding the faces smooth with a hammer or by filing with a coarse file. Then heat to the proper temperature as follows:

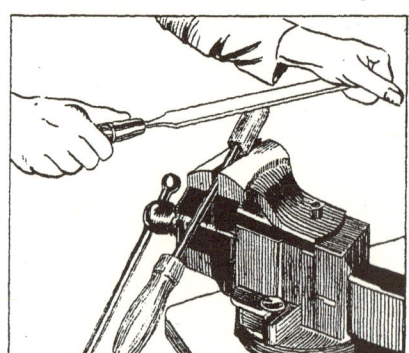

Heating With Gasoline Blow Torch

Best for general farm soldering, since heat is steady and easily controlled, and can be carried to the roof or the tractor or wherever needed. It is also useful for pre-heating large pieces, thawing frozen pipes, burning off old paint, etc.

Heating Over Open Burners

Most farm homes have open flame stoves burning gas, bottled gas, gasoline, kerosene, etc. Not satisfactory to heat soldering copper directly over these, but the use of a cover or hood heats quickly

and turning down the flame holds the heat steady as desired. The hood may be simply a bread pan with about a dozen layers of sheet asbestos pasted on the inside and then fastened with tin washers and nails or rivets, and with holes cut in each end for inserting the soldering copper and holding it up off the burner. Two bricks laid a short distance apart, a third on top and another across the back also can be used.

Heating in Coal or Wood Fires

Hard to hold the soldering copper at the proper heat and keep the point clean in a stove, furnace, or other coal or wood fire. However, this method works fine if a short piece of pipe closed at one end or an old wagon skein is put into a hot fire and the soldering copper put inside. Can be heated nicely in a forge after the coal has been coked.

Failure to secure the proper heat for the soldering copper is the stumbling block for most beginners, since successful soldering cannot be done if the point is a few degrees too cool and the tinning will be burned off if the proper heat is exceeded very much. The expert

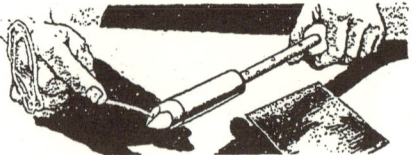

solderer can tell the proper heat almost at a glance, but the beginner should go by these tests:

1. Rub the hot copper on a large crystal of salammoniac. If hot enough, it will give off clouds of pungent smoke.

VARIETY

IN
G.W.G. WOOL SHIRTS

Your dealer will receive a fair share of G.W.G. WOOL SHIRTS.

There will be checks and plaids, in harmonizing colors . . . and you'll find the fitting comfortable—the fabric pure wool—the details smoothly finished.

Ask your dealer for a gay G.W.G. WOOL SHIRT . . . it's another dandy G.W.G. garment for hunting, sports or work.

G.W.G. WESTWOOL JACKETS

HEAVY DUTY MACKINAW JACKETS IN BRIGHT COLORS

TYPE I
Vancouver Stag-- with extra warmth in double yoke and double sleeves.

TYPE II
Cruiser coat with double yoke, and game pocket across back.

TYPE III
Button-fitted jacket, double yoke for extra warmth. Tightens at side band.

TYPE IV
Zippered jacket that is tailored to neat lines and comfortable fit. Zipper breast pocket.

Exclusive G.W.G. pure, warm WESTWOOL is the fabric.

FOR THE MAN OF ACTION...
TWEED PANTS

- FULL CUFF
- FIVE POCKETS
- REINFORCED SEAMS
- REINFORCED BELT LOOPS

GWG *Red Strap* OVERALLS

STAPLED BUTTONS

FAMOUS RED STRAP

SNOBACK DENIM

REINFORCED SEAMS

For children too!

Your first impression of G.W.G. RED STRAP OVERALLS will be, "Good-looking" . . . but that's only part of the story.

G.W.G. RED STRAP OVERALLS are tailored from Snoback Denim—a fabric exclusive with G.W.G.—that is made from tightly twisted strands of cotton, woven to give extra-long wear.

G.W.G. RED STRAP OVERALLS have smooth seams, and big pockets, and they're cut to give maximum ease for your work.

G.W.G. RED STRAP OVERALLS are not plentiful yet, but when you CAN get them, the quality is always tops!

2. If hot enough, the point should melt solder almost instantly.

3. If hot enough, the point rubbed on a pine stick should smoke freely.

4. If too hot, the tinning will be burned off the point.

5. It is much better to have it a little too hot than too cool, and the beginner will do well to keep the heat as high as possible and still have the tinning on the point bright or **tinged with yellow.**

An electric soldering copper is very convenient where electric power is available, and is especially good for electric and radio work, where the space is often limited and excess heat is likely to cause damage to insulation.

Soldering Fluxes

Some kind of flux is necessary with nearly all farm soldering to keep the surfaces free from oxides and to make the solder flow more freely. The most common is zinc chloride, which can be used with tin, copper, brass, steel, iron,

and is about the only flux which will give good results with steel. It can be secured from any tinner, or can be made by diluting commercial hydrochloric acid with an equal amount of water in a wide-mouthed glass jar. Small pieces of scrap zinc, such as can be cut from old dry cells, are then added slowly until there is no further action of the acid. This can be applied with a small brush or swab cut from an old inner tube. It is quite corrosive and should not be used on electrical work, and any joint on which it is used should be thoroughly washed.

For electrical connections and radio work, a *non-corrosive soldering paste or special liquid flux* should be used. These are on the market or can be made by stirring zinc chloride crystals in water until no more will dissolve. To five parts of this solution add one part of glycerine and four parts denatured alcohol. A *non-corrosive paste flux* can be made by rubbing together equal parts of zinc chloride crystals and vaseline. These non-corrosive fluxes are also suitable for general soldering work, and can be as bright dips for the hot, tinned soldering copper.

Commercial hydrochloric acid diluted with an equal volume of water should be used as a flux for soldering zinc or galvanized iron. *Powdered resin* is the favorite flux for bright tin, and is also helpful in difficult soldering with almost any metal. Self fluxing wire solders with either a rosin or a cut-acid core are handy to use on light jobs and in close quarters, such as electrical and radio work. *Aluminum soldering* requires special treatment and will be discussed in a later article.

Soldering a Seam

Surfaces which are to be soldered must be as clean as possible, with all grease, paint, dust, rust, and other foreign material removed as completely as possible. Many beginners do not realize the importance of thorough cleaning and fail on that account. File, scrape, or grind the surface until the bright metal shows before trying to solder. Emery paper or steel wool is good for removing scale and rust, while soft surfaces may be scraped with a knife or a tinner's scraper. When the work is clean, apply just enough flux to be sure both surfaces to be joined are wet.

Remove the soldering copper from the fire, quickly test it for proper heat by rubbing one or two faces on the salammoniac crystal, dip it for an instant into a jar of salammoniac or zinc chloride cleaner, touch it to a bar of solder, and quickly put one face flat against the joint, holding the work so the solder can run into the joint. Start at one end of the joint and hold it stationary until the work becomes hot enough for the solder to run freely, then draw the point down the seam only as fast as the solder will follow. When the solder on the point is used up, pick up a new charge or better hold the bar of solder to the point as it is moved along. When the soldering copper becomes so cool that the solder does not flow freely, it must be reheated. Two well tinned coppers heating at the same time will save time if much soldering is to be done. When done the seam should be completely filled, with no excess solder, and should then be scrubbed with a solution of soap and washing soda in water to remove excess flux.

Sweated Joints

Pieces of sheet metal are "sweated" together, and this is the usual method of patching a large hole in a pan. The

Thousands of farm wives throughout Canada agree that HABACURE is the most efficient, economical meat-curing compound. You buy HABACURE ready-to-use—nothing to add or mix.

Cure delicious hams, and bacon, also beef, lamb, mutton and fish right in your own home.

Each pound smokes, cures, flavors ten to fourteen pounds of meat. It is safe, sure and thrifty. Buy from your dealers today.

5-pound drum $1.00
10-pound drum $1.75

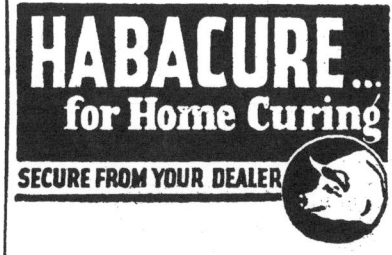

two surfaces are cleaned and fluxed in the usual way, then each one "tinned" with a coating of solder. They are then placed together and reheated by holding the hot copper on the patch while pressing down with a stick. When the solder flows freely again, the heat is removed and the patch held down until the solder hardens.

Soldering Enameled Ware

Holes in enameled ware can be soldered in the usual way after the enamel has been removed by grinding or otherwise so as to expose a ring of clean metal. Zinc chloride flux should be used, and all excess scrubbed off after the job is completed.

Comments

Failure to pick up solder means that the soldering copper is dirty or not hot enough. Slow running solder means that the copper is not hot enough or has not been held on the surface long enough. Rough uneven solder means not enough heat. Patchy adhering means grease or dirt on the surface. Failure of the solder to run into the seam means poor cleaning, too little heat, or not enough flux to keep down the oxide.

—I. W. Dickerson.

Methods of Welding

There are several methods of welding. Those which are most adaptable for farm use are:

1. Forge welding.
2. Oxy-acetylene welding.
3. Electric arc welding.

Forge welding — This is the familiar blacksmith method by which pieces of metal are heated to melting point and then immediately forced together by means of hammering.

Oxy-acetylene welding — Metal is heated to melting or fusion temperature by an intensely hot flame. This very hot flame is created by the ignition of a controlled mixture of two gases, namely, oxygen and acetylene. No pressure is used to unite the heated metal. The molten metal is allowed to run together and solidify.

Forge Welding

The conventional blacksmith's forge, anvil, hammer, tongs and quenching pail or tank are the essential tools for forge welding. Special blacksmith's bituminous coal is most commonly used to supply heat. This is usually obtainable through any local coal dealer.

Welds are made in different ways. The five principal methods are:

1. Butt weld.   4. Lap weld.
2. Scarf weld.   5. Jump weld.
3. Cleft weld.

In order to supply enough metal to make a good weld, the ends to be welded are first made thicker as shown in the sketch. The parts are then heated again to white heat sufficient to cause the surfaces to melt. The ends are then hammered together.

The fire for welding should be fairly deep. A deep fire prevents too much air passing through the fire. This enables most of the oxygen to be consumed before it reaches the parts to be heated.

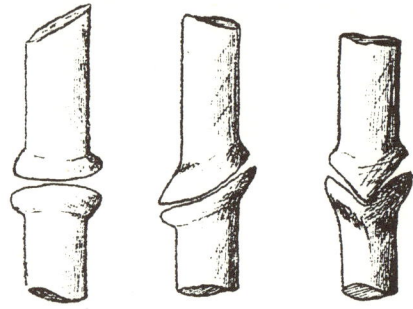

BUTT WELD   SCARF WELD   CLEFT WELD

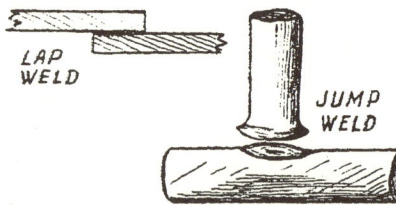

LAP WELD

JUMP WELD

Too much oxygen causes the metal to scale. Flux is used to coat the metal to exclude the air. The blower should supply air with a pressure of about one-quarter pound to the square inch.

Flux is not usually required for welding ordinary wrought iron, unless it is thin. Flux, however, is required for welding steel. A 50-50 mixture of borax and sand, plus about 25 per cent of iron filings is a satisfactory flux for steel. Good welding compounds can also be purchased. Steel is first heated to a yellow color. Then the flux is sprinkled on to the heated parts. The parts are then hammered together lightly for the first few blows, followed by heavier blows as soon as the two parts have stuck together.

Some practice is required to become efficient in welding. If one can obtain a few lessons from an experienced blacksmith, or from an agricultural engineering school, so much the better.

Oxy-Acetylene Welding

The oxy-acetylene torch can be used for welding many kinds of metals. It is usually more suitable for commercial work, or where electric power service is not available.

An oxy-acetylene welding outfit consists of an oxygen tank, an acetylene tank and a torch. The torch is equipped with a handle, a tube for the oxygen, another tube for the acetylene. A needle valve is used to control the flow of each of the two gases. A head and tip unite the gases to form the desired arc. Oxygen is supplied to the torch at somewhat higher pressure than the acetylene. The pressure of both gases is controlled by regular valves on the tanks. Various size tips, together with the controlled pressure of both gases, determine the size and shape of the arc and the intensity of the heat.

The thickness of the metal to be welded determines the size of tip to be used. The construction and types of nozzles vary. Some are straight, while others are bent. Manufacturers supply various sizes of tips. Charts are also supplied by manufacturers indicating the proper sizes of tips, welding rods, etc., to be used for welding various kinds of metals. The beginner is advised to start to learn welding with strips of wrought iron or steel about ⅛-inch thick. Welding rod of filler metal is not usually required for this light work. As proficiency is attained, thicker metals may be welded using filler metal. When welding, the torch is moved across the joint in semi-circular or zig-zag motion. This tends to avoid overheating and at the same time causes the metal to flow together. The edges of thin metal of ⅛-inch, or less, are made square preparatory to welding. Edges of thicker metal up to ⅜ inches in thickness should be levelled on one side at about 45 per cent to a depth of ¼-inch. Metal, which is ½-inch or more thicker, is preferably levelled on both sides.

Welding Rods — Generally speaking, the welding rod should be of the same kind of metal as the parts to be welded. When welding metals of two different kinds with different melting points, the welding rod should have a melting point equal to that of the lower.

Welding Cast Iron — Cast iron should be pre-heated to cherry red before welding. The pre-heated area should extend to a considerable distance on each side of the joint. As soon as the metal is cherry red, apply a little scaling powder. As soon as the metal begins to run, the cast iron welding rod should be applied.

Welding Steel — A mild steel welding rod is used for welding steel of more than ⅛-inch in thickness. The edges of the steel parts are heated until they become fused together. As soon as the metal begins to run, the welding rod is used to apply more metal. Flux is not required for low carbon steel. Flux or welding compound is required for the higher carbon steel to prevent oxidation.

As the art of welding has become more than simply melting or fusing metals together, it is advisable for the beginner to become more acquainted with the fundamentals of welding operations. Instruction from a local experienced, practical welder, or through some of the University Farm Short Course Schools during the winter months, will reward those who are especially anxious to become efficient.

Electric Arc Welding

Metal is heated and fused together by means of the heat liberated by an electric arc. The arc is formed between the end of a metallic or carbon rod and the metal parts to be welded. The metallic rod or electrode supplies additional metal to the parts being fused together, as well as being used to create the electric arc. Carbon rods are used to create the electric arc only, and additional metal is supplied by means of a separate welding rod similar in some respects to oxy-acetylene welding.

Electric arc welding outfits are of two main general types. These are classified according to the kind of electric current used, namely:

1. Alternating current.
2. Direct current.

The alternating current type uses an A.C. current from an electric power service line, and is equipped with a transformer for supplying electric current for welding. The direct current type consists of a direct current generator driven by either an electric motor or a gas engine. Its chief advantage is that it can be made portable for use in the field when it is powered with a gas engine.

The transformer type is considered most desirable for farm use where an electric power service is available. It is less expensive and has sufficient capacity to meet most of the farm requirements. The electric power line service must have sufficient voltage and a suitable size service transformer to equal the total load of all electric farm power and lighting.

A 250 ampere welder, together with the average farm electric light and power load, requires a service transformer with 5 to 10 K.V.A. capacity. If, in addition to the above, there are some of the large electric appliances, such as electric stove, refrigerator and hot water heater, then a 15 K.V.A. service transformer may be required. The ordinary light and small domestic electric appliances require a 1½ to 5 K.V.A. Before installing any large electric equipment, the company which furnishes the electric power supply should be consulted.

Capacity — The minimum amperage capacity of electric welders for average farm repair work is 200 amps. This capacity requires a 220 voltage service. Welding outfits of this size create a heavy demand on a rural electric power line, especially if used continuously for a considerable period of time. Some welders are now equipped with a power factor correction unit, which reduces this demand. Most farm electric supply lines are single phase, which consists of two lines. The three phase supply consists of three wires. The single phase line limits the load that can be carried without too much interference with lights and other electric farm equipment.

Masks, Goggles and Gloves—The electric rays from electric welding arcs are dangerous. Eyes and skin are especially susceptible to the irritating effects of electric rays. Special masks, goggles and gloves for welding requirements can be purchased. Never attempt to do any welding, even for a very short time without protecting the eyes, face, hands and arms.

Process of Electric Arc Welding— Metals are fused together by means of an electric arc. The electric arc is created by placing an electrode in contact with the metal and then moving it

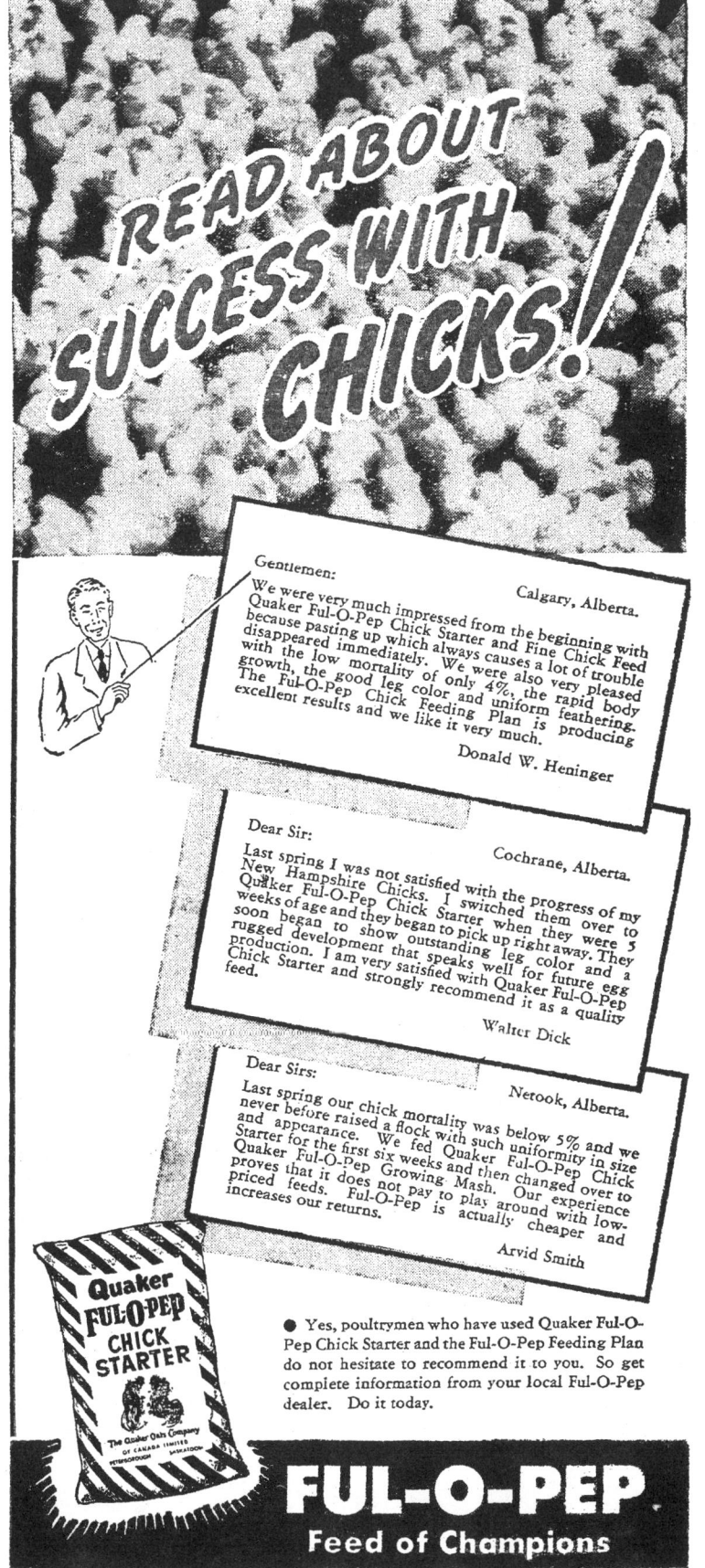

away from the metal a very short distance. The heat of the arc thus created varies from 6,300 to 7,200 degrees Fahrenheit. This intense heat causes metals to melt or become fused together. In actual practice one wire of an electric circuit is attached to the part to be welded by means of a clamp. The other wire is connected to a hand clamp, which holds an electrode or welding rod. When the electric current passes across a small gap between the end of the electrode or welding rod and the metal parts to be welded, the electric arc thus created melts the metals so that they run together and become fused.

The electric arc current required to fuse metals varies with the different kinds of metals and thicknesses. Expansion and contraction of different metals also vary. Uniting two different kinds of metal or two metals of different thicknesses becomes a more complicated problem that requires more skill.

Different Methods of Electric Welding—There are many methods of welding metals, depending largely on size or thickness, shape and kind of metal to be joined. Special welding machines are required for certain methods, as for instance "spot welding." The method most likely to be used on a farm can be classed as "butt welding." This method is used when two pieces of metal of the same thickness are to be joined together end to end. Similarly broken castings may be welded together. The parts are held in position by means of electrode clamps so that they may become fused together as the current passes through.

Metals Welded by the Electric Arc

Wrought iron, low carbon steel and steel castings can be welded with the electric arc. A great deal of the above metals are used for constructing farm equipment. Cast iron is more difficult to weld. Pre-heating is often necessary to prevent cracking by contraction in cooling. Brass, bronze and aluminum are more difficult to weld.

Welding Rods—Mild steel rods or electrodes are suitable for welding wrought iron, low and high carbon steel and occasionally malleable cast iron. Copper-aluminum-iron electrodes are most suitable for brazing cast iron and malleable iron.

The size of welding rods varies in diameter. They may be obtained from 1/16-inch up to 1/4-inch. The size of welding rod to use depends on the kind of work to be done and the amount of current used. Manufacturers of welding outfits and rods supply a chart or table which indicates size and kind of electrodes and the voltage and amperage of electric current for welding various metals.

Some welding rods are coated to prevent the air contacting the molten metal in order to effect a better weld.

Carbon electrodes are used for heavy, thick welding jobs where it is necessary to heat large surfaces and melt considerable metal and filling in large gaps in castings.

For Beginners—Any young farmer, with average mechanical ability and a desire to learn, can become efficient in electric arc welding for most farm needs. If a special course in welding can be obtainable by attending a farm mechanic short course as sponsored by some commercial companies, the beginner is more certain of obtaining more skill than if he attempted to acquire knowledge by himself. Accidents are more frequent where beginners attempt to acquire knowledge entirely by their own efforts. It requires considerable time to become familiar with the essential principles of welding. Knowledge and experience are best obtained systematically and in progressive steps. This means that the beginner should start learning with simple jobs of welding. More difficult jobs can be attempted as skill is obtained.

Grinder—Wood Block Frame

A block of hardwood is the special feature of this easily made bench grinder. The block is 6x4½x3 inches when planed and finished. It is cut to shape and two 11/16-inch holes bored for Model "T" Ford spindle bushings, as shown in the sketch. A 9 x 6 x ¾-inch hardwood base is glued and screwed on the bottom. Oil holes are drilled at the top after 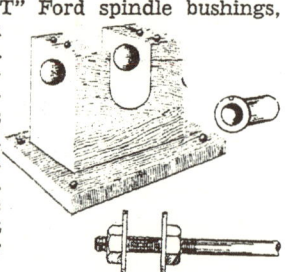 the bushings are set in the block. A slot in the flange of the bushing, together with a roundhead wood screw fastened to the block, will keep the bushings from turning. The shaft is made from an 11-inch length of ½-inch cold rolled shafting. About two inches of each end of the shaft is threaded and the emery wheels are kept tight by means of a washer and nut on each side of the wheel. End play is taken up by means of washers. The 2-inch "V" pulley is purchased at most hardware stores at very reasonable cost.—H.J.K.

Grinder with Self-aligning Bearings

A simple bench grinder for sharpening small tools can be made with easily replaceable self-aligning bearings  mounted on an iron "U" shaped bracket. The bracket should be made of 3x⅜-inch flat iron. A piece of old wagon tire will serve the purpose. An 11/16-in. hole is drilled close to the top of each bracket arm to fit a Model "T" Ford spindle bushing. Care should be taken to drill these holes exactly opposite each other. The bushings should not be tight in the hole so that they can be self-aligning. A small hole or slot in the flange of the bushing and a pin or screw in the bracket arm will prevent the bushing from turning in the bracket. The spindle is made from a piece of ½-inch steel shafting 12 inches long, about 2 inches of each end should be threaded. If a lefthand thread and nut for one end can be used so much the better, as then the grind stones can both be kept tight.

Old Piston for Grinder

Old pistons have long been useful to many home workshop enthusiasts in providing a ready made bearing and base assembly, such as shown in the sketch. One or two holes are drilled in the head so that it can be bolted to a base or bench. The side of the piston is cut away, as shown, to provide access for the drive belt to a 2 or 2½-inch "V" pulley. The drive shaft must be of the same diameter as the bearings. The ends are machined down to suit the size of  the holes in the emery wheels and threaded. One end should have a right-hand thread and the opposite end a lefthand thread. Hexagon steel nuts and washers hold the emery wheel on the shaft. A local machine shop or garage can machine a piece of round cold rolled steel to make the shaft at little cost, since it is a simple operation. Washers or spacers made of metal tubing or pipe are placed between the "V" pulley and the bearings to eliminate end play.

Pipe Fittings Grinder Stand

One-inch pipe fittings consisting of one "Tee," two short nipples and a flange are used to construct the stand. These are connected as shown in the sketch. One nipple is used to connect the "Tee" to the flange. The other nipple is cut in halves. Each half is screwed into the "Tee" to provide short housing for bronze or brass bushings used for the shaft bearings. Discarded brass or bronze bushings of suitable size, from other machines or automobile parts, can usually be found for such a purpose. One bushing is inserted in the end of 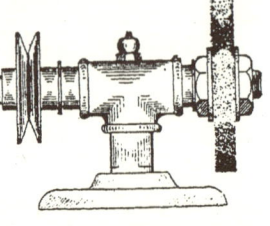 the halved nipples. A Zerk grease nip-

FORD TRUCKS BRING ECONOMY TO THE FARM

1-TON EXPRESS... 122-INCH WHEELBASE
Powered by the famous Ford V-8 engine with 4-speed transmission standard equipment. Tires, 7.50-17, 8-ply. Full-floating rear axle. Body sides are heavy-gauge steel with panels stamped outward for increased rigidity. Body measures 8 feet long by 4½ feet wide by 22 inches high to top of flareboards.

The economy of the Ford 1-ton truck has been proven over and over again on Canadian farms. Because of its ruggedness and stamina, it stays on the job and out of the repair shop. New advancements in its powerful Ford V-8 engine assure you of even greater saving in gas and oil.

Farmers and business men who look for high level performance with low operating economy have kept Ford first in truck sales across Canada year after year. Keep in touch with your dealer for up-to-date information on deliveries.

FORD AND MONARCH DIVISION
FORD MOTOR COMPANY OF CANADA, LIMITED

COMPLETE RANGE FROM ½-TON TO 3-TON IN VARIOUS WHEELBASES

MORE FORD TRUCKS SOLD IN CANADA THAN ANY OTHER MAKE

ple in fitted on the top of the "Tee" for lubricating the shaft. The shaft should be about ¾ inches in diameter or to fit the bushings. One end of the shaft is threaded and fitted with nuts and washers to hold the emery wheel. A 2½ or 3-inch "V" belt pulley is set-screwed on the opposite end with a wear washer between the hub of the pulley and the bearing. The pulley is adjusted on the shaft to take up any end play and is securely fastened with two safety set screws set at right angles to each other. —H.J.K.

Treadle for Emery Wheel

The old treadle power is better than hand crank power, since it leaves both hands free to do the work required. The handle of the crank may be removed. A piece of 1¼ x ¼-inch flat iron is fastened to crank with a bolt in place of the crank handle. 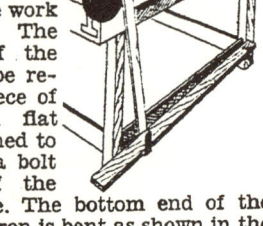 The bottom end of the piece of flat iron is bent as shown in the sketch. A length of 1x3-inch hardwood is used for the treadle.

Anvil Bench

A solid block of wood about 20 inches high and about 12x16 inches square makes the most suitable support for a blacksmith's anvil. Such large pieces of timber are not readily obtainable in treeless areas. A solid bench can be built up with 2x6-inch lumber for legs and cross braces. A piece of 2x12-inch plank, with corners cut out to fit the legs, is used for the top. Pockets and racks to hold 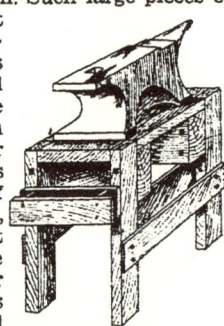 tongs and other tools are provided as shown in the sketch. The height of the bench should be made to suit the user. Usually it is from 20 to 24 inches high. The dimensions of the top are 12x20 inches.

Saw Gauge

A handy gauge for a cross cut saw can be quickly and easily made with two short pieces of strap iron and a piece of light angle iron about 15 inches

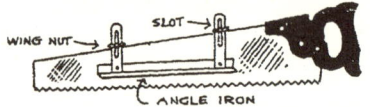

long. Two small holes are drilled through the saw near the back edge. Small stove bolts with wing nuts hold it in position. The slots in the strap iron provide room for adjustment to any depth.—Paul Tremblay, St. Paul, Alta.

Circular Saw Set

I made this circular saw set from an old Model T Ford iron monkey wrench. 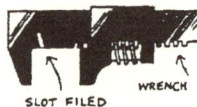 A slot is filed between the jaws to fit over the saw tooth and the handle pressed down until the inner jaw hits against the blade of the saw. As this jaw is adjustable, any set can be put on the saw. The end of the wrench may have to be heated and tempered to make it tougher and prevent the slot from spreading. I have used this set for 15 years and find it works well.—Harry Cooper, Glenora, Man.

Workbench and Wood Vise

WITH these drawings before him the farmer will have no difficulty in framing a workbench with a wood vise attached. The idea followed in constructing the wood vise is to have the outer jaw always perpendicular when gripping the work. The frame arrange-

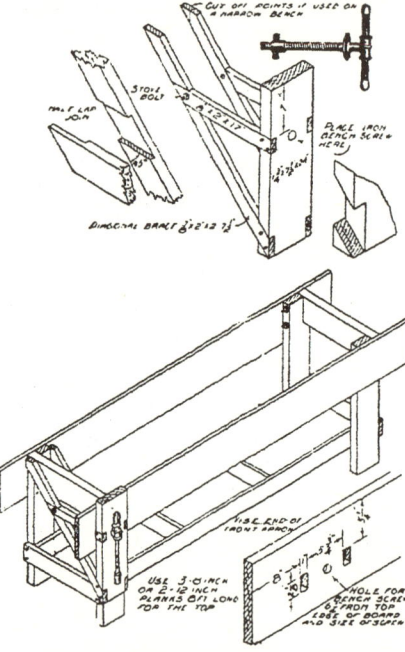

ment slides along under the bench top to accomplish this object. The iron screw normally is available in any good hardware store.

Post Hangers and Drop Hangers

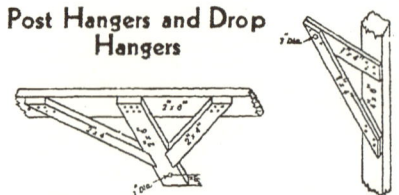

In putting a line shaft in the workshop you may need one or both of these two types of hangers. On the left is a drop hanger, which is fastened to an overhead joist. The main pieces are 2x6, while the braces are of 2x4 and are let in to the main piece. The hole is the size of the shaft and is three inches from the end.

The post hanger is attached, like a bracket, to a stud. Two pieces of 1x4 will do with the hole bored through them where they lap.

Converted Breast Drill

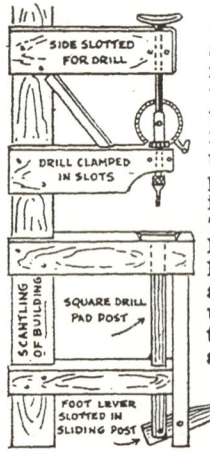

This is a light drill which I constructed and used many years ago. It will be noted that the drill is fixed rigidly in place and that the work is pushed up against it by the foot lever. The pressure is in line with the drilled hole. In drilling, say a ½-inch hole, first use a ¼-inch drill then a ⅜-inch one and finish with one of the desired size.—John H. Foreman, Arden, Man.

Miter Box is Easy to Make

On jobs of sawing small-sized lumber, molding, etc., where diagonal ends are desired, the best and quickest way to make accurate cuts is with a miter box. 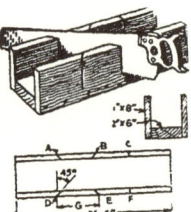 The figure shows a simple box that you can make in a short time, from a piece of 1x8 inch lumber 5 feet long cut in the centre to form the sides and a piece of 2x6 inch lumber 30 inches long to form the base. Assemble the box with wood screws, measure for cuts as shown and follow the marks carefully when sawing.

Handy Shop Drill

An ordinary hand brace can be made to save a lot of labor in drilling holes in iron as shown. The easiest way to make the shallow holes to hold the round brace head is to burn them with a red-

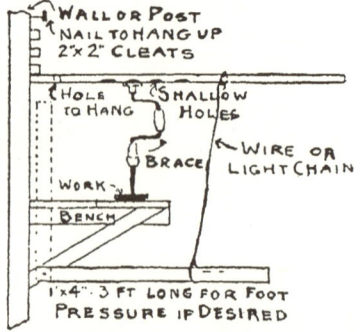

hot iron door knob or large pipe cap. A good sized clamp is desirable to hold metal pieces for drilling.—I.W.D.

Combination Farm Workshop and Garage

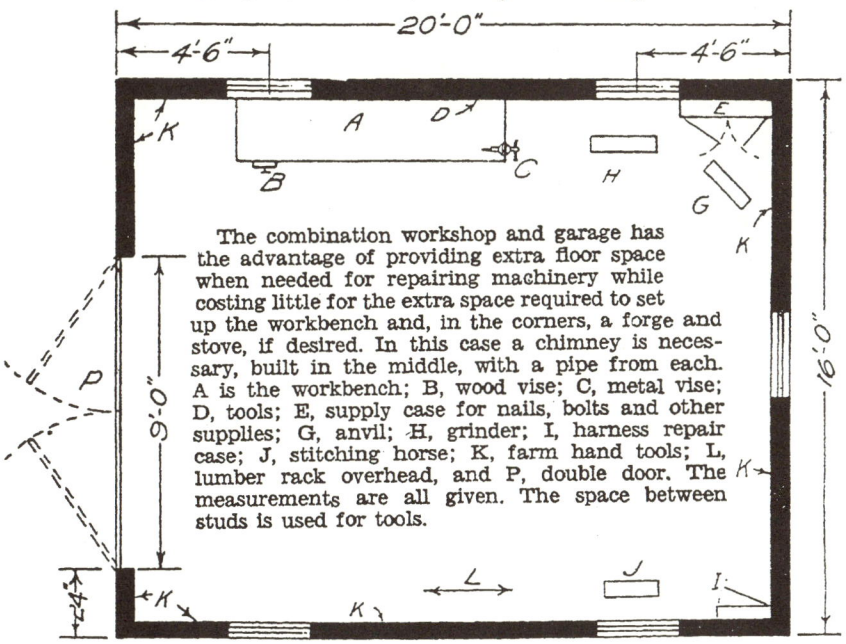

The combination workshop and garage has the advantage of providing extra floor space when needed for repairing machinery while costing little for the extra space required to set up the workbench and, in the corners, a forge and stove, if desired. In this case a chimney is necessary, built in the middle, with a pipe from each. A is the workbench; B, wood vise; C, metal vise; D, tools; E, supply case for nails, bolts and other supplies; G, anvil; H, grinder; I, harness repair case; J, stitching horse; K, farm hand tools; L, lumber rack overhead, and P, double door. The measurements are all given. The space between studs is used for tools.

Cheap Press Drill

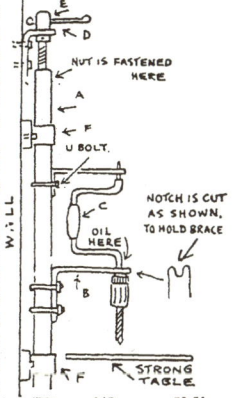

All this press drill cost me was a thorough search of the junk pile and some careful work. The parts required are: A 3-foot length of 1½-inch pipe (A); a brace with the head removed (C); two pieces of heavy steel (B and D), which can be cut from the corners of an old plow frame; a screw and nut (E) with a sliding handle; and two guides (FF) to hold the pipe and allow it to slide up and down freely. The unit is bolted to a stout stud. The steel bar (D) must be made very secure.

Handy Sheet Metal Bender

This handy sheet metal clamp will be found very convenient for making square bends in sheet metal for use on self feeders, roofing, gutters, etc. It is simply one two-inch angle iron, three to four feet long fastened on the edge of a somewhat longer 2x8 by heavy flat headed screws placed both in the top and edge of the board. A second two-inch angle operates on the first one by using two ½x3½ inch machine bolts with the heads below the board and covered with hardwood blocks with square depressions to keep the bolts from turning. The metal is placed as at B projecting to the line of the desired bend, and the nuts drawn down tight. The bend can then be easily made with the hammer or mallet.—I.W.D.

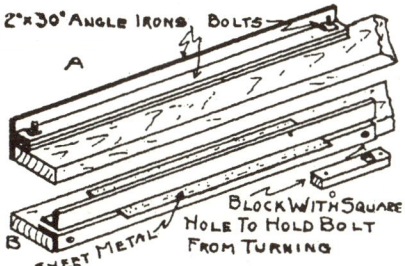

High Speed Drill

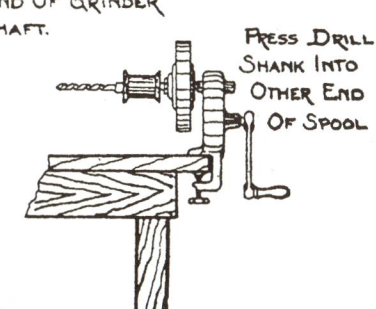

This shows how to convert an ordinary hand grinder into a high speed wood drill by using a common thread spool. Square one end of the spool hole by pressing the bit shank into it a short distance. Turn the opposite end on to the protruding threaded grinder shaft. This is especially useful to model makers and shops without electric power. While it would not be suitable for heavy work, it does give the high speed necessary for fine, smooth, accurate jobs.

All progressive hatchery men use

"Double Duty Egg Shell Maker"

2 PRODUCTS IN 1

1. Is the better NATURAL EGG SHELL MAKER WITH LESS WASTE
2. A better FOOD GRINDER

"Double Duty Egg Shell Maker"

Contains 96% Calcium Carbonate; all the birds need, and is hard enough for grinding purposes (31,000 lbs. crushing strength).

stays in the gizzard
—No Waste—
Economical to use

PRICE $1.25 PER 100 lbs.

Double Duty Products
including

"BAKN-MAKR"

for profitable pigs, are backed by 20 years of satisfactory service

Write for full information to:

Sewing Horse

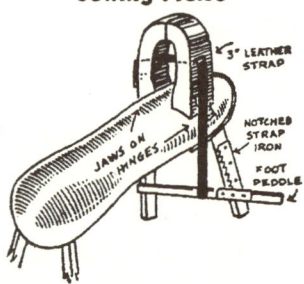

This sewing horse will hold anything from a thin strap to a horse collar and it comes in very handy when getting the harness ready for work. The jaws are fastened on the block with hinges for a wide opening. A 3-inch strap is fastened to one jaw and passed through the other and down to the foot pedal which pulls the jaws together when pressed down. A notched piece of iron is fastened to the leg to hold the pedal. The diagram shows the seat made of a piece of log but it can also be made from a piece of good planking.

Clamp for Thin-Wall Pipes

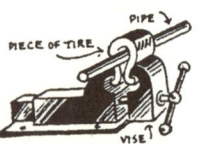

A clamp to be used with a vice for holding sheet metal pipes can be easily made by cutting a section out of an old tire and gripping it in the vise as shown. It will hold the delicate pipe without bending it out of shape.—A.F., B.C.

Stitching Horse

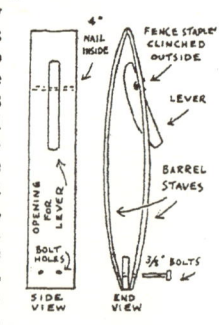

This handy stitching device is made from two barrel staves. The two ⅜-inch bolts at the bottom should be tightened until the staves fit tight on the 1x4 between, giving pressure on the jaws. In one stave there is a slot into which a lever is hinged. A 4-inch nail is fastened on with staples, forming a pivot. The lever spreads the jaws for inserting the work.

Stitching Clamp for Harness

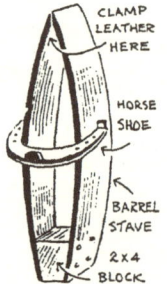

A stitching clamp for the repair of harness can be made from two barrel staves. When using the clamp hold it between the knees and your shoes. You will find that it is worth while to make such a holder even for a few jobs in harness sewing.—W. Kalbfleisch.

Saw Vise

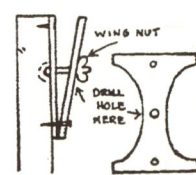

A simple but efficient saw vise can be made from the bridges of an old stove. Drill a hole a little above the centre in both sections. Fasten one to the post with a screw nail through the top hole, first inserting a bolt, the head of which is recessed in the post. Now put the second section on the bolt and fasten with a wing nut. Then fasten the two pieces to the post at the bottom with a long screw with a washer between. This brings the tops together tightly on the saw when the vise is tightened.—R. C. Willett, Cochin, Sask.

Clamp For Carpenter Work

For this clamp use two pieces of 2x4 about five or six feet long. Bore holes four inches apart the full length of both pieces. Make four bolt hooks of half-inch iron, two of which are made just long enough to fasten through the holes in one bar, the other about eight

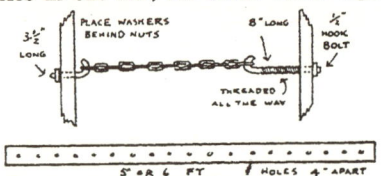

inches long and threaded most of their length. To connect the bars use six-foot trace chains, which can easily be adjusted to the depth of the work; but any kind can be used, even pieces of wire. Use washers under the nuts. Hardwood bars are best if you have them.

For Smooth Jaw Vise

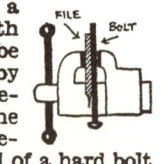

When the jaws of a vise become worn smooth a substantial hold may be secured on the work by inserting an old file between the work and the jaw. This is especially useful when burring the end of a hard bolt.

A Handy Soldering Vise

A pair of magnets will serve as a handy vise for holding small articles for

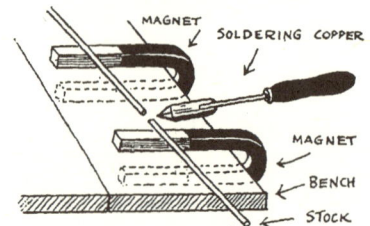

soldering. They will hold the articles to be soldered without absorbing the heat as is the case when using a heavy vise. They also prevent burning the workbench when soldering is done directly on the wood. They will hold only steel or iron.—A. S. Wurz, jr., Rockyford, Alberta.

Convenient Home-made Anvil

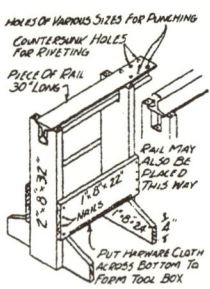

A short piece of railroad or street car rail, which can be bought very cheaply from section foremen or junk dealers, makes a very satisfactory farm anvil when mounted as shown. It is fine for straightening, flattening, bending either square or rounded shapes, punching, riveting, knocking off and replacing mower sections, etc. The method of mounting shown makes an outfit which can be moved from place to place, or it can be mounted on a concrete base if preferred. The framework will be very much more rigid if the joints are brushed with waterproof glue before being put together.

Knife Rack

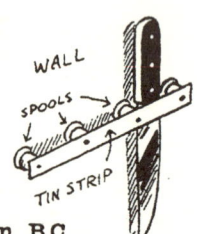

A knife rack can be made with four or five spools and a strip of tin a foot or so long. The nails hold the tin against the spools and the spools against the wall.—Leonard A. Atrill, Seaton Station, B.C.

Vise from Auto Jack

A vise for holding boards while planing and for general woodwork can be improvised from pieces of lumber and an old screw type auto jack. The jaws have a spacer between them at the bottom and the outer one is hinged as shown. The screw is received at the back by the sprocket, which has been squared to form a nut and is kept from turning by passing a piece of strap iron and nailing it down at both sides. A coiled spring pushes the jaws open. For fine work the strips of old rubber tire can be put on the jaws or short pieces of angle iron can be slipped over the ends when metal objects are being held. A

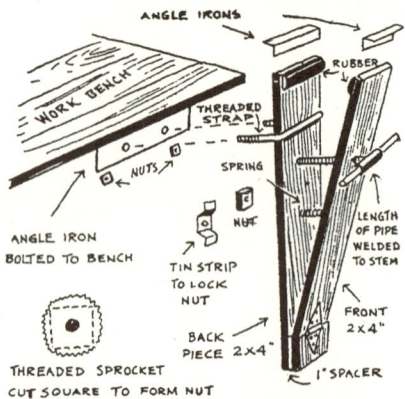

short piece of pipe is welded to the stem to receive the handle.

Handy Open-top Sawhorse

No one can do efficient repair or construction work without at least two good sawhorses. Open-top sawhorses can easily be made out of scrap material around every shop, are light and easily handled, and are very convenient for the many repair jobs around the farm. The chief advantages of the open top are for ripping short boards and for holding a stick

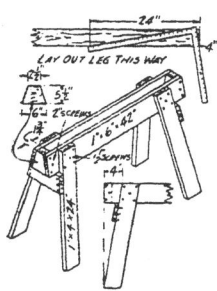

or a handle in a vertical position where a vise is not available. The sawhorse will be at least twice as strong and rigid if all joints are brushed with waterproof glue (which can be bought in any hardware store) just before they are put together; and will last much longer if given two coats of good paint. If intended for use with heavy timbers and poles, the legs should be 2x4's and the cross braces of 1x8-inch material.

Portable Workbench

On the farm there is a lot of repair work that cannot be taken indoors to the workbench. The sketch shows a simple way of taking the workbench to the job. This portable workbench is sturdily constructed of old pieces of lumber and two old plow wheels or

other similar wheels that may be available. A box is built into one end to hold the necessary tools while a drawer to hold bolts, nails, screws, etc., may be placed at the other end if desired. The side rails extend at the end opposite the wheels to serve as handles when the bench is being moved. The whole structure should be rigidly built and the wheels must fit tightly to secure a firm surface.—J.A.S.

Swinging Bench Drawer

This style of a bench drawer, first published in a mechanics magazine, has many advantages. It can be swung completely out where the contents can be gotten at without difficulty and it is not in danger of dropping, as a drawer is when it is pulled too far

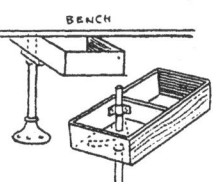

out. The support of the bench is part of an old axle housing. The hole in the bottom of the drawer must fit the support snugly and should be reinforced by an extra ply of lumber screwed on along with glue. The upper drawer support can be made of a piece of 2-inch stuff with the proper size of a hole bored through it and also fastened in place with screw nails and glue.

Re-shaping Grindstone

If the grindstone has worn out of shape or has a piece broken out of it, take two pieces of 2x4 about 18 inches long, nail them together at one end with a piece of soft lumber leaving clearance enough for the grindstone to turn freely between them. Bore two small holes exactly opposite each

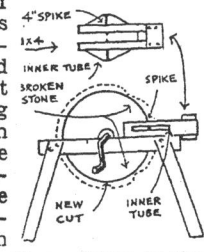

other through the 2x4's as shown in the sketch so that they will be inside the broken part of the stone. The holes should be just large enough so that a 4-inch spike will pass through them freely. Then nail or wire the device on the frame with a 2x4 on each side of the stone. Put a 4-inch spike in each hole and over them tack pieces of inner tubing to give pressure. Start turning the grindstone without using water and soon the nails will shear off the stone to a true circle.—Daniel Harris, Edgeworth, Sask.

Shelf for Small Parts

Lengths of eave trough, nailed between the studs of the garage or work shop provide convenient shelves for bolts, nuts or small parts. With the curved side of the trough outward it is easy to pick out

the parts wanted.—Edwin Unger, Mayfair, Sask.

Empty Oil Can Cabinet

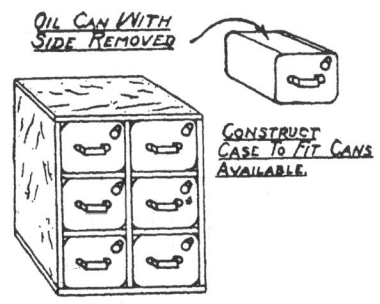

The diagram shows how to make a convenient cabinet out of empty oil cans for storing nails, screws, bolts, and other small articles about the shop and household. The top side is cut out and the cut edges turned in and hammered

---

FARM ANIMALS USE UP SALT

JUST AS YOUR TRACTOR USES UP OIL

You wouldn't think of running a car or tractor that was short of oil...

But do you realize that all farm animals USE UP SALT just as farm machinery uses up oil? And that an animal's *salt balance* can only be maintained by a *daily salt* ration?

Cows put one ounce of salt in every 3 gallons of milk. Horses lose one ounce of salt in every 14 pounds of sweat. Sheep put salt into their fleece and hens put salt into their eggs. In addition all animals need salt to build bone and muscle and retain stamina and vigour.

No animal, unless it is salt starved, will take more salt than it actually requires. In addition, therefore, to a basic daily ration of one pound of "Windsor" Iodized Stock Salt with each 100 pounds of feed, keep "Windsor" Salt Blocks handy so each animal can fill out its own salt requirement.

Plenty of salt means healthier, more productive farm stock.

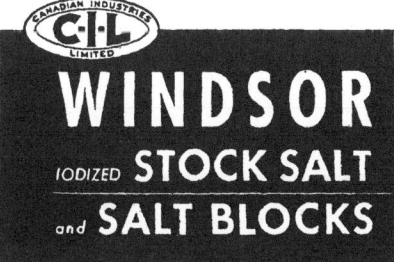

Ta·pat·co
IMPLEMENT and TRACTOR SEAT CUSHIONS

Made by the Makers of

**WALKER
—Trainmen—
OVERALLS**

SANFORIZED - SHRUNK and REGULAR FINISH
DENIMS and DRILLS

Ask Your Dealer
THE AMERICAN PAD & TEXTILE CO., Chatham, Ont.

down flat to avoid danger of cutting the hands. The handles on the cans are used for pulls.

Easily Made Tool Rack

The diagram shows a handy rack on the work bench. It is made from two boards four inches wide and four feet long, nailed to three 1x2-inch spacers as shown, with holes of different sizes bored through the top board. This rack keeps the tools off the bench and available at all times. Drawers to hold small materials could be arranged under the lower board, or better still they could be put into small screw top glass jars,

with the caps fastened with nails or screws to the under side of the lower board.—I. W. Dickerson.

Bench Catch-all

Here is a diagram of a handy catch-all under the work bench into which to brush sawdust, shavings, chips, and so

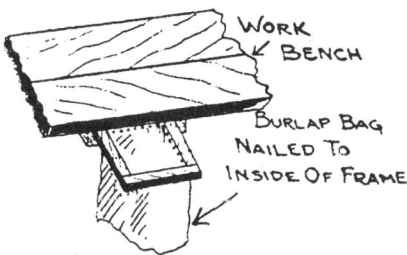

on. It is simply a frame made of three-eighths inch material tacked and glued together, with an old sack fastened on the inside. This slides in and out on cleats.—I.W.D.

Sharpening Twist Drills

Straighten bent drills by tapping lightly with a hammer on a straight surface, as a bent drill does not bear equally on the lips and is likely to break.

Grind the two cutting lips of the same length and at an angle of 59 degrees, as shown in the diagram. Keep a cup of water handy and dip the drill frequently to prevent overheating and destroying the temper. The blade of the T-bevel set at five inches and three inches on the square as shown gives the proper angle of 59 degrees. If this is laid off on heavy sheet metal and a templet cut out, it will help the beginner very much in getting the proper shape to the drill point. Check one lip for shape and mark the centre point, and see that the other lip has the same angle and that the centre is on the

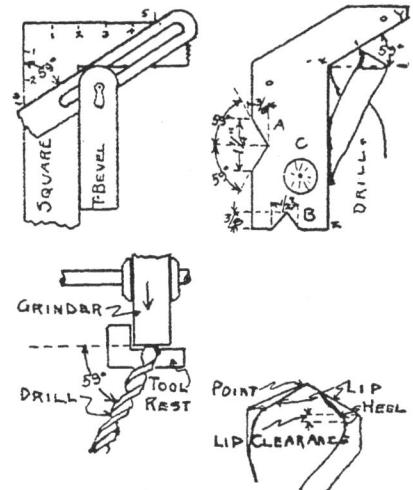

same mark. Clamping the templet to the tool rest with the edge XY parallel to the grinding surface and holding the drill along the line XZ will give the proper lip angle. The ⅜ by 1¼ inch notch A on the side of the templet also gives the exact shape of the drill point. The ⅜ by ½ inch notch B gives the angle of 70 degrees for grinding a cold chisel for ordinary use.

Proper lip clearance, or the angle between the cutting lip and heel, is also very important to the proper operation of the drill. This should be about 10 or 12 degrees, but can best be checked by setting the drill point in the drilled countersink, and seeing that only the two cutting lips touch the templet, while the heels both clear slightly. The lip clearance angle should be rounded rather than straight, and is secured by holding the cutting lip to the grinder and then turning it slightly to the right.

Homemade Forge

This homemade forge is recommended by the North Dakota Extension Service. A wooden bench with a two-inch top is built. A bull wheel from an old binder

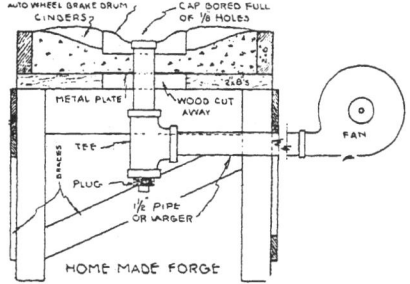

with the spokes cut out, makes a good outer rim. Two-inch pipe would be none too large. A wooden fan speeded up with a wheel and belt can be put together. The illustration gives all the details that are necessary.

Screw-driver from Bit Shank

Most farm work shops have a wood bit which is done for. A good screw-driver can be made  by cutting off the auger part, leaving only the round shank which can be ground to the proper shape with an emery. A screw-driver that is driven with a brace does its work much easier than the ordinary screw-driver. — Norman Yates, Grenfell, Sask.

Sharpening Tiller Discs

This is how I built a very satisfactory device for sharpening my tiller discs. Do not place the pivot (disc bearing) directly behind the emery wheel, but about two inches toward the most convenient side. For mounting use a 2x6 or 2x8 and on this use a piece of 2x6 which is fastened to the mount by two hinges placed side by side. To this 2x6 an old ball race is securely fastened, and by means of a nut or collar to centre the disc, a concave washer and a tightener bolt and nut, the whole assembly is put together so that the weight falls on the grinder. This pressure may be regulated by moving the disc pivot closer to or farther from the hinges. I find that about six inches from the back of the hinges to the pivot about right for a 22-inch disc.

Put on the disc and tilt by means of the wedges until the edge lies flat on the centre of the grinder, and adjust until about half an inch is brightened when the disc is sharp. The grinding wheel will keep the disc revolving and there is no danger of it becoming over heated. The hinges allow an even pressure all around. When the edge of the disc is ground down to about the thickness of the back of a thin table knife take disc off pivot, turn it over and finish from the inside by hand.

Chisel for Cutting Drums

To make a chisel that will cut oil drums, take an old 12-inch file and shape it as shown. Make it a little thicker at the end than elsewhere to give it clearance. Heat to a dull red and cool in old oil. This chisel will be found handy for cutting any thick sheet iron.—H. Fuller, McCreary, Man.

Simple Bench Vise

It is usual to nail a bit of inch stuff with a V-shaped notch in it on the work bench to hold stock while planing it. This device has that one beaten all hollow. For holding stuff while it is being planed or worked down with the

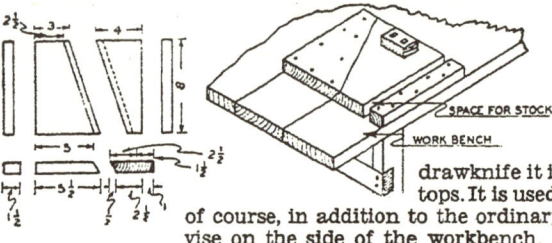

drawknife it is tops. It is used, of course, in addition to the ordinary vise on the side of the workbench.

Wooden Hack Saw Frame

This frame is best made of ply wood as there is less danger of splitting. Two or three pieces of thin wood can be glued together and the same result

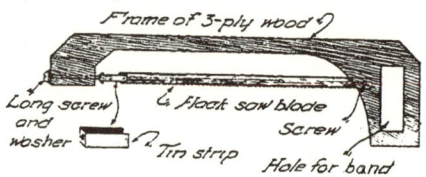

achieved. A long screw or stove bolt is used to put the tension on the blade. The proper length of the handle is worked out from the length of the blade. Be sure and have the clearance great enough to take any work which you may have to cut.—Dale Van Horn.

How to Sharpen Auger Bits

Nearly all farmers have auger bits for boring in wood, but few know how to sharpen them so they will do good work. An auger bit has three working parts—the lips which cut the bottom of the hole; the spurs, which cut the sides of the hole; and the screw or threaded point, which forces the bit into the wood.

An auger bit file is the most convenient for sharpening, but a small taper file works very well. The lip is sharpened by resting the point on a board and then filing only on the upper sur-

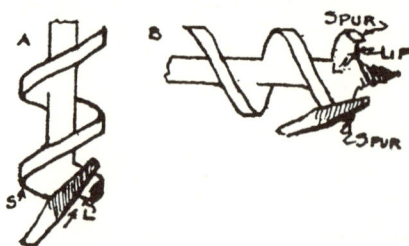

face of the lip. The spurs are sharpened by filing them on the inside only. Any filing on the outside will make the bit cut too small a hole and cause the twist to bind. If the screw is rusted or badly worn, the bit will work better if the threads are cleaned out and sharpened with a small taper file.

Hand Box for Tools

In the old days carpenters used a hand basket for carrying the tools they needed on the job. A box is handier and can be made from stuff picked up around the place. The sides are best of thin material. For carpenters' tools it

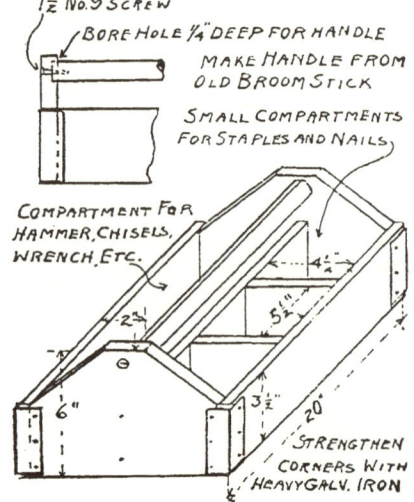

is best made long enough to take a saw though this is not absolutely necessary. The box shown is 20 inches over all, and has corners reinforced with sheet iron. The sides are 3½ inches deep and the ends at the highest point six inches. The width is about 10 inches. One side is divided into compartments for nails and staples.—I.W.D.

Planes and Chisels

A smoothing plane bit should be at right angles to its length, but rounded about 1-64 inch at the corners to prevent ridges on the planed surface, as shown. Next set the T-bevel at an angle

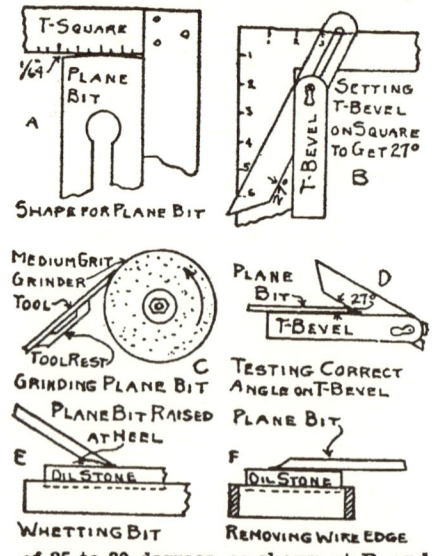

of 25 to 30 degrees, as shown at B, and grind the bit on a medium grit-grinding wheel. Test the angle with the T-bevel D until a slight wire edge can be felt.

Dip the edge in water occasionally, as overheating as shown by a bluish color will ruin the temper. The bit should then be whetted on a good oil stone, lifting it slightly at the heel E so it touches for about 1-32 back from the edge. The bit should then be rubbed flat on the stone F to remove any remaining wire edge. A plane bit should be whetted several times before it needs regrinding, and less sharpening will be needed if a scraper and brush are first used on boards to remove grit.

Chisels should be sharpened in exactly the same way as plane bits, except that the corners are not rounded.

Homemade Cold Chisel

A cold chisel that may be used on rough work can be made from a discarded magneto. Heat the magneto in a forge and straighten it. Then reheat and bring one end to a blunt taper. Bury the iron in wood ashes so that it will cool slowly and then grind the tip to the regular chisel shape.

Board Divided Into Equal Spaces

Suppose you have a board five and one-half inches wide that you want to rip into equal strips of equal width. Of course, you can figure out the width of the strips mathematically, but there's a much easier way of doing the job. Just lay a ruler across the board at an angle, with the figure eight at one edge and the figure one at the other. The inch marks automatically divide the space into eight equal parts. Better keep this dividing trick in mind as you'll be sure to need it sooner or later in your shop work.

Fitting and Filing Farm Saws

To file a saw properly you must first have in mind the proper shape to which the teeth should be filed. The following instructions are taken in part, from a Cornell bulletin.

The first operation is not filing but jointing. To joint a handsaw or rip saw put it firmly in the saw clamp with the teeth up. Then take a small piece of three-quarter inch board, saw six inches long and two inches wide, and plane one edge until it is exactly at right angles to the side. Then take a flat file and hold it firmly on this edge. Hold the block against the side of the saw, with a file projecting out to cover the teeth and rub back and forward until they are brought to the same height. Do not file more than is absolutely necessary. It may take more than one jointing and filing to bring a badly used saw back into good condition.

The next operation is setting. For properly setting a saw, one of the sawsets on the market should be used. Just set enough to give the blade nice clearance. It will require more set if it is to be used on wet or green lumber.

Fig. 1 shows the proper shape of the teeth of a ripsaw. At the bottom is shown the set, which is here slightly exaggerated. The front or cutting faces of the teeth are at right angles to a line along the points of the teeth. This may be tested with a square as shown.

The proper position of the three-cornered file is also shown. The file is

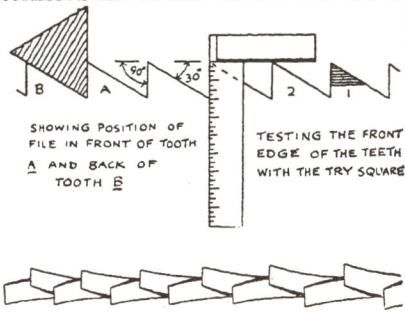

Fig. 1. Ripsaw teeth, showing angles and cutting edges.

held level and straight across, at right angles to the saw blade. The file is pressed down into the gullet and files the front of one tooth and the back of the other at the same time. Be careful to put the pressure on so as to keep the points of the teeth evenly spaced. The teeth should all be the same size and shape. File all the teeth from the same side and work from the handle toward the small end of the saw. Each tooth is brought to a sharp edge as you go along; that is, the final touches are applied to the front side of each tooth.

Some file every other tooth from one side of the saw and then turn the saw around in the clamp and finish the rest from the other side. This has to be done with a handsaw, but is not necessary with a ripsaw.

Fitting a Handsaw

In fitting a handsaw, as it is generally called, though it is a crosscut handsaw,

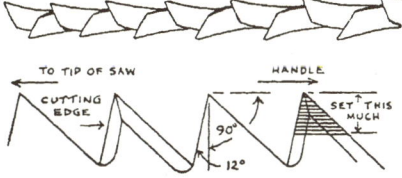

Fig. 2. Crosscut handsaw teeth, showing angles and cutting points.

it is first jointed and then set. The teeth are filed so that each alternate one cuts the grain of the wood on one side of the saw kerf, and the other one on the opposite side. They have therefore to be filed at a combination of

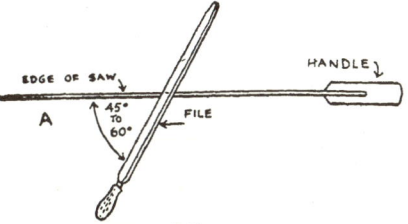

Fig. 3. Position of file across a crosscut saw.

angles. Fig. 2 shows these angles. Note that the handsaw tooth hasn't a cut-

DOUBLE CHECK
AGAINST ROOFING TROUBLE

● "Storm King" Roofing laps into place ... not just a few inches of lap, but a full 19" lap over the roll below. This means a DOUBLE THICKNESS of roofing over the entire surface and a triple thickness along the edges. And that is only one feature of this patented "Storm King" Roofing. In addition it provides:

Completely concealed nailing ... no trouble from nails rusting or pulling loose.

All laps cemented down ... no buckling or open laps to catch wind, rain and snow.

Fire-safe ... mineral surface ... in Red, Green or Black.

Easy to apply and low in cost.

Better insulation against heat, cold and weather.

Ask your dealer for

STORM KING*
*TRADE MARK REG'D

ROLL ROOFING
THE BARRETT COMPANY LIMITED
MONTREAL TORONTO WINNIPEG VANCOUVER

ting edge, like a ripsaw, but a cutting point. Study closely the shape of the teeth in Fig. 2 and get a general idea of the angles.

The angle at which the file is held across the saw blade is shown in Fig. 3. It is also held with the point higher than the handle. Half the teeth are filed from one side of the saw and the other half from the other. It is when doing the last half that each tooth is brought to the fine cutting point. Be careful as you go along to keep all the teeth as near as possible the same size. This takes care and practice.

If your saw has become badly out of condition with teeth of various sizes and gullets of various depth, Fig. 4 gives some hints on how to make the corrections. Of the four teeth shown, tooth No. 1 has just been touched up and is the right shape. Tooth No. 2 has been very much longer and much of the tip has been filed away in jointing, leaving a large, flat surface. Tooth No. 3, due to poor filing, has been left smaller than the others and No. 4 is larger than any of the others.

Fig. 4. Correcting the shape of the teeth of a poorly filed saw after jointing.

To file the teeth properly, No. 1 is left as it is, No. 2 is brought to a point by filing against the front edge only; the back of No. 3 is filed with the same strokes of the file as used for the front of No. 2; and No. 4 is brought to a point by filing the front edge only. The teeth are thus brought to the same size and shape.

Circular Saws

In fitting a circular saw the first thing to do is to true it up if it is out of round. This is done by holding a piece of emery or grindstone or a flat file fastened to a board, squarely across the points as the saw is rotated by hand. Rotate the saw backward. This operation is carried on until the saw is not

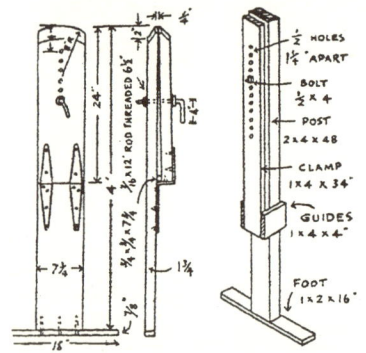

Fig. 5. Two types of circular saw clamps.

only round but also the points of the teeth are all the same distance from the centre, so that each tooth will do its work.

The saw may be set by using a standard circular set or a hammer-and-anvil set. For the latter a piece of railroad iron may be used. From the square end of the iron, file a bevel which runs back one-quarter of an inch, and is 1-32nd of an inch deep at the end. The setting is done by two men, one holding the saw and the other using a heavy hammer. A piece of cardboard or thin piece of wood is placed on the rail under the saw opposite the hammer to keep the edges of saw tooth from coming in contact with the steel rail. The tooth is struck firmly, with the face of the hammer parallel to the bevelled surface.

A simple gauge, to measure the amount of set, is easily made by taking a small piece of hardwood and putting four small screw nails in it. By placing this on a flat surface one of the screw nails is adjusted to clear between a 64th and a 32nd of an inch. Then this can be used on the saw to gauge the amount of set on each tooth.

A circular saw clamp is necessary for filing and Fig. 5 shows two types that can be made. They are simple and need no explanation. When filing they can be leaned against the workbench.

Fig. 6 shows one type of cordwood saw tooth. The front edge of the tooth is in line with the centre of the saw. The teeth are filed on a bevel and the bevel recommended is 105 degrees to the side of the saw. This bevel can be

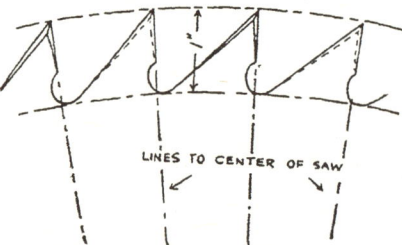

Fig. 6. Gullet type of cordwood circular saw

found by placing a T-bevel on a steel square, Fig. 7, at 1¾ inches on the tongue and six inches on the blade. It is then applied to the tooth of the saw as shown in Fig. 8.

Both the front and the back of the tooth are filed from the same side with a flat file, using long, light, even strokes. When every other tooth of the saw has been filed from one side, the saw is reversed in the clamp and the other teeth filed the same way. For properly spacing the teeth of a saw that have been badly

Fig. 7. Getting the bevel.

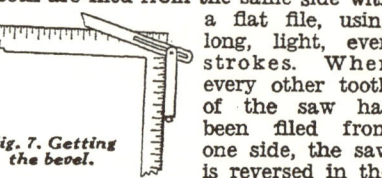

Fig. 8. Applying the T-bevel to cordwood saw tooth.

filed the same principles are used as have been described for conditioning handsaws.

Circular Ripsaws

A gullet-toothed circular ripsaw is shown in Fig. 9. It is jointed and set according to the directions already given for cordwood saws. In gumming, the gullet is ground toward a tooth seventh or eighth beyond the one being ground as shown by the dotted line at the top. Another type of ripsaw tooth is shown in Fig. 10.

In filing a circular rip saw the fronts of the teeth are filed straight across the saw at an angle as shown by the lines A, B, C and D, which are tangent to a circle drawn half way between

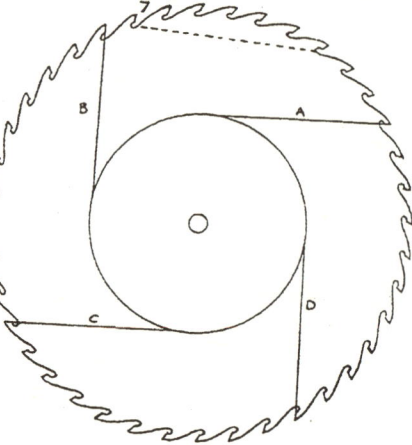

Fig. 9. Circular ripsaw of gullet type.

the centre and the rim. Each tooth is bevelled slightly on the back.

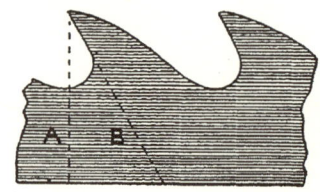

Fig. 10. Common type of ripsaw tooth.

Take one precaution. Never let the edge of the file touch the gullet. Notches interfere with the discharge of the sawdust by the saw.

A Lead Hammer

This lead hammer is very useful for pounding the threaded end of a bolt without damaging it. Take a piece of ½-inch round iron rod about 10 inches

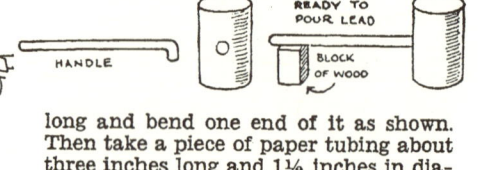

long and bend one end of it as shown. Then take a piece of paper tubing about three inches long and 1½ inches in diameter. Cut a hole on the side of it to take the handle and put the handle in place. Then fill up with melted lead; when it is cool strip off the paper and there you have your hammer.—Alex. Wilson, Star City, Sask.

CHART FOR CALCULATING PULLEY SIZES AND BELT SPEED

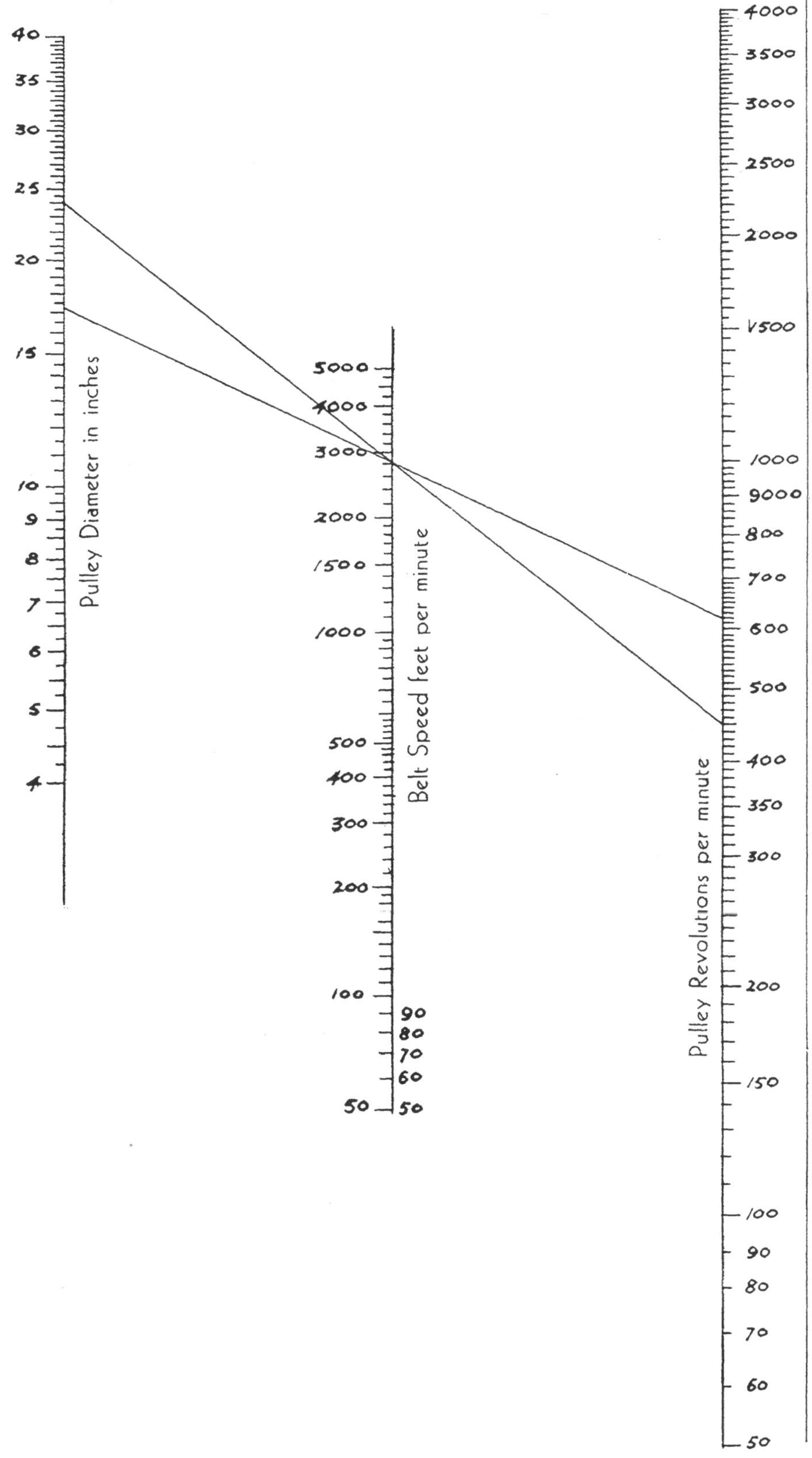

SECTION 2.

Engines, Implements, Heavy Equipment

The Farm Engine

The farm is using more and more engines in mechanizing various farm operations. The stationary engine in pumping water, cleaning grain, running line shaft in shop, generating electricity, compressing air, and running the washing machine; the tractor, combine, automobile, truck, and hay baler engines as well as the four cylinder engine skid mounted for wood sawing, feed grinding, and general utility work are some of the engines used on the farm.

Types of Engines:

The farm engines can be classified as four stroke cycle (4 cycle) engines except for some of the two stroke cycle engines on washing machines.

The farm engines are mostly gasoline engines. Some of the tractor engines are Diesel engines.

The four stroke gasoline engine requires four strokes of the piston to complete the cycle of operation. The four strokes require two revolutions of the crankshaft. The piston travels from head centre to crank centre during the suction stroke, from crank to head centre during the compression stroke with ignition occurring along the end of the compression stroke 15° to 20° before head centre, from head centre to crank centre during the power stroke, and from crank to head centre during the exhaust stroke.

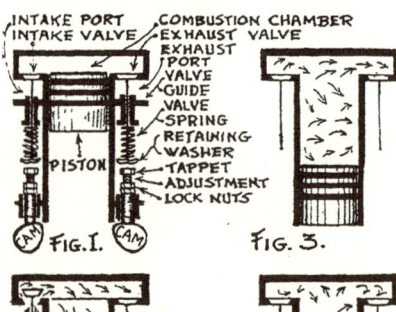

Suction Stroke: (Fig. 2)

On the suction stroke air is forced through the air strainer, carburetor, manifold and into the engine by the difference in pressure between atmospheric pressure, 14.7 pounds per square inch, and the partial vacuum produced by the piston travel in the cylinder. Fuel is taken into the engine by the air stream passing by the nozzle in the carburetor. The carburetor adjustment regulates the ratio of fuel to air so that an explosive mixture is always present. The throttle of the carburetor regulates the air and fuel to the engine to vary the performance of the engine from idle, acceleration, part throttle, or full load.

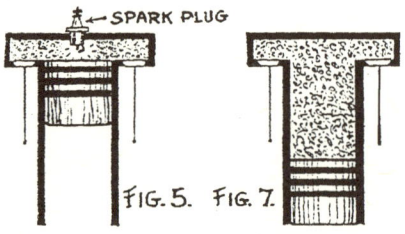

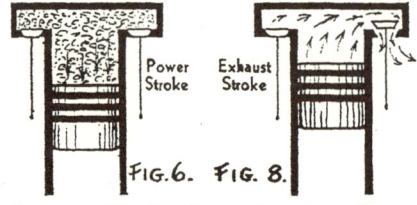

Compression Stroke: (Figs. 3 and 4)

The compression stroke of the engine compresses fuel and air into the combustion chamber to ratios 1:4 for low compression engines, 1:5 for medium, and 1:7 for high compression engines. The most economical gasoline for these engines is low octane No. 2 for the low and medium compression engines, and high octane No. 1 gasoline for the high compression engine.

Ignition occurs 15 to 25 degrees before head centre on the compression stroke.

Power Stroke: (Figs. 5 and 6)

Combustion takes place during some 35° of crank travel 10° before head centre and 25° after head centre where the temperature of combustion increases from 200° Fahr. to 4,700° Fahr. maximum. The increase in temperature of the gases in the combustion chamber of the engine increases the pressure from 110 pounds per square inch to 350 pounds pressure. The high pressure causes the power of the engine through the piston, connecting rod, and crankshaft to the flywheel, clutch, etc.

The Exhaust Stroke: (Figs. 7 and 8)

The exhaust valve starts to open 45° before crank centre so that the valve will be fully open during the travel of the piston on the exhaust stroke when the by-products of combustion are scavenged from the cylinder.

Combustion:

Before Combustion
Air 1,200 cubic feet.
Gasoline 6.25 pounds.

Gasoline consists of:
5.26 pounds of carbon.
0.99 pounds of hydrogen.

After Combustion

Exhaust gas	1,350 cubic feet
Nitrogen	775 cubic feet
Carbon dioxide	10.0 pounds
Carbon monoxide	5.9 pounds
Hydrogen	0.2 pounds
Oxygen	0.2 pounds
Water	6.9 pounds

The exhaust gas resulting from burning 6.25 pounds of gasoline contains 6.9 pounds of water. The water vapor must be carried out of the engine and the exhaust system at a temperature well above the dew point (120° Fahr.) or rusting and corrosion will take place. The water vapor of the exhaust gas is only visible during the winter. Any procedure which can be followed to reduce the time during which the engine operates at a temperature below 150° Fahr. will reduce engine wear and tend to improve lubrication conditions.

The carbon monoxide and water vapor of the exhaust gases are the cause of the trouble. Carbon monoxide is a poisonous gas responsible for many casualties. Care must be taken not to run an engine in a closed garage without ample ventilation.

The water vapor (steam) must be carried out of the engine in a superheated condition. Water sludges are formed in quantity where engines are operated in cold weather, particularly where the engine is worn so that "blow-by" past the piston occurs. The water of the exhaust gas condenses and mixes with the oil forming sludge to such an extent that the engine may seize and cause serious damage.

Top Lubrication:

The remedy is to warm the engine up quickly so that lubrication and piston fit will be effected quickly. Then too, the use of top lubrication provided by a pint of engine oil S.A.E. 10 or 20 to 6 to 8 gallons of gasoline will provide protection for the top end of the engine while the engine is warming up. Research work on engine wear when using top oil showed a reduction of top ring and cylinder wear of 4 to 1 for rings and 3 to 1 for cylinders where top oiling was employed.

Engine Efficiency:

Gasoline contains 23,000 B.T.U. per pound, which is released as the gasoline is oxidized or burned in the engine. The engine is designed of materials which resist heat making high temperature operation possible. The valves are of steel alloy designed to resist exhaust temperature of 1,300° Fahr. The exhaust gas temperature varies from 900° Fahr. part load to 1,250° Fahr. full load.

Aluminum or cast iron pistons operate at 750° Fahr. at the top centre to 300° Fahr. at the skirt. Lubricating oil is selected to resist heat up to 700° Fahr. where it begins to crack and break down. The pistons are cooled through the cylinders and cooling water and through the lubricating oil flowing over the inside surface of the piston head and skirt. It is necessary to carry away 29 per cent of the heat of the fuel through the cooling system to keep the pressure of the pump and the resistance of the nozzle so that combustion takes place with a minimum of ignition lag. The highly heated turbulent air heats, ignites, and oxidizes the fuel releasing the B.T.U. content of the Diesel fuel which causes expansion and power transmission from the engine similar to the gasoline engine.

The exhaust gases contain no carbon monoxide because of combustion with an excess of oxygen. The temperature

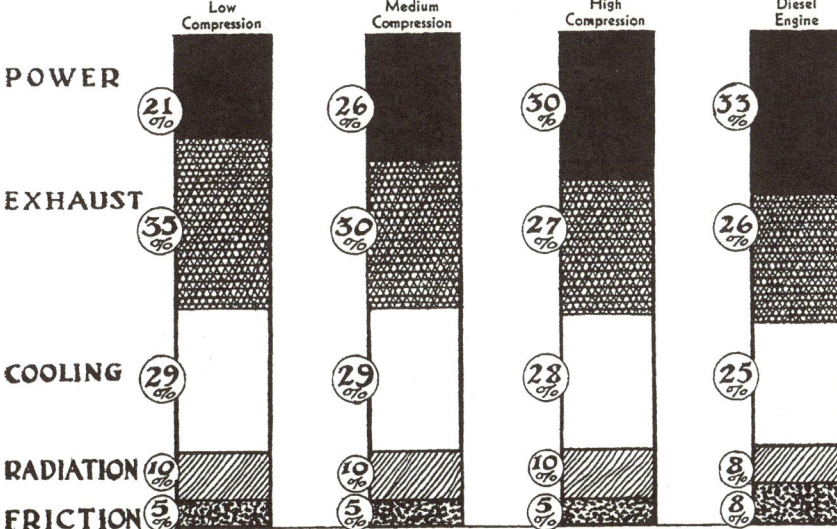

pistons, rings, lubricating oil and valves at safe operating temperature. Thirty-five per cent of the heat of the fuel is carried out with the exhaust gases. The exhaust gases must be hot enough to carry the water vapor out of the engine in a superheated condition to prevent water condensation in the exhaust pipe and muffler. Ten per cent of the heat of the fuel radiates from the working parts of the engine while five per cent is dissipated in friction of the working parts of the engine, leaving 21 per cent of the heat of the fuel for power from the engine. Efficiency and power improves with the high compression engine.

The Diesel Engine:

The Diesel engine is the most efficient of all the internal combustion engines. The Diesel cycle consists of four or two strokes where air is drawn in on a suction stroke and compressed on the compression stroke to a ratio of 15 to 1, compression pressure 550 pounds per square inch, during which the air is heated to 1,200° Fahr.

Diesel fuel, fluid at all operating temperatures but non volatile, clean (free from dirt), and possessing lubricating properties to lubricate the pump, is injected by pump and nozzle into the highly compressed and heated air. The fuel is broken up into a fine spray by

of the exhaust gas is lower than the exhaust from the gasoline engine, being 400° Fahr. to 800° Fahr. maximum. Water vapor must also be carried from the Diesel engine in a superheated condition to prevent condensation.

The Diesel engine presents real economy in the field of medium to large size tractor, the truck and bus engine, and in large power plant and marine engines.

The table on the next page indicates a comparison of fuel cost and efficiency for the different engines in the farm tractor.

Engine Lubrication:

Engines are lubricated so that the working parts will operate cool and in good condition for long periods of time. Engine wear is caused by contamination in lubricating oil of dirt, gum, water sludge, and oxidation sludge, and also from corrosion caused by acidity and the presence of water in the exhaust gas and oil. The use of cracked gasolines and Diesel fuel in engines have caused more severe operating conditions for engines requiring special cleansing and acid neutralizing properties of a lubricating oil. The result of research and development in the field of lubrication has developed the heavy duty oil or compounded oil for use in both the Diesel engine and the gasoline engine in heavy duty service.

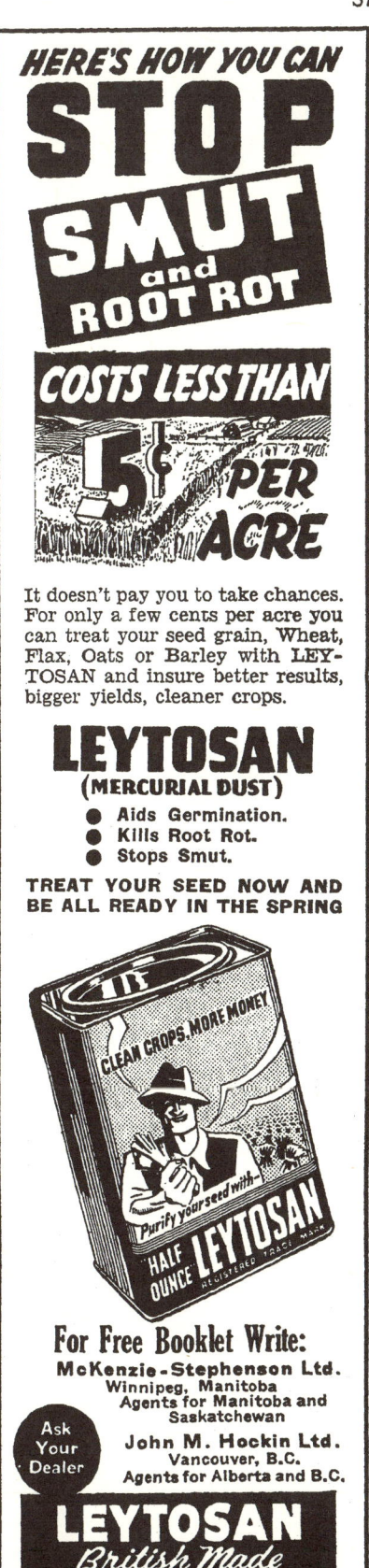

Compounded "Heavy Duty" Oils:

A considerable amount of engine damage has resulted from the dirt which loosens from the inside of an engine and plugs up the engine lubrication system resulting from changing from the straight mineral oils to compounded "heavy duty" oils. During the war the armed forces changed to heavy duty oils and found that the following procedure was satisfactory for cleaning an engine during the change-over period.

The straight mineral oil was drained when the engine was hot. The first filling of heavy duty oil was operated in the engine for 15 minutes under reasonably high speed engine service, after which the oil was drained. The second filling was then operated under normal engine service for two hours and drained. The third filling was then put into normal service for ten hours and drained. At the fourth filling the oil was found to be reasonably clean so that the normal draining periods could be followed.

It is particularly important that if an engine, even though it is assumed to be clean, has been lubricated with straight mineral lubricating oil and has been changed over to heavy duty lubricating oil, every precaution be taken to prevent engine damage due to the accumulation of quantities of dirt, carbon, and oxidation accumulations, which are loosened from the inside of the engine block.

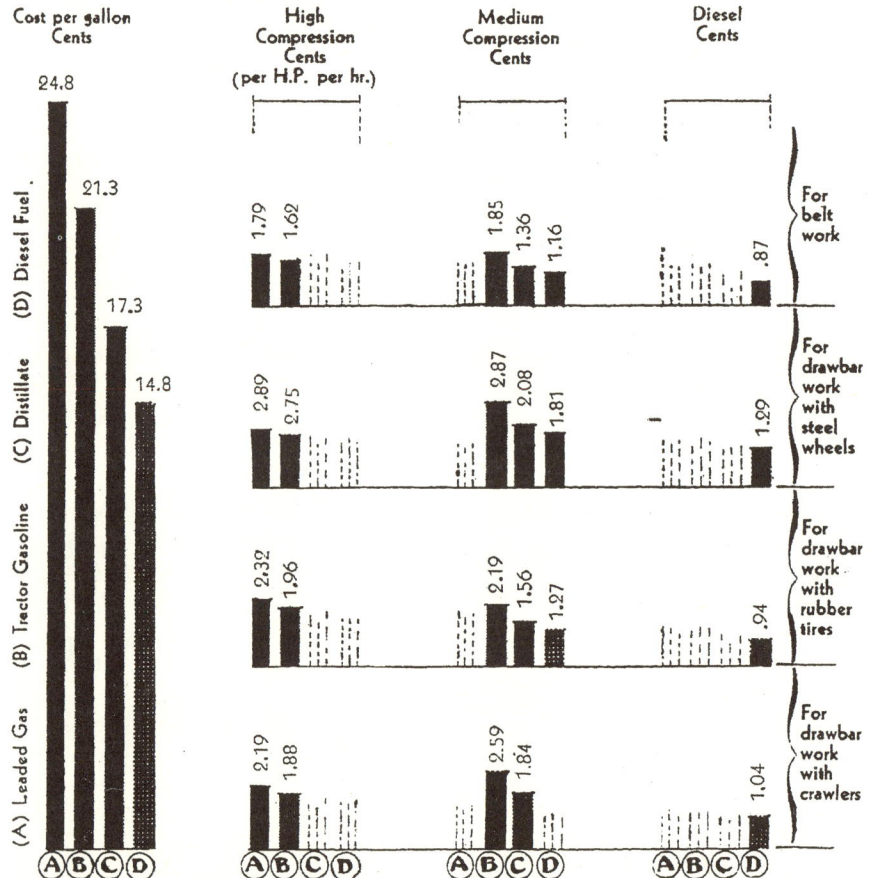

Acidity:

Engine oils become acid during operation due to the presence of combustion gases passing by the pistons which are absorbed by the unstable fractions of the lubricating oil. The by-products of combustion consist primarily of carbon dioxide and water vapor. There may be traces of sulphur and lead. The acidity during normal summer operation is not injurious due to the fact that there is very little blowby and the water vapor is carried out of the crank case by the normal crank case ventilation. Engines fitted with cast iron pistons and babbitt bearings operate even in the winter time where oils become quite acid without damage. Engines fitted with aluminum pistons and precision type alloy bearings are subject to damaging corrosion when operating under cold temperatures. The corrosion takes place somewhat as follows: The acidity of the acid and the water sludges which form from the condensed exhaust gas vapors form an electrolyte which stimulates electrolytic corrosion between any two dissimilar metals such as aluminum and iron, or lead and copper. Consequently any tractors, trucks, or automobiles which are operated during the fall, winter, and spring where the operating temperatures may not normally be high, should be so operated that the temperature of operation will be quickly brought up to normal and maintained at normal. The oil should be drained much more frequently than during summer operation in order to remove the accumulations in the crank case which are causing corrosion.—*Prof. Evan A. Hardy, University of Saskatchewan.*

Maintenance and Overhaul of the Farm Tractor

DAILY SERVICE

Cooling System

The cooling system should be inspected for water level at frequent intervals or whenever the engine is filled with fuel. Use soft water if available.

Fuel

Keep the pump and container clean. Any precautions for keeping the fuel clean will assist greatly in efficient operation. A pint of clean oil to five gallons of gasoline is a good mixture

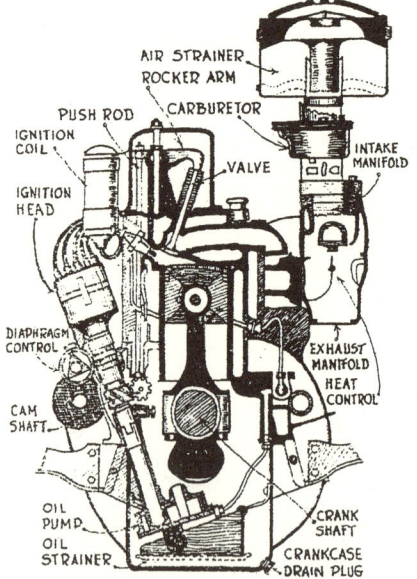

Valve-in-Head Type Engine

to use when starting, for cylinder and ring lubrication. Tractors mounted on rubber should be equipped with a short length of chain dragging on the ground to prevent the possibility of fire when filling the fuel tank.

Engine Oil

Engine oil level must be inspected every time the engine is refueled either by checking oil level gauge or cocks. When adding engine oil put in only sufficient to maintain the proper level. That is, do not put more than the prescribed amount of oil in the crank case.

Air Cleaner

The air cleaner must be kept clean. If of the cartridge type, it should be washed out with kerosene and dipped in oil at proper intervals—every half day if conditions are sufficiently dusty to warrant it. The oil bath air cleaner needs less attention although the level of the oil should be inspected regularly and maintained. All connections between the air cleaner and the carburetor should be inspected regularly to prevent leakage.

Greasing

The steering assembly and rear wheel seals should be inspected whenever the fuel tank is refilled. The fan and the other parts need attention as outlined in the instruction book. Frequent attention should be given to inspecting, tightening or repairing any disorders which are noticed during operation.

Starting

When starting, the engine should be warmed up as quickly as possible by the use of the radiator curtain or shutters. The engine should be hot before the distillate is turned on, and should be kept hot whenever distillate is used.

WEEKLY SERVICE

Oil Filter

The oil filter should be serviced as provided for in the instruction book. It is particularly important that the oil filters be clean and free to function. All types need careful attention, and this applies to the breather filter as well. Be sure that after cleaning, the filter housing is wiped clean. When the element becomes worn it should be replaced.

Engine Oil

The time of draining crank case oil varies with the fuel, type of oil filter, and condition of the engine. It should be drained as recommended and always drained when the engine is hot. Drain oil at least every 250 hours.

Air Cleaners

The oil bath air cleaner should be inspected at regular intervals and the oil changed as recommended by the manufacturer, considering operating conditions such as dust, etc. Use the grade of oil recommended. Other grades will not function as well.

Tire Pressure

Air pressure in tires varies from 12 to 20 pounds per square inch depending upon the size of the tire and the total weight upon the tire. Keep the air pressure up so that practically no side wall flexing of the tire takes place. To control slippage increased air pressure should be balanced by increased total weight.

PERIODIC "TUNE-UP"

The modern tractor is so refined as to materials and construction that the annual overhaul common to the tractors of the early thirties or earlier is no longer necessary. The periodic tuneup has developed in its place. Tractors operating less than 600 hours per year will probably require this tune-up only once every two years. Tractors operating more than 600 hours per year will require the tune-up every year.

Compression

The loss of compression pressure indicates the need of a periodic tune-up. Compression pressure may be readily tested on the farm by pulling against the compression with a hand crank. The operator can easily tell whether the compression is even and adequate. Where the compression is weak for one or more cylinders, a tune-up is necessary.

Cylinder Head

The cylinder head should be removed and all carbon cleaned from the combustion chamber.

Valves

Valves should be cleaned of all carbon and lead deposits, when leaded fuels have been used. The seating of the valves should be inspected and those showing uniform seating should only be cleaned and rubbed in with fine compound. Warped or burned valves or badly worn valves should be refaced or replaced and the valve seats reseated. When reassembling, valve clearances should be as recommended. Where clearances are recommended as hot setting, they should be approximately double for cold setting and checked for the hot setting after running on load for at least two hours. The valve clearances should be checked at least twice a year to insure proper valve seating.

Gaskets

All gaskets should be carefully inspected and replaced when necessary. They should only be re-used when found to be in excellent shape.

Spark Plugs

The spark plugs should be removed and cleaned and where leaded gasoline has been used this should be done with a sand blast cleaner. Before reassembly, the spark plug gap should be adjusted as recommended.

Timing

The timing of the magneto or ignition head must be checked with the timing marks on the engine and ad-

MORE *work in* LESS *time with*

SARGENT HYDRAULIC LOADERS

SENIOR MODEL

Fits all standard row crop tractors.

JUNIOR MODEL

Fits Farmall B and John Deere H Tractors.

Sargent Attachments Available:
HAY SWEEP
BULLDOZER
SNOW SCOOP

"K" MECHANICAL FENCER

Fits all standard row crop tractors.

The **CUTHBERT** Co. Ltd.

167 Grain Exchange Bldg., Winnipeg

Jamieson Farm Equipment Co., Winnipeg
The Cuthbert Co. Ltd., Yorkton.
Wm. Cozart & Son Ltd., Calgary.

$4 COMPLETE

PAY ONLY $1 A MO.

for AUTO MECHANICS

AUDELS AUTO GUIDE explains in detail with illustrations and diagrams for better service the whole subject of auto mechanics: Basic principles—Construction—Operation—Service—Repair. Handy Size, Easily understood. NEW FLUID DRIVE, HYDRAULIC SHIFT AND DIESEL ENGINES FULLY COVERED. Over 1700 pages—1540 illustrations showing inside views of modern cars, trucks and buses with instructions for service jobs. This standard book saves time, money and worry for operators and mechanics.

Step up your own skill with the facts and figures of your trade. Audels Mechanics Guides contain Practical Inside Trade Information in a handy form. Fully Illustrated and Easy to Understand. Highly Endorsed. Check the book you want for 7 days' Free Examination.

Send No Money. Nothing to pay postman.

---- CUT HERE ----

MAIL ORDER

AUDEL, Publishers, 49 W. 23 St., NEW YORK 10, N. Y.

Please send me postpaid for FREE EXAMINATION books marked (x) below. If I decide to keep them I agree to mail $1 in 7 Days on each book ordered and further mail $1 monthly on each book until I have paid price. Otherwise, I will return them.

- ☐ AUTOMOBILE GUIDE, 1700 Pages . . . . $4.
- ☐ REFRIGERATION & Air Conditioning, 1280 Pgs. 4.
- ☐ OIL BURNER GUIDE, 384 Pages . . . . . 1.
- ☐ POWER PLANT ENGINEERS Guide, 1500 Pages. 4.
- ☐ PUMPS, Hydraulics & Air Compressors, 1658 Pgs. 4.
- ☐ WELDERS GUIDE, 400 Pages . . . . . . 1.
- ☐ BLUE PRINT READING, 416 Pages . . . . 2.
- ☐ SHEET METAL WORKERS Handy Book, 388 Pgs. 4.
- ☐ SHEET METAL PATTERN LAYOUTS, 1100 Pgs. 4.
- ☐ AIRCRAFT WORKER, 240 Pages . . . . . 1.
- ☐ MATHEMATICS & CALCULATIONS, 700 Pgs. . 2.
- ☐ MACHINISTS Handy Book, 1600 Pages . . . 4.
- ☐ MECHANICAL Dictionary, 968 Pages . . . . 4.
- ☐ DIESEL ENGINE MANUAL, 400 Pages . . . 2.
- ☐ MARINE ENGINEERS Handy Book, 1280 Pages 4.
- ☐ SHIPFITTERS Handy Book, 272 Pages . . . 1.
- ☐ MECHANICAL DRAWING COURSE, 160 Pages 1.
- ☐ MECHANICAL DRAWING & DESIGN, 480 Pgs. 2.
- ☐ MILLWRIGHTS & Mechanics Guide, 1200 Pgs. 4.
- ☐ CARPENTERS & Builders Guides (4 vols.) . . 6.
- ☐ PLUMBERS & Steamfitters Guides (4 vols.) . 6.
- ☐ MASONS & Builders Guides (4 vols.) . . . 6.
- ☐ MASTER PAINTER & DECORATOR, 320 Pgs. . 2.
- ☐ GARDENERS & GROWERS GUIDES (4 vols.) . 6.
- ☐ ENGINEERS and Mechanics Guides
   Nos. 1, 2, 3, 4, 5, 6, 7 and 8 complete . . . 12.
- ☐ Answers on Practical ENGINEERING . . . . 1.
- ☐ ENGINEERS & FIREMANS EXAMINATIONS . 1.
- ☐ ELECTRICIANS EXAMINATIONS, 250 Pages . 1.
- ☐ WIRING DIAGRAMS, 210 Pages . . . . . 1.
- ☐ ELECTRICAL DICTIONARY, 9000 Terms . . 2.
- ☐ ELECTRICAL POWER CALCULATIONS, 425 Pgs. 2.
- ☐ HANDY BOOK OF ELECTRICITY, 1340 Pages . 4.
- ☐ RADIOMANS GUIDE, 914 Pages . . . . . 4.
- ☐ ELECTRONIC DEVICES, 216 Pages . . . . 2.
- ☐ ELECTRIC LIBRARY, 12 vol., 7000 Pgs., $1.50 vol.

Name _____

Address _____

Occupation _____

Employed by _____ Guide

---

justed as necessary. Late ignition will cause excessive fuel consumption and loss of power.

General Cleaning

General cleaning and tightening completes the periodic check up for the tractor. Many tractors operate up to 5,000 hours under this system of periodic tune-ups before requiring a major overhaul.

MINOR OVERHAUL

Oil Consumption

Where continued operation results in an abnormal rate of wear and inefficient operation, the engine is worn to an extent where it will require either a minor or a major overhaul. If oil consumption has gradually increased from a pint to two quarts in twelve hours an overhaul is required.

Piston Rings

Excessive oil consumption may be controlled by the installation of new piston rings if the cylinders have worn uniformly. If the cylinders are badly out of round new rings will seldom control oil consumption and a major overhaul will be necessary. Occasionally rings become stuck in the piston grooves causing a sudden increase in oil consumption which may be remedied by freeing the pistons and rings from carbon.

Fitting New Piston Rings

After pulling out the piston—meanwhile making sure the bearing cap and shims are marked so that they may be reassembled in the same position—remove the rings and scrape all the carbon from the ring grooves, making sure that the oil drain beneath the oil ring is clean. (The oil ring scrapes excess oil from the cylinder wall which passes through the slots in the oil ring, then through the piston to the oil sump). Clean all carbon from the piston, also from the cylinder head. For the first replacement of rings use standard width and standard diameter rings. (It is questionable whether rings of oversize diameter should ever be used, but if they are used, .003-inch to .004-inch clearances per inch of diameter at the gap of the ring must be provided for all compression rings and from .002-inch to .0025-inch for the oil rings. Piston ring gap measurements must always be taken at the bottom of the ring travel). With alloy rings, instructions supplied with the rings are to be followed. (New rings can be placed on the piston quite readily by using three or four thin strips of tin one-half inch wide).

Piston Lands Badly Worn

If the clearance between the ring and the land of the piston is .010-inch or more a slightly wider ring should be installed. Secure the next wider ring and have a competent machinist widen the ring groove to fit the ring. Rings should have .001-inch to .0025-inch side clearances in the groove. When installing over-width rings, the ridge formed (by wear) at the top of the ring travel must be removed with a suitable reamer.

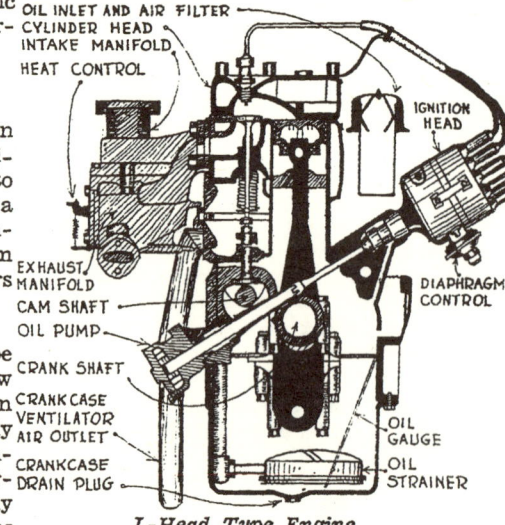

L-Head Type Engine

Clearance Between Piston and Cylinder

When clearance between the cylinder and piston is more than .012-inch or where .015-inch oversized rings are required, not only will the engine be noisy, but it will be practically impossible to control oil consumption with new rings. Reboring of the cylinders will be required, or new sleeves, thus a major overhaul.

MAJOR OVERHAUL

When considering a major overhaul of the tractor there are three possibilities.

1. The engine may be overhauled at home, if there are the proper tools available and the operator is a good mechanic.

2. The engine may be taken in to a dealer or mechanic who is fitted for overhauling.

3. Where the engine has an industrial motor it may be taken in to the branch or dealer and exchanged for a completely rebuilt engine.

Reboring the Cylinder

To rebore a cylinder it is necessary to take the cylinder block off the tractor. The reboring and honing is done with specialized equipment for the purpose. Proper size pistons and rings and if necessary, new piston pins in the pistons and piston pin bushings in the small end of the rods must be properly fitted. The reconditioned cylinder may then be reassembled on the tractor.

Replacing Sleeves and Pistons

Where the engine is equipped with removable sleeves it is necessary to either have a sleeve puller or one may be made out of a three-quarter or one inch round rod threaded at one end and long enough to reach from the bottom of the sleeve to the top stud bolts above the cylinder block. Below the sleeves place on the bottom end of the rod a heavy flat plate long enough to reach across the bottom of the sleeve. A hard wood block should be placed on the top end of the cylinder block or stud bolts so that the sleeve is free to be loosened. Assemble this inside the sleeve and tighten up the nut on the top end. If the sleeves are difficult to loosen, it may be necessary to use a hammer or short heavy bar from underneath the tractor. After the sleeves have been removed, clean out the water jacket thoroughly and remove the old rubber rings from the block at the bottom of the sleeves.

New sleeves and pistons come in matched pairs and they must be assembled in the tractor in the matched sets. All grease must be washed from the sleeve and the piston underneath the rings in the ring grooves. The old pistons must be removed from the connecting rod and new ones put on. Care must be taken when the new piston pin is driven through the piston and the connecting rod to see that the set screw hole in the pin matches with the hole in the boss of the piston. The set screws which are used to hold the piston pin in the piston should turn with very little effort. Otherwise, the boss in the piston may be cracked. The set screws must be locked in place. Wherever cotter keys are used for this purpose they should be of such length that the ends may be spread and turned firmly into position around the set screw but still not long enough to scrape the cylinder wall. Before the sleeves are replaced, a small quantity of light alemite grease should be placed on the outside sleeve where it makes contact with the cylinder block. Use a hardwood block on top of the sleeve to drive it into position. See that the gaps in the rings are alternately spaced on the piston and that there is a quantity of light oil placed on the piston before it is inserted in the sleeve. The sleeves are usually bell-mouthed at the bottom to make entry of the piston and rings easy. Where this is not the case, a ring compressor will be required. (Most of the above instructions with regard to pistons and rings will be true whether the engine has been rebored or equipped with new sleeves).

Bearing Adjustment

Engines may be fitted with either of two types of bearings: (1) copper lead alloy bearings (precision bearings) or (2) babbitt bearings.

Precision Bearings

The precision bearing shell is only .072-inch thick with the bearing material only .020-inch thick. They are interchangeable and are put in without scraping or fitting. The crank shaft is machined to very close measurements so when the shells are assembled they fit with .003-inch clearance for the oil film.

Precision bearings must be replaced when worn for they are not fitted with shims. When badly worn, the crank shaft must be ground down to uniform dimensions and fitted with undersized bearing shells. The tools and equipment necessary to install precision bearings is not available in the ordinary farm shop. They should be fitted and assembled in competent service or machine shops.

Babbitt Bearings

Almost all connecting rod or main bearings of this type are equipped with two heavy shims and several thin shims in each half. To adjust the bearing remove the cap, noting the marks and remove one thin shim from each side (these shims are usually .003-inch thick). Replace the cap as removed and tighten the bearing. Continue to remove shims in this fashion until the bearing binds on the shaft. Then add a thin shim on each side and tighten (the bearing should always have a .003-inch clearance to provide for lubrication—if the bearing is too tight excessive wear will result. If too loose it will be noisy and hammer the crank shaft, also causing excessive wear).

The crank pins on the crank shaft will gradually wear out of round, although they will operate fairly satisfactorily if not more than .005-inch out of round. To maintain a minimum of wear, keep the bearings properly adjusted and use an oil of the proper weight and viscosity for each season.

When the crank shaft is out of round, the bearing adjustment should, of course, be made at the tightest point on the crank pin. If the shims get rough on one side due to the shell rocking a little or due to ordinary wear, the shim should either be filed smooth or turned over to prevent the shell from rocking.

If the shells are rocking in the caps, it will likely be necessary to replace them. When putting in new shells, allow from .004-inch to .012-inch side clearance. When the bearings have been tightened several times, so that all thin shims have been removed, then it is necessary to replace one of the heavy shims from each side with a thinner shim. (These thinner shims are quite often called "half shims" and may often be purchased from the tractor companies). Thin shims may then be

100's of Jobs
Made easier

SKOTCH WOOD JOINERS:

Perfect in the repairing and making of furniture, toys, screens, and windows, etc. Skotch Fasteners are designed specially to make rigid and permanent wood joints; so simple to handle anyone can use them, and with professional results. There's always a job for Skotch Wood Joiners.

SUPERTITE STEEL SHEETS

Save your money — protect your grains and livestock by lining your henhouse, stable and other constructions with durable Supertite Steel Sheets. Large sheets for easy application, maximum width 26 inches, maximum length 28 inches. Selected sheets for each shipment all one size as above. Prices per 100 square feet are as follows: Manitoba, $3.75; Saskatchewan, $4.25; British Columbia, $4.50. Immediate delivery. Freight paid on orders of 300 square feet or more.

SASHMASTER FASTENERS:

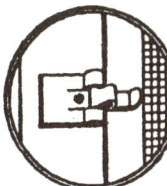

One flip of the finger and Sashmaster Fasteners secure your door, window or removable panel. The strong steel spring holds securely—eliminates all rattles and drafts. Sashmaster can also be installed as a burglar proof window lock. More people today are using Sashmaster Fasteners.

SUPERIOR PRODUCTS
LIMITED
8 NELSON ST.
Sarnia   Ontario

used to fill up the gap. When one bearing has been adjusted as outlined above loosen it and adjust the next. After all bearings have been adjusted tighten them all.

Main Bearings

It is not necessary to adjust main bearings as often as connecting rod bearings. The clearance of the main bearings may be determined by lifting the shaft by a bar over a block and where clearance is noticeable the bearing should be taken up. The same procedure as outlined above should be followed in tightening the main bearings.

When a bearing burns out, do not install a new bearing until the cause has been found. That is, the oil lines and drilled passage ways in the crank shaft should be checked for obstructions, etc. Engines equipped with ball bearings on the crank shaft cannot be adjusted for wear. They should, however be replaced when there is more than .021-inch clearance between the ball and the cone. To change these bearings the motor must be lifted from the chassis and the crank shaft and fly wheel removed and taken to a competent service man who will remove the old bearings and instal new ones. When installing in the tractor, make sure that the opening in the outer race is installed at the top.

After the bearings have been adjusted and set up, all nuts should be keyed on the main and connecting rod bearings. When putting in the new cotter keys, use as large a cotter key as possible and make sure that the key is the correct length so that one-half can be turned up along side the nut and the other half across the bottom side of the nut. The keys must be driven in solid and the ends placed tightly against the nut so that they will not shake loose.

End Play

Excessive end play is caused either by the wear of the main bearings or by wear on the thrust flange. (End play of the crank shaft should be .004-inch to .008-inch). This situation may usually be remedied by replacing the main bearings or by having the thrust flange built up.

Crankshaft Overhaul

When the crankshaft is out of round to such an extent that it is no longer possible to keep the bearings tight, then it is necessary to replace or recondition the crankshaft and the bearings. This should not be necessary until after the tractor has been run for a number of seasons.

It is usually preferable that the crankshafts and bearings be replaced with standard sized parts because too much material may be removed in making the crankshaft journals round and also due to the fact that new undersized bearings have to be made or purchased.

Most machine shops are, however, equipped to do regrinding of crankshafts and pouring new shells or bearings. Whenever considering crankshaft reconditioning, the exchange policy of the company should be examined. The advice of tractor companies is based upon experience and should be given every consideration.

FROZEN BLOCK

If an engine is in good condition except for the block, which may have been cracked due to freezing, a competent welder may be able to braze or weld the crack. The water jacket will return to shape as soon as the block is heated. The block may also be patched by drilling and tapping 3/16-inch or ¼-inch holes back from and all around the crack. A patch of fairly stiff sheet iron should be cut and drilled to match the holes in the water jacket. A rubber gasket fitted between the block and the patch will make a tight join if the studs are carefully and uniformly tightened. This system, however, should not be followed unless the block is badly worn and not worth dismantling or a competent welder is not available.

Protecting Radiator in Winter

This is what I attached to our tractor and find it very successful in winter. I took a blind from an old car and fastened it at the bottom of the radiator. When starting the engine on a cold morning, draw the blind up. The engine will heat up much more rapidly. Then, when it is warm, the blind can be lowered again.—J. W. Nichol, Onanole, Man.

Fertilizer Distributor

This fertilizer distributor is made of wood, sheet metal, some 1x¾-inch flat iron and two wood or metal wheels about 8 to 10 inches in diameter. Hardwood about ¾ inches thick is used for the two ends of the hopper. A piece of 28 gauge galvanized sheet iron 27x20 inches is used for the sides of the hopper. The top edges of the metal sides are folded twice to make the hopper rigid and prevent the raw edges from cutting the hands. The sheet metal is also folded over the hardwood ends and fastened with ¾-inch round head screws about 2 inches apart. A hole is bored in each of the hardwood ends to fit a ½-inch bushing such as a Ford Model T brass spindle bushing. The hole for the bushing is bored 2 inches (on centre) from the bottom. A piece of 1x1-inch hardwood or iron 16 inches long is used as a feed device. A ½-inch hole is drilled into each end of the feed bar to a depth of 2½ to 3 inches. Two ½-inch diameter steel axles, about 8 or 9 inches long, are fitted to the feed bar and secured with 3/16-inch stove bolts or cotter pins after they are assembled in the hopper.

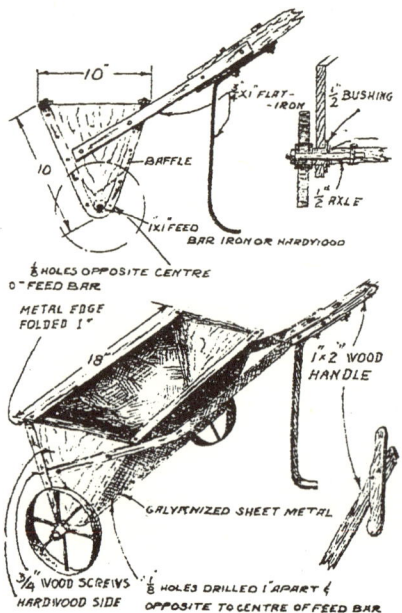

The wood handle and flat iron braces are bolted to the hopper at an angle suited to the user. A flat iron rest is bolted to the handle to keep the hopper upright so as to prevent the fertilizer falling out when the machine is standing alone. Fertilizer outlet holes are drilled one inch apart and in a straight line and immediately opposite the centre of the feed on a horizontal plane. The holes are ⅛-inch in diameter. A baffle board or metal plate is fastened inside the hopper as shown in the sketch. This is necessary to prevent the fertilizer from becoming packed at the bottom by the weight of the fertilizer above. Both drive wheels may be fixed to the shaft with set screws where there is not much turning. Where there is a great deal of turning, one of the drive wheels can be left loose so that it can revolve freely on its axle and facilitate turning. In operation the feed bar revolves and the square edges act like small paddles, which pile the fertilizer over the feed outlet holes, through which it falls to the ground. The holes may be made slightly larger to apply more fertilizer to suit local requirements. A slide adjustment can be added to vary the size of the openings. This, however, is not shown in the sketch in order to make the construction as simple as possible. If a heavier application of fertilizer is required, the machine can be pushed over small areas two or three times with ease in very little time. For applying fertilizer alongside of drill rows, the feed openings not required can be closed temporarily by bolting on a strip of metal or wood on the outside of the openings, using two or three of the non-wanted feed holes for the small bolts to hold the covering strips.

Use and Adjustment of Tillage Equipment

Moldboard Plow:

The moldboard plow is used in the more humid areas of Western Canada where weed control cannot be accom-

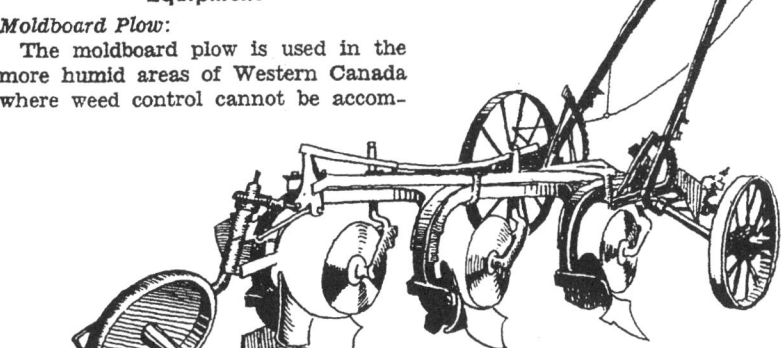

plished by surface tillage. The 14-inch bottom general purpose or high speed bottoms where no difficulty is experienced with scouring are most suitable. The high speed bottom turns the furrow slice with less pulverizing than the quick turn bottoms at the speeds of our modern tractors (3½ to 4½ miles per hour). The 14-inch bottom will turn a shallow furrow slice (3½ inches to 4½ inches) much more uniformly than the wider bottoms with less pulverizing. Shallow plowing is safe, efficient, and economical plowing which is essential in a *sound* agricultural program.

Adjustments of the Moldboard Plow:

Plow bottom alignment, colter, rear furrow wheel, and hitch adjustments are the important adjustments of the moldboard plow for efficient operation.

Plow Bottom Alignment:

The bottoms of a plow are mounted on beams parallel with the beams and with each other. Bottoms should be shimmed back into alignment when found to be bent or out of line.

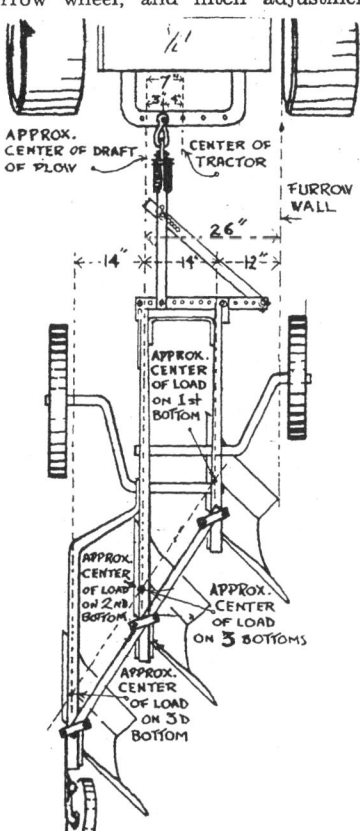

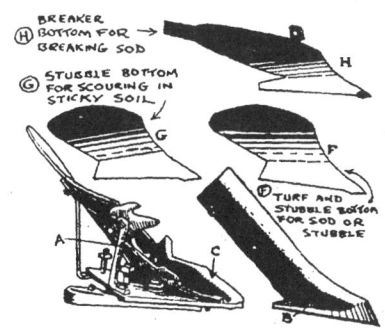

A—Share Holding Clamp.
B—Stud Rivet in Share.
C—Frog Supports Share.

Colter Adjustment:

Colters must be set back of the point of the plow some four inches and outside the shin of the plow so that the plow will clear the furrow wall. The colter should be set only deep enough to cut through the stubble and trash on top and hold against the soil being raised by the point of the share.

Rear Wheel Adjustment:

The rear wheel on either the horse or engine gang plow must support the rear bottom of the plow. The wheel should be set to carry the side thrust and the weight of the plow. The rotation of the wheel should be straight along the furrow wall with a slight lead to the furrow wall.

Hitch Adjustment:

The hitch of the plow must be so ad-

EATON'S *Acme* GRINDER

Note These Important "Acme" Features

- Balanced pulley assembly reduces vibration.
- Two supporting bearings for sliding shaft and two sets S.K.F. deep grooved ball-bearings totally enclosed in the pulley.

Balanced Assembly Showing Oil Reservoir and S.K.F. Bearings

- Burrs can open 1-inch instantly if foreign materials get between the plates, and then instantly reset.
- Requires lubricating in one place only, when first put to work, and thereafter not more than once a year.

Additional Reasons For Its Popularity

High capacity Acme auger ensures a uniform grind by direct flow feed. Depending on power, grain and fineness of crush, direct flow will feed more than 200 bushels per hour to burrs.

Double box edge galvanized hopper holds about five bushels of oats.

Ten-inch, double-sided, grinding burrs are made of a very hard, high grade alloy of white iron.

Quick release lever separates plates and shuts off grain feed in single operation. Once the correct setting has been made there's no need to adjust burrs each time grinder is used.

See Farm Machinery pages in your EATON's General Catalogues for full description and price.

THE T. EATON C? LIMITED
WINNIPEG   CANADA

a CRANE WATER SYSTEM will bring you...

COMFORT IN THE BATHROOM...

CONVENIENCE IN THE KITCHEN...

BIGGER PRODUCTION ON THE FARM..

If you do not enjoy the advantages of running water on your farm, now's the time to plan for the installation of a complete Crane water system, including an efficient Crane pump, sparkling, easily cleaned Crane plumbing fixtures in the bathroom, kitchen and laundry, and Crane piping throughout. Not only will running water add to your comfort and convenience in the house, but in the barns and feed lot, it will reduce labor and help increase production.

Planning a water system calls for an expert's help. And there is one nearby . . . your Plumbing and Heating Contractor. Why not visit him the next time you're in town? He will help you develop a system to exactly fit your needs.

Obviously, the demand of farm owners is so great that everyone cannot be supplied with a new water system at once. But now is the time to get your plans ready— to consider the type of pump you need and decide on a bathroom and kitchen arrangement that will provide you with the maximum comfort and convenience as soon as equipment becomes available.

CRANE
AND ITS SUBSIDIARIES

CRANE Limited   **WARDEN KING** Limited   **CANADIAN POTTERIES** Limited   **PORT HOPE SANITARY** Manufacturing Co. Limited

Plumbing Fixtures • Heating Equipment • Valves • Fittings • Piping

CRANE LIMITED, 1170 Beaver Hall Square, Montreal, 2, Canada — 18 Branches in Canada and Newfoundland

NATION WIDE REPRESENTATION THROUGH PLUMBING AND HEATING CONTRACTORS EVERYWHERE

justed that the team or tractor will draw the plow without pulling up or down on the front of the plow. This adjustment is particularly important. The horizontal hitch must be adjusted so that the front bottom will cut the proper width with either the team or tractor operating normally in the furrow.

The One Way Disc:

The one way disc has entirely replaced the disc plow and is rapidly replacing the moldboard plow where weeds can be controlled with the machine. The one way disc operates at considerably less

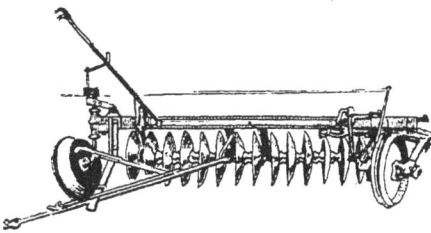

power per acre than the moldboard plow, produces quality tillage, leaving the stubble and trash in and on the surface of the soil and the soil in a level lumpy condition when moderate speed is used. High speed tends to pulverize excessively and should be avoided with the one way disc.

Size of Disc:

There are three general sizes of discs used. The 22-inch disc with 2½-inch or 3-inch concavity. The 24-inch disc with 3¼-inch concavity and the 26-inch disc with 3½-inch concavity. These discs are spaced at 8-inch, 9-inch, and 10-inch spacing. The 22-inch discs with 2½-inch concavity are best in clay soils inclined to be wet and sticky. The straighter discs tend to run clear where the discs with more concavity will tend to clog.

Discs with shallow concavity are more desirable for all soils except in stony land where the blades are apt to contact stone and no serious difficulty is experienced in clogging. The discs with deep concavity should be used in stony land, for the increase in concavity strengthens the disc so that it offers greater resistance to contact with stone with minimum damage.

The wider spacing with fewer discs will provide better penetration in hard soil for a given weight of one way disc. The one way disc is adjustable so that the weeds will be cut cleanly at all desired depths of operation.

Adjustments:

There are three important adjustments on the one way disc.

(a) Frame. The frame must be level so that the frame supporting the hitch will be low. The frame may be levelled by raising the back upon the rear wheel and lowering the frame over the land wheel.

(b) Wheels. The rear and front furrow wheels must be mounted so that the flange operates in the bottom of the furrow with the tire sloping up the furrow wall. The front furrow wheel should run straight ahead, while the rear furrow wheel should be set with lead away from the furrow sufficiently to hold the land wheel straight. The rear furrow wheel frequently requires wheel weights to provide traction to hold the one way disc straight.

(c) Hitch. The hitch of the one way disc must be low so that there will be no pull down by the tractor when checked by a straight line from the bottom of the centre disc to the drawbar. This adjustment is most important; where there is insufficient adjustment to make this possible either the hitch of the one way disc must be extended down or the hitch of the tractor must be raised by adjustment or offset hitch. The horizontal adjustment must be set so that the front disc will cut the proper width of cut.

Rubber Tired One Way Disc:

The one way disc mounted on rubber tires saves from 30% to 10% of the power required to operate a steel wheeled one way disc. The greatest saving is made in the spring when the soil is soft and the one way disc is being used for seeding. The least saving is made when summerfallowing in hard ground where the most of the weight of the machine is carried on the discs for penetration.

Seeding with the One Way Disc:

The seeding attachment for the one way disc has been used almost universally in areas where the one way disc is used. Some difficulty has been experienced in obtaining depth control. Much grain has been seeded too deep, while seed has also been scattered from the bottom of the furrow to the surface of the soil where the seed spouts have been slightly short or have been in misalignment. It is necessary that all seed spouts be uniformly adjusted so that the seed will be uniformly placed in the bottom of the furrow. The hitch must be low on the one way disc so that there is no pull down upon the front of the machine. It has been found necessary to seed heavier to obtain good stands when seeding with the one way disc than with the drill.

Packing One Way Disc Seeding:

Grain seeded with the one way disc is covered with very loose soil. Where wire worm damage is serious it is necessary to pack the soil either when seeding or shortly after in order to control wire worm damage by producing a firm

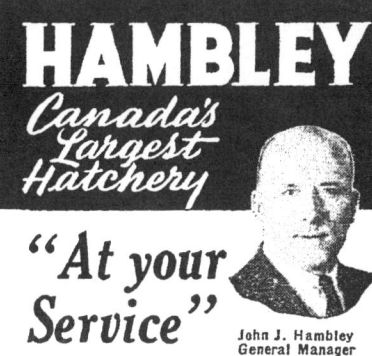

HAMBLEY
Canada's Largest Hatchery

"At your Service"
John J. Hambley
General Manager

When you order chicks three things are very important to you. You must be satisfied that you are going to get GOOD QUALITY chicks from government approved blood tested flocks. You must have confidence that the hatchery from which you buy is DEPENDABLE with a well-won reputation for supplying chicks that are good egg producers; finally you must be able to count on prompt delivery at the time you want your chicks.

When you order HAMBLEY ELECTRIC CHICKS you can be sure you are going to receive GOOD QUALITY CHICKS from a firm that has an established reputation and that is completely equipped to meet every demand for service.

HAMBLEY ELECTRIC CHICKS all come from government approved pullorum tested flocks. HAMBLEY ELECTRIC CHICKS have been ordered year after year since 1927 by thousands of poultry raisers throughout western Canada. Millions of HAMBLEY ELECTRIC CHICKS each season are proving their dependability in egg production by giving flock owners early season volume when egg prices are higher. With 12 hatcheries located at central points in the western provinces you are assured of QUICK SERVICE from a point nearby, eliminating delay and inconvenience.

MAKE MORE MONEY from your poultry. HAMBLEY chick buyers are successful. Order from your nearest HAMBLEY branch. Catalog and price list sent on request.

WINNIPEG REGINA SASKATOON HAMBLEY Electric HATCHERIES BRANDON CALGARY EDMONTON
PORTAGE BOISSEVAIN DAUPHIN ABBOTSFORD, B.C. SWAN LAKE PORT ARTHUR

HAMBLEY'S CHICK ZONE
The Life Saver for Baby Chicks
Don't wait till wings droop and birds get listless and pale. Be prepared. Hundreds of satisfied poultry raisers find Chick-Zone a real Life Saver for Baby Chicks. Just add a teaspoonful of Chick-Zone in the first quart of drinking water.

CHICK ZONE PRICES
Prepaid
12 oz. 75c
6 oz. 40c

Collect
40 oz. ..1.00
½ gal. .1.50
1 gal. ..2.75

Send for your supply today. Have it ready when chicks arrive.

Order from nearest HAMBLEY HATCHERY

seed bed. Subsurface type packers are best, consisting of either the V shaped or the blade type packers.

Packers must be loaded so that penetration will result so that the soil will be packed around the seed. Seed placed uniformly in a firm moist seed bed will germinate more uniformly than when a loose seed bed at irregular depth.

The One Way Disc Harrow:

The one way disc harrow is a machine which has been designed and built by farmers. The 3½-foot disc harrow sections consisting of six discs on 6½-inch centres are mounted independently upon a frame mounted on wheels and drawn on an angle similar to the one way disc. Each gang flexes up and

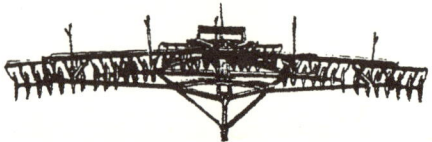

down with variations in the field. The one way disc harrow has been tested and found to operate very well. It cuts cleanly through the stubble and weeds. The soil is disced and placed in a smooth uniform land with no ridging. The power per acre is less than the one way disc by about one third. Its use has increased the tillage per acre per unit to almost double that of the one way disc. The machines run from 3½ to 4 inches deep which is more desirable than deeper tillage. Where the one way disc harrow has been mounted on rubber, depth control and draft per acre has been still further improved and reduced.

Seeding with the One Way Disc Harrow:

There have been only comparatively few one way disc harrows built with the seeder box and attachments but those which were, seeded quite satisfactorily. It was felt that the shallow seeding caused earlier maturity than where seeded deeper with the one way disc.

The Field Cultivator:

The field cultivator is a valuable implement for land tillage for Western Canada. The power per acre is less than for the one way disc. The cultivator must be maintained with sharp clean shovels (sweeps) so that the weeds and root systems may be cut without operating too deeply. The three gang cultivator will operate in combine stubble and one wayed land without clogging. The cultivator drag bars and standards need to be straight so that the shovels are held rigidly in line so that all weeds are cut. The cultivator may be operated at high speed without pulverizing the soil or affecting the quality of work. In stony land slower speeds are advisable.

Blade Weeders:

The heavy trash and stubble required to protect the soil from washing or blowing encourages the building and use of wide sweep cultivators or blade weeders. The blade weeders are heavily constructed, fitted with either the straight or V blades to suit the type of soil being worked. Blade weeders have ample capacity for straw and stubble with no clogging. The blades pass through the soil without much soil disturbance probably not enough for a sure weed kill. The blade weeder must be loaded to hold the blades into the ground and must be supported by the wheels to hold the blades for shallow operation. Soils which scour the blade are most ideal for blade operation. Soils which are inclined to be dry are cultivated with the blade weeder more successfully than the moist soft soils.

The Rod and Wire Weeders:

The rod and wire weeders pull weeds which have been missed when using

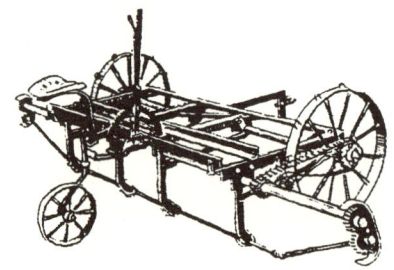

the cultivator or disc harrow. The land should be reasonably uniform and in good tilth for best operation.

The wire weeder works well in stony land which is in good tilth and free from heavy stubble.

Disc Harrow:

The disc harrow has been used a great deal for spring and fall shallow tillage and far too often for summerfallow. The disc harrow pulverizes and ridges

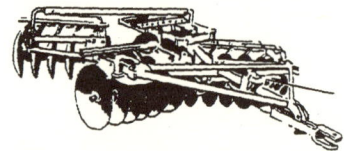

the soil badly, particularly when the soil is dry. The use of the disc harrow should be carefully related to the amount of stubble and condition of the soil so that soil drifting will not be aggravated by its use. The disc harrow is quite apt to be replaced by the one way disc harrow which is a somewhat heavier machine but is one which makes a much more complete cut and leaves the land level.

Where double discing is practised so that the land is left in a level condition and where the soil is in good condition to work, the disc harrow is an economical implement to use.

Spike Tooth Harrow:

The spike tooth harrows, either rigid or flexible, are efficient implements for the destruction of weeds. It is import-

ant that the teeth be sharp and that the soil be in a semi-moist condition so that the land will be lumpy after being harrowed. The speed of harrowing will vary with the moisture content of the soil and type of soil. Where the soil is dry slow speeds are desirable. Where the soil is moist, high speeds are necessary to tear the weeds from the lumps.

The Rotary Harrow:

The rotary harrow is designed with clearance for heavy stubble. It is ideal

to draw behind the one way disc when seeding. All fine weeds which are germinated are loosened and exposed so that wilting results.

Spring Tooth Harrow:

The spring tooth harrow is frequently a desirable implement to draw in heavy clay soil summerfallowed for the first tillage in spring before seeding. The

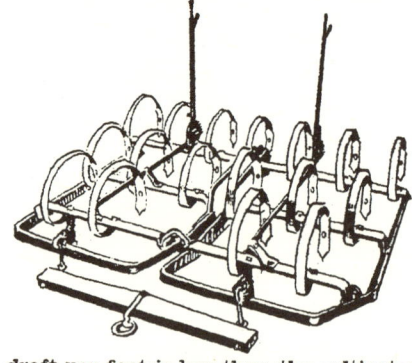

draft per foot is less than the cultivator and the work is deep enough to bring moist clay lumps to the surface to control soil drifting and also destroy small weeds.—*Professor Evan A. Hardy, University of Saskatchewan.*

Tillage and Western Agriculture

There is some doubt of the function of tilling of the soil. Is the soil tilled to produce a seed bed, to control weeds, or to conserve moisture? As a matter of fact, all three are accomplished. Where the rainfall is limited and the yield is definitely related to the rainfall, moisture conservation is essential. Weeds are controlled because weeds in growing use more moisture than the crop.

Research at the Swift Current Experimental Station indicates that where the rainfall was 19.54-inches the possible moisture saved by preventing all weed growth was 5.1-inches, but that where the first cultivation was delayed until May 15, 4.5-inches was conserved, if on June 15, only 3.6 inches was conserved, and when weeds were allowed to grow so that the first operation was July 15, only 1.9-inches of moisture was conserved. (Early summerfallowing pays.)

It was also observed that where the weeds were controlled there was no difference in moisture conserved for the use of any combination of machines.

Average of Four Years:

The moisture conserved in the field was measured to a depth of four feet. When the precipitation was high, the moisture penetrated below this depth.

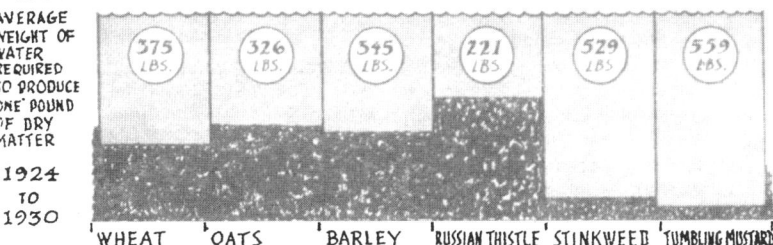

AVERAGE WEIGHT OF WATER REQUIRED TO PRODUCE ONE POUND OF DRY MATTER 1924 TO 1930

WHEAT 375 LBS. | OATS 326 LBS. | BARLEY 545 LBS. | RUSSIAN THISTLE 221 LBS. | STINKWEED 529 LBS. | TUMBLING MUSTARD 559 LBS.

Depth of Tillage:

The depth of tillage best for spring work, summerfallow, or fall tillage depends upon the type and size of weeds to be controlled. Annual weeds must be controlled by shallow tillage which plants weed seeds close to the surface of the soil to stimulate rapid germination and growth. Perennial weeds may need turning up at a greater depth in order to destroy more of the root systems mechanically, thus leaving fewer roots for smothering by the black fallow. Deep tillage is not effective in weed control.

The Conservation of Moisture Under Different Fallow Treatments and in Standard Fallow Tanks

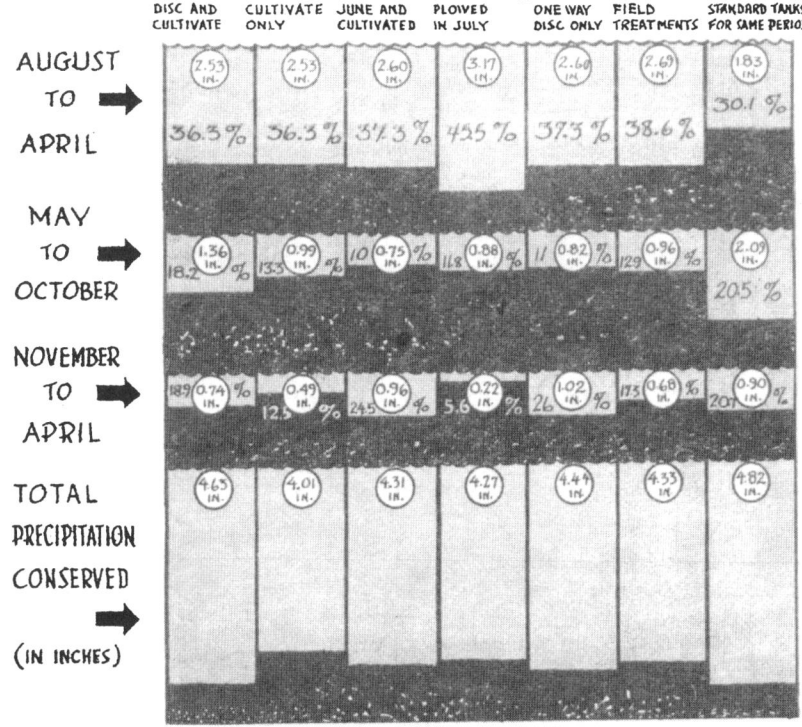

"THAT FRESH"
VICTORIA
GREENMELK CHICK STARTER

Rich in Vitamins and Minerals

STARTS CHICKS into MONEY-MAKING PULLETS

VICTORIA EGG MASH

more than pays for itself in

EXTRA Eggs-EXTRA Profits

VICTORIA HI-HATCH BALANCER

Mixed with Home-Grown Grains
for
High-Quality, Hatchable Eggs,
Strong, Healthy Chicks

"THERE'S A VICTORIA FEED FOR EVERY NEED"

Ask your Victoria dealer or McCabe elevator agent

VICTORIA
FEEDS - BALANCERS - MINERALS

Manufactured Fresh Daily by

McCABE GRAIN COMPANY LTD.
WINNIPEG - REGINA - EDMONTON

In Western Canada where the frost penetrates the soil to a depth of from five to six feet, the frost heaves and expands the soil particles through the action of freezing of water suspended in the subsoil. The frost opens the sub-

soil so that no hard pan ever develops, thus eliminating the need for loosening the subsoil for water absorption or breaking subsoil for root penetration.

Root System Study:

Research on the development of root systems in soils in Saskatchewan has proven that the expansion of root systems is not effected by the depth of tillage, that the frost has opened the subsoil so that free expansion of root systems takes place to four and five feet deep.

Consequently, if subsoiling is not necessary for soil tilth and root development and if weed and moisture control is best where soil is tilled at a shallow depth, all tillage in Western Canada should be as shallow as possible to control the weeds at hand.

Power per Acre of Tillage Machines at 3½ Miles per Hour:

The graph below indicates the horsepower hours per acre required to operate the various tillage machines in clay loam soil under average conditions. The average tractor requires one-fifth of a gallon of fuel to develop one horsepower hour under field conditions. The power variations may be of value in choosing the most economical implement to use where the quality of the tillage will control the weeds and leave the soil in condition to resist wind and water erosion.

Soil Drifting:

Tillage most suitable for soil drifting control is of the type which will leave stubble and trash on the surface and will mix with the surface soil. Decomposition of the stubble and trash is much faster when close to the surface, exposed to the heat, moisture, and bacterial action in the soil.

Tillage most suitable for soil drifting should be lumpy, with a minimum of pulverizing. High speeds of operation are the cause of finely pulverized soils which are subject to blowing, more than the implement being used. Thorough shallow tillage at speeds best suited to the moisture condition of the soil and the implement to cut all weed growth and yet leave the soil in a lumpy condition well mixed with the stubble and trash close to the surface, will resist the action of the wind with minimum drifting.— *Prof. Evan A. Hardy, University of Saskatchewan.*

Fix Up Binder and Combine Canvas

Get out the binder and combine canvases and put them in first class condition now, so you won't have to stop for repairs during the harvest rush. If the front edge is frayed, loosen the front ends of the slats, cut strips of canvas about a foot wide and fold it in the middle lengthwise, then slip the folded strip over the front of the old canvas, and tack and rivet it solidly to each slat. The spaces between the slats can be stuck down on both sides with ordinary rubber cement or with flexible fabric cement. If the flap at the end of the canvas has whipped so short that it lets straw get inside, put on an extra piece by tacking it under the first slat and cementing it to the old flap, and then fasten on shoe strings or thongs so the flap can be tied down to prevent any whipping. Any holes or tears in the canvas can be patched by cementing on a piece.—I. W. Dickerson.

Homemade Corn Cutter

A wheeled type of corn cutter is shown here. The main frame is made of two 2x4's about eight feet long. They are set on edge and tapered together at the front. The platform is made of inch

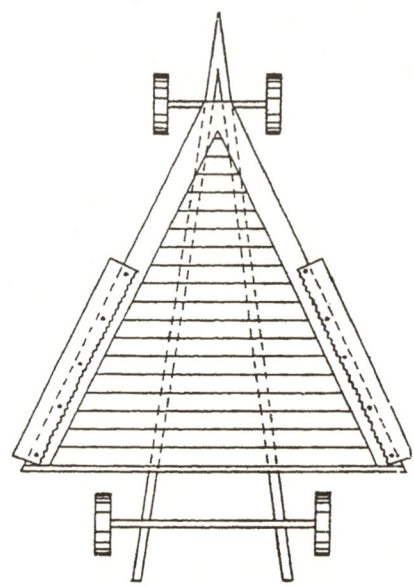

material. A 1x6 hardwood strip 4 ft. 8 ins. long is put at the rear and also 1x6 hardwood cutting V boards the same length at the sides.

The cutting blades are thin pieces of steel. In some places old cross cut saws are used, with the backs ground sharp, but they are not available on the prairie. The frame should be hung from the axles so as just to clear the ground.

Boulder Remover

If you have a medium sized rock that you want moved to the fencerow this device saves the back breaking job of loading it on to a stone boat and is

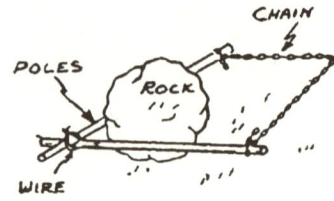

easier to draw than if it is simply dragged with a chain. The members have to be of good strong seasoned wood. Either horses or tractor can be used for the motive power.

Packer after One-Way

Last spring I bought a tractor and tiller combine but I decided I needed a packer behind it, so I got busy and made one. The frame is from an old drill with a through axle, with a second axle for the second row of wheels. Discarded plow wheels were used and they are spaced nine inches apart, ten in the front row and 11 in the back row. The wheels in the back row are spaced in

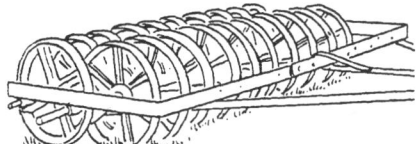

between the front wheels. The axles are 18 inches apart and the width of the frame is 8 feet 6 inches. This is an excellent packer and did not clog when pulled behind the One-way.

Rod Weeder

This rod weeder is easy and inexpensive to build. It consists of two round rods, one inch in diameter, attached to sled-like runners, one at the front and the other at the rear. When in operation only one rod is in the ground at a time. By shifting his weight back and forth the operator can put one or the other into the ground as he pleases. By this means it is possible to clear the rods of any accumulation of trash. When hitched to a tractor, this implement can be operated by one man, by the use of a tight cable attached to the front and rear of the implement and a 100-pound weight to slide back and forth with a rope.

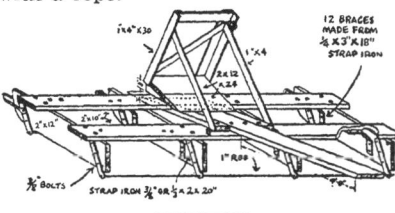

Bait, Seed or Fertilizer Spreader
Cone Hopper Type for Wagon or Truck

The cone-hopper wagon or truck type of spreader has been especially designed at the experimental station, Swift Current, to make possible the construction of a simple spreader from stock materials when suitable discarded parts are not available. Tradesmen may also find this type suitable for local manufacture.

The cone-hopper spreader may be constructed to fit on a wagon or truck box. The machine obtains its power from one of the rear wheels through a belt, preferably round or V-shaped, or a piece of sash-cord to serve as a belt. This is connected directly with the spreader shaft by grooved pulleys, thus avoiding gears. A convenient belt tightener also serves as a clutch. The cone-shaped hopper feeds bait, etc., automatically to the rotating spreader table below. The amount of bait delivered is

MASSEY-HARRIS SERVICE HELPS YOU IN BUSY SEASONS

Through its network of branches and dealers, the Massey-Harris organization gives prompt and dependable service to users of Massey-Harris equipment.

This service is especially helpful during rush seasons—it means so much then in the saving of valuable time and in preventing annoying delays to be able to get needed parts quickly. Use Massey-Harris equipment and buy genuine Massey-Harris parts—they are high in quality and are guaranteed to fit.

Early ordering of repair parts helps you and helps your dealer

MASSEY-HARRIS
COMPANY, LIMITED

Winnipeg - Brandon - Regina - Saskatoon - Yorkton
Swift Current - Calgary - Edmonton - Vancouver
Toronto - Montreal - Moncton

700,000 Copies

1947 T & T SEED DIGEST

EARLY IN THE ARCTIC
FIRST IN YOUR GARDEN

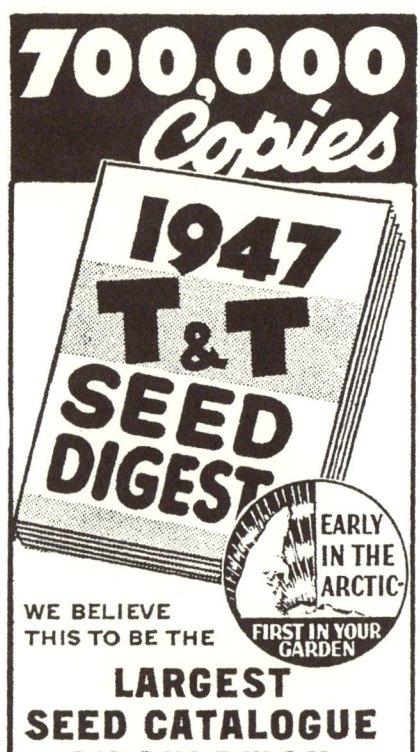

WE BELIEVE THIS TO BE THE

LARGEST SEED CATALOGUE CIRCULATION

EVER MAILED IN THE NORTHWEST

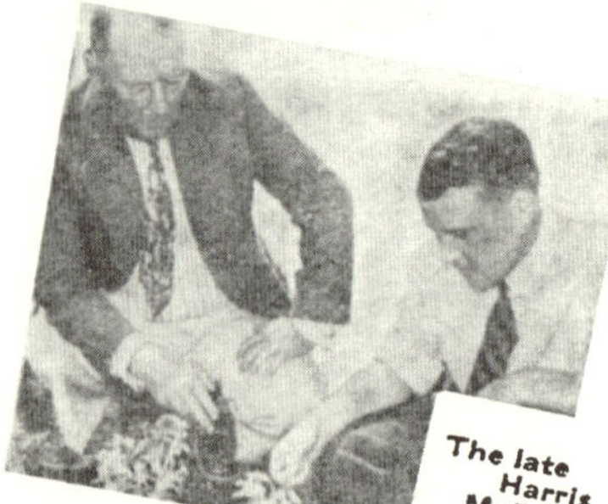

The late Harris McFayden and our Jerry Twomey.

Harris McFayden Was Right . . .

93,000 Home Gardens

by planting T & T Seeds last year showed that they still believe in his fair deal policy. Mr. McFayden before his untimely death in 1941, believed and expressed in his business policy, that the returns from large volume, low priced seed, should go into Plant Breeding and development work for the Benefit of Northern Agriculture rather than into the coffers of large corporate trusts. It is to keep these ideals alive that we dedicate T & T SEED DIGEST to the memory of the late Harris McFayden, Canada's Greatest Seedsman.

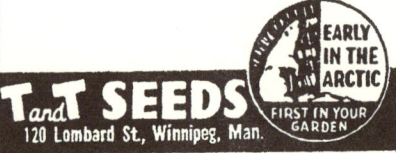

T and T SEEDS
120 Lombard St., Winnipeg, Man.

EARLY IN THE ARCTIC
FIRST IN YOUR GARDEN

regulated by means of a wing-nut on the top.

Parts numbered in the figures are as follows:—

1. Sheet metal cone-shaped hopper, about 36 inches in diameter across the top.
2. Supports for hopper, flat iron, 1½ by ¼ inch.
3. Support for feed control rod, 1½ by ¼-inch flat iron.
4. Lower support for feed control rod, 1½ by ¼-inch flat iron.
5. Feed control rod, ½-inch round iron, threaded at top end, fork-shaped

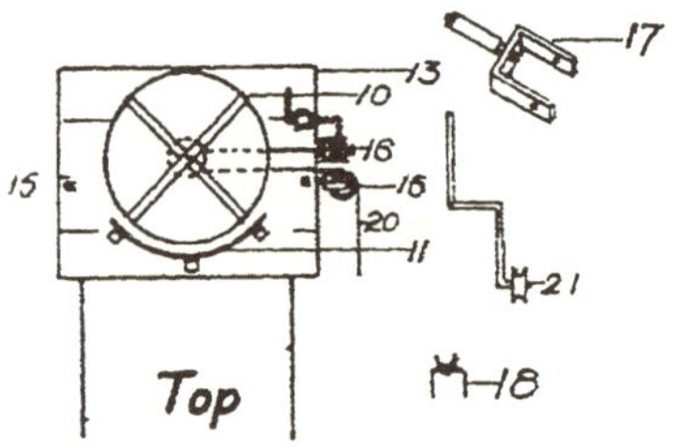

at lower end to connect to spreader table shaft below, and secured by a loose-fitting split pin or bolt.

6. Metal cone and two ¼-inch round iron agitator arms attached to the feed control rod 5.
7. Wing-nut at top end of feed control rod to regulate the size of feed opening between the metal cone 6 and the sides of the hopper.
8. Agitator made of two ¼-inch round iron arms attached to opposite sides of a collar. Collar fitted with a set-screw to enable agitator to be set in proper position on the feed control rod 5.
9. Sheet metal spout (4-inch rainwater 45-degree elbow) may be turned by means of a handle and clamped in any position so that the bait may be dropped on any part of the table.
10. Spreader table, 30 inches in diameter, consisting of four blades made from 4- by 1¼-inch lumber fastened to a wood disk, the top of which is covered with sheet metal. A hub with a bore equal to the size of the propeller shaft is required. The hub may be secured to the shaft by two set-screws at right angles to each other or, better still, it

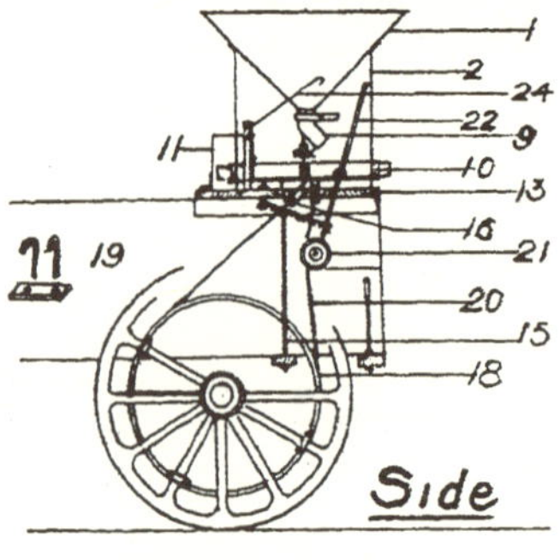

may be keyed and set-screwed. Rubber belting is fastened to the tips of the blades. The pieces of belting should be of sufficient length to just contact the shield No. 11.

11. Shield of sheet metal at back of the fan blades, about 10 inches high and fastened to wood platform No. 13 by pieces of 2- by 3/16-inch strap iron bent to suit.
12. Grooved pulley about 4 to 6 inches in diameter fastened to spreader table propeller shaft.
13. Wood platform constructed of 2-inch lumber. Hopper and spreader table are mounted on this platform. Platform may be moved back or forward to secure the right tension of the belt. It is secured in position by tightening the tie rods.
14. Brackets to support the spreader table propeller shaft. Made from 2- by ¼-inch flat iron. Also holds a bushing or bearing for the lower end of the propeller shaft.
15. Tie rod of ½-inch round iron connected to wood platform on top of the box and a 2-by-4 cleat below the bottom of the box to hold platform in position.
16. Grooved guide pulleys about 4 inches in diameter.
17. Swivel bracket for grooved pulley No. 16. Brackets made from 1½- by 3/16-inch flat iron. Swivel formed with piece of ½-inch pipe over a round iron stem.
18. Pulley or circular band iron drive wheel clamped to wagon wheel. May be constructed from discarded wagon or drill wheel tire and reduced to 36 inches

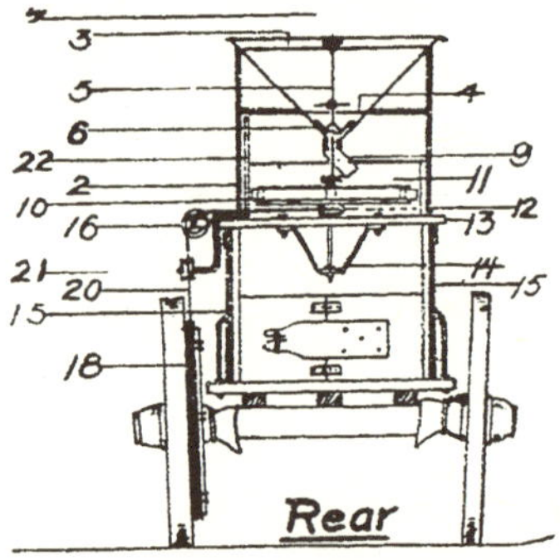

in diameter. A groove for the drive rope or belt is provided by splitting a piece of ¾-inch rubber hose riveted around the outside of the drive-wheel. (See small sketch No. 18.) A 14- to 16-inch pulley may be used on truck wheels.

19. Hooks of ½-inch round iron and a plate made from 1½-inch by 5/16 flat iron will serve to clamp circular iron band drive to wagon wheel.
20. Rubber "V" belting or ½-inch window sash cord.
21. Grooved tightener-pulley 4 inches in diameter.
22. Crank arm from tightener-pulley made from ⅝-inch or ¾-inch round iron. Crank arm is pivoted on the wood platform with a plain strap iron bearing or keeper. A coil spring fastened between the platform and crank arm provides tension for the belt.
23. Support for ⅜-inch iron hook.
24. Hook of ⅜-inch round iron holds tightener-pulley in slack position in order to move wagon or truck without operating the bait spreader.

Header Barge

This header barge was designed at the Dominion Experimental Station at Swift Current in 1933, and has been built and operated successfully by many farmers since that time.

The special feature of this header barge is the automatic discharge of the stack. With one movement of a lever, the dump bottom is tilted by the weight of the stack. At the same time the rear gate is lifted and the stack slides out.

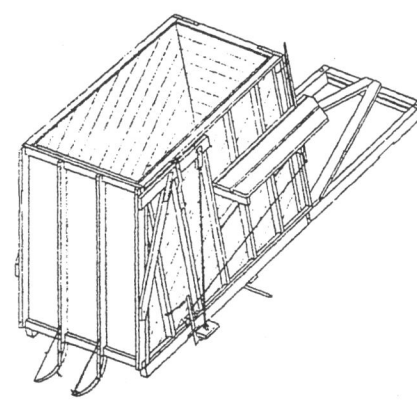

As soon as the stack is discharged, the rear gate falls back to its former place and lifts the dump bottom back to its normal loading position. Thus every movement becomes automatic after the first and only manual operation of the trip lever.

Suggestions to Barge Builders

1. The successful barge must be simple in design—a minimum of working parts.

2. The type of barge should be such that it will prevent any loss of time when cutting.

3. The barge must be inexpensive to construct.

4. The size of the barge box must be made to suit threshing methods.

5. The size of wheels and power available to pull the barge will influence the size of the stack or box.

6. The stacks should be as high as possible. Width not over seven feet and as long as it is convenient to build the box.

7. Floor and sides of the box should be lined with smooth lumber or tin to make dumping easier.

8. The tilting floor, pivoted on the main wheel axle is the most satisfactory type of floor construction.

9. The solid frame and sides with tilting floor is cheapest to construct.

10. The rear gate hinged on the top and opened by the floor makes an automatic, time saving barge for cutting without stopping when discharging the stack.

11. Pieces off old machinery or granaries, etc., should be used wherever possible in constructing the barge to cut down costs.

12. Old tractor, swather, header or other large wheels are the most suitable for use on the rear of the barge.

13. Any low wagon, truck or implement forecarriage wheels are suitable for front barge wheels.

14. The larger the rear wheels, the lighter the draft of the barge.

15. The wider the face of the barge wheels, the lighter will be the draft.

16. The higher the rear wheels, the greater will be the slope of the floor and the more positive the discharge of the stack.

17. The barge box should be wider at the rear than at the front—usually 6 inches wider at the back.

18. The type of cutting device should be decided upon or secured before starting to build the barge.

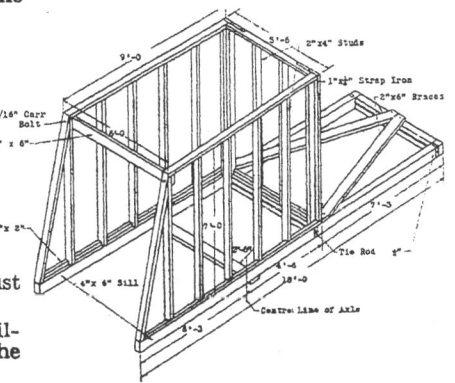

REAR AND SIDE VIEW OF BARGE FRAME

19. Elevators mounted on the barge and run from the rear wheels of the barge are the most satisfactory type of loading devices.

20. End delivery swathers, headers, binders or detachable combine table may be used with the barge which has a permanent elevator.

21. Elevators attached to the barge make it possible to build a higher stack.

22. The higher the stack, the more grain will be under one top—thus reducing the chances of spoilage.

23. Elevators mounted on the cutting device and high enough to deliver into the barge box are frequently used.

24. Delivery of the cut

HALLMAN SHUR-SHOCK
PRODUCTS
and
HANDY FARM TOOLS

Hallman Shur-Shock Fencers offer every livestock producer the most modern, efficient and economical method of stock control on today's market; available with all accessories in two models, the Hallman Senior Model, and the Hallman Standard Model. Simplify your fencing problems by ordering a Shur-Shock Fencer today.

This all steel, handy grip hog holder will last indefinitely. With one hand, one man can easily hold the largest hog for ringing, vaccinating, ear tagging, worming, etc. One end for holding large hogs and the smaller end for pigs.

Hallman Shur-Shock Livestock Prodder will solve your shipping problems by conveniently and effectively controlling stubborn animals. Sturdily built, two feet long, sure-grip handle.

JACKALL JACK & FENCE STRETCHER

This Jack is indispensable for many jobs around the farm or shop. Pulling and hoisting can be done up to three tons. Jacks from 4½-inches to the top of the beam. Lifts and lowers easily with a strong lever action. All steel construction —no castings. Capacity three tons. Three sizes. 32-inch, 48-inch, 64-inch. Weight 26, 31, 36 lbs.

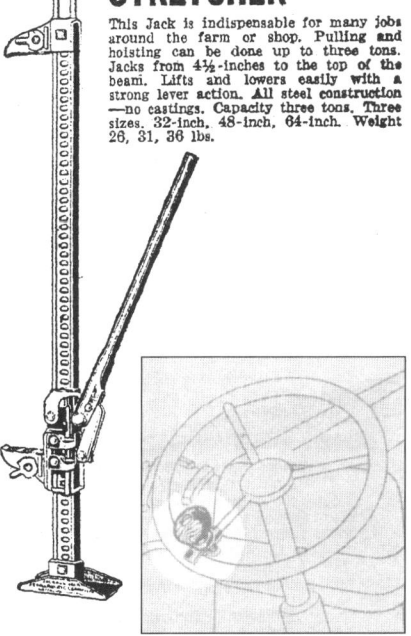

Handle your tractor or truck the easy way with a Hallman Heavy Duty Wheel Spinner. You will wonder how you ever managed without one. Attaches to steering wheel easily.

J.C. HALLMAN MFG. CO. LIMITED
WATERLOO   ONTARIO

grain over the front of the box has been found to be the most satisfactory with the best "keeping" stacks as a result.

Suggested Bill of Material

The list of materials used in constructing this barge is suggested in order to guide the builder in gathering enough suitable material. The kind,

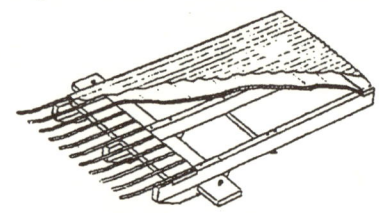

sizes and quality of material used by the farmer for his own barge will undoubtedly be determined by his own pocketbook and the amount of scrap material from old buildings, old discarded machines and other places which are usually found in every district.

Lumber Required

9 pcs. 2-in. x 4-in. 8 ft. long
3 pcs. 2-in. x 4-in. 12 ft. long
9 pcs. 2-in. x 4-in. 14 ft. long
2 pcs. 2-in. x 4-in. 16 ft. long
2 pcs. 2-in. x 6-in. 14 ft. long
1 pc. 2-in. x 6-in. 10 ft. long Oak or Fir
1 pc. 2-in. x 8-in. 12 ft. long (Platform)
1 pc. 3-in. x 6-in. 10 ft. long
1 pc. 3-in. x 6-in. 18 ft. long
2 pcs. 4-in. x 6-in. 18 ft. long
1 pc. 4-in. x 6-in. 10 ft. long
280 board feet Shiplap—14 foot lengths.

Other Material Required

Frame:
Flat iron for reinforcing corners.
Two tie rods—½-inch round iron, 5 ft. 7-in. and 5 ft. 3 in. long.

Running Gear:
Two rear wheels about 48 ins. in diam. and a 9-in. face.
Steel axle to suit wheels, about 2 ins. in diam.
Two front wheels, small trucks off some farm implement suggested.
2⅝-in. U-bolts 2 ins. x 12 ins.

Gate:
Two brackets, four pieces 3 ins. x ¼ ins. x 2 ft. 4 ins. flat iron.
Four pieces flat iron, 2 ins. x ½ ins. x 1 ft. 4 ins.
Two pieces flat iron 1½ ins. x ⅛-in. x 3 ft. 8 ins.
Two pieces flat iron 1½ ins. by ⅛ in. x 1 ft. 4 ins.
Two pieces round iron, ⅝-in. diam. 2 ft. long.
Two 10-in. clevises—2 pieces, 1¼-in. x ¼-inch x 2 ft.
Two ½-in. rods, 6 ft. long.

Dumping Floor:
Eleven teeth—11 pieces 1 in. x ¼-inch x 2 ft. 11-ins. flat iron.
Three pieces—3-in. x ⅛-in. x 5 ft. flat iron for skids.
Two pieces angle iron for trip catches.
3½-in. U-bolts, 2½-ins. x 10-ins.
Two ⅝-in. eye bolts.
One piece of pipe about 5 ft. to fit over axle.
Two sheets of corrugated galvanized iron, each 2 ft. 6 ins. x 9 ft. 0 in.
Sheet iron may be used by making minor changes in the floor construction.

Tripping Device:
One lever—5 ft. long.
Two arms—7 ins. long.
One shaft—¾-in. round iron 6 ft. long.
Two trip-hooks—two pieces flat iron 2 ins. x 1-inch x 2 ft. 0 in.
Two rods or cables—¼-in. diam.

Header for the Barge

While push-headers and binders with elevating equipment can be hitched to a barge, the most suitable heading machine for the barge is the swather with an end delivery. The design of swathing machines varies. Some swathers deliver cut grain over the end of the table clear of the ground wheels. Others deliver the grain on the ground on the inner side of the ground wheels. The extra end delivery type is more readily equipped with an elevator to deliver the grain over the side of the barge. See Fig. (1). Since only the barge floor

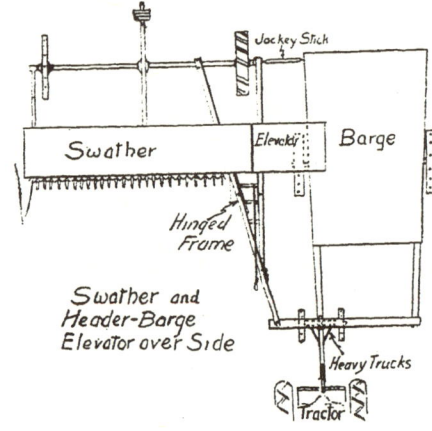

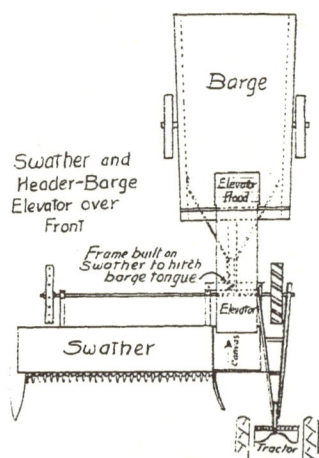

dumps and as the sides do not tilt, the dump floor type of barge is quite suitable for the side delivery elevator. Swathers which deliver grain on to the ground on the inner side of the drive or bull wheel are easily equipped with an elevator to carry the grain over the front of the barge. See Fig. (2). Dimensions are not shown in the sketches. It is expected that the builder will use his own ingenuity in adapting such cutting machines as he can obtain to be used as a header attachment for the barge.

Portable Grain Elevator
Drag or Flight Type

This type of elevator may be found on many of the old threshing machines. Such elevators are best driven from the top to prevent buckling of the loaded flights. When driven from the bottom, the flight chains must be kept tight at all times to avoid buckling. Wheels can be used instead of skids: Skids on the ground, however, prevent undue vibration and permit easier shovelling into the hopper. The hopper is supported on its own framework rather than on the leg. Wood with sheet metal lining makes the strongest and most suitable hopper. The framework supports both the counter shaft drive and the elevator leg. Heavy gauge sheet metal makes the best slide feed gate. An 18 to 20 foot elevator, with 6x2½-inch flights, will elevate 100 to 200 bushels per hour using 1½ to 2 horse-power.

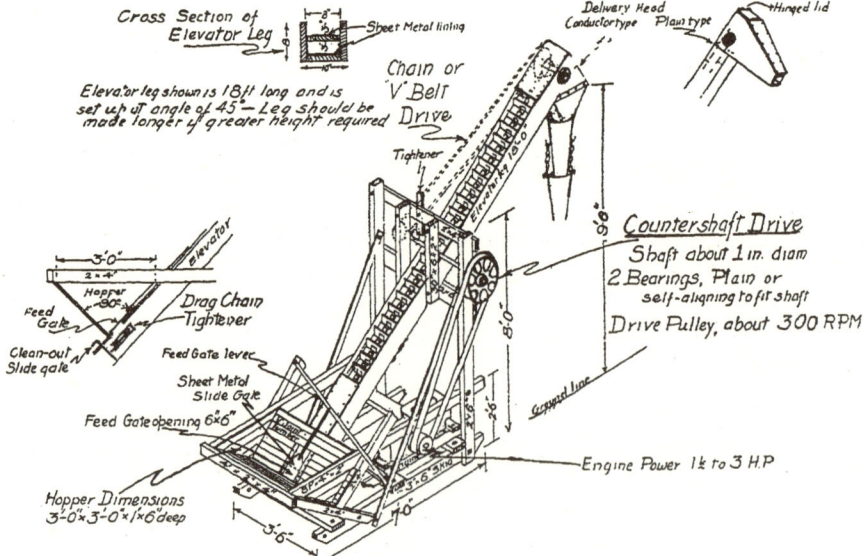

Seed De-Awner and Dresser

Awns of barley and oats and appendages of other seeds may be removed with a combined auger and peg machine. Seed to be treated is placed in a hopper. The seed passes through a 4½-inch diameter hole into the auger housing. The combined auger and steel pegs revolve 300 to 500 r.p.m. depending on the kind of seed to be processed. As the auger and pegs revolve, the seed is moved along so that both the auger and the pegs break off the awns, etc. Three pegs arranged in a spiral at the end of the auger shaft complete the de-awning process and force the seed upwards and out of the 4½-inch opening in the end of the housing. When the awns, etc., are difficult to remove, the opening in the end of the auger can be partly closed by means of a slide with a sloping cut-off edge. When the outlet hole is partly closed the auger then compresses the seed in order to force the seed through the reduced outlet. The revolving auger and pegs working in the closer packed seed are then able to break off the tougher awns, etc. The slide at the outlet is the only control required, and this is used solely to meet the varying needs of the different seeds. The machine is cleaned out by turning it upside down and sloping it downward towards the outlet end. The auger is revolved by hand to complete the cleaning.

A 6 or 7-inch auger, such as found on many combine harvester threshers, is used. The auger shaft should be about 44 inches long. The auger flight is 32 inches long so that 6 inches of the auger shaft protrude at each end. The auger shaft is drilled with ½-inch holes to receive ½-inch round steel pegs or pins as shown in the sketch. These are brazed or welted in the holes. The length of the pegs should be equal to the diameter of the auger. The auger housing is fitted with a hardwood disc 1 to 1½ inches thick at each end. Plain iron

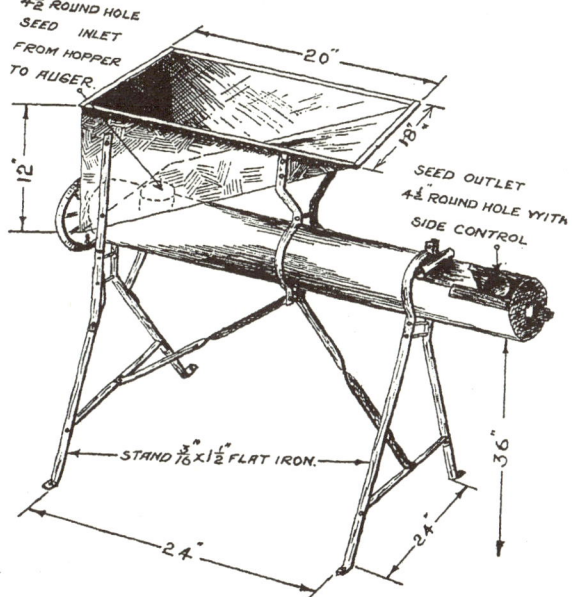

or brass bushings in the ends serve as bearings. The auger housing is fastened to the wood discs with ¾-inch wood screws. The 4½-inch round openings at each end are cut with a cold chisel using a round fence post for a block. The opening is then filed smooth. Note inlet opening is on top of the auger housing, while the outlet opening is placed a little on one side so that the seed will fall only on one side of the auger housing. Do not make the outlet opening in the bottom of the auger housing, as that will reduce the efficiency of the augers and pegs in the de-awning process. The slide and guides are made of about 22-inch gauge sheet metal. The drive pulley may be of the popular "V" belt type or just a plain kind. Speed of the drive pulley can be from 300 to 500 r.p.m. Use as large size diameter pulley as possible. The legs and braces are made of 1½x3/16-inch flat iron. This can be bent cold to the desired shape. Drill 5/16-inch holes as required and use 5/16-inch machine bolts and washers for assembling.

FETHERSTONHAUGH & KENT

REGISTERED PATENT ATTORNEYS

CANADA AND UNITED STATES

422 C.P.R. BUILDING, WINNIPEG

CECIL C. KENT,
Consulting Engineer
Fellow, the Patent Institute of Canada

F. B. FETHERSTONHAUGH, K.C.
Of Osgoode Hall (1890-1945)

A Better Battery...
FOR Every FARM NEED

| HOME RADIO BATTERIES | ELECTRIC FENCE BATTERIES | LANTERN BATTERIES | IGNITION BATTERIES | TELEPHONE BATTERIES |

RAY-O-VAC (CANADA) LIMITED - WINNIPEG
MANUFACTURERS OF EVERY TYPE OF DRY BATTERY

**FOR ECONOMY - LONG LIFE - BETTER PERFORMANCE -
INSIST ON RAY-O-VAC BATTERIES**

HOUSEHOLD CEMENT — For mending china, glass, crockery; repairing wood, metal, fabrics, leather articles, furniture, books, toys, shoes. Excellent for model making.

PLASTIC WOOD — For repairing wooden articles, filling cracks and holes. Can be nailed, screwed, sawed and sanded. Takes paint and varnish.

Wood Splitter

This is the wood splitter that J. H. Cooper, of Neelin, Man., rigged up. He got a pulley from an elevator. It has a 12 inch face. To the rim on one side is riveted a wedge, faced with two leaves of an old auto spring, brought to a sharp edge in front. The splitting block is made as shown, with a notched iron plate. The belt from a 3½ h.p. engine runs on the other side of the wheel. The stick of wood is held on its side and

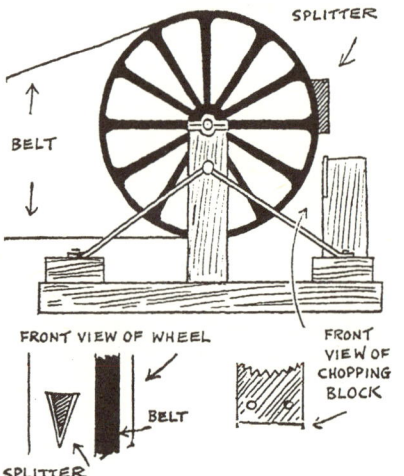

pushed in so that the wedge shaped splitter comes down on it. If the splitter sticks, it simply throws the belt and no harm is done. As fast as a man can feed the blocks to the machine, it will split them.

Power Sweep for Hay or Stooks

The time to make a power sweep is not when it is needed but in some slack time, if any, before the crop is ready. This design has simplicity resolved into its simplest elements and at the same time it works. It was designed by S. L. Tallman and Prof. G. L. Shanks, and published by the Manitoba Extension Branch.

As shown, the push arms (1) are run inside the front wheels of the tractor. The rear of the push arms butt against the drawbar frame or carry to the rear of the drawbar where they are bolted. Or they may be carried outside the wheels directly back to a crossbar anchored securely to the drawbar. The crossbar (2) is suspended to the tractor frame carrying the rear of the sweep eight inches from the ground. Chain (3) on the push arms is fastened to the tractor frame to prevent the sweep from swinging while the other chain (9) is used to lift the sweep off the ground while moving. The sweep runs on the ground while working. The hinges are heavier than can be bought in the store and are made by the blacksmith. The unloading gate, at (16) is hinged with shoes (19) shaped to slide over the ground with a forward motion, but hook into the ground when you back the tractor, so that the gate starts the load to move off the sweep. The sweep operates most satisfactorily with a rubber tired tractor moving at 6 to 7 m.p.h.

This is the bill of materials:

Item No.	Details
1—Push Arms:	4—2x8 (8' to 10')
2—Cross Bar:	1—2x8 (4' to 5')
3—Logging Chain:	Approximately ⅜"x7'
4—Tie Bar:	2—2x4—6'
5—Hinges:	2—¼"x4"—7"
6—Rear Crossbeam:	2—2x8—12'
7—Front Crossbeam:	1—2x6—10'
8—Teeth:	10—2x4—10' clear fir
9—Lift Chain:	Approximately ¼"—15'
10—Top Cross Piece:	1—2x4—12'
11—Middle Cross Piece:	1—1x8—12'
12—Upright:	4—2x4—4½'
13—Side Rail:	2—2x4—7'
14—Side Rail:	2—2x4—9'
15—Diagonal Brace:	2—2x4—5½'
16—Connecting Straps:	4—¼"x1½"—18"
17—Gate Uprights:	2—2x4—4'
18—Gate Boards:	2—1x8—11'
19—Gate Shoes:	2—⅜"x3"—18"
20—Diagonal Braces:	2—¼"x2"—2½'.

Bolts: For convenient removal and adjustment it is recommended that only machine bolts with washers on both sides be used. 39—⅜x4½"; 24—⅜x6"; 16—¼x3½"; 6—⅜x2¼" (for shoes). Lag Screws: 12—⁵⁄₁₆x5". Washers: ½ pound ⅜".

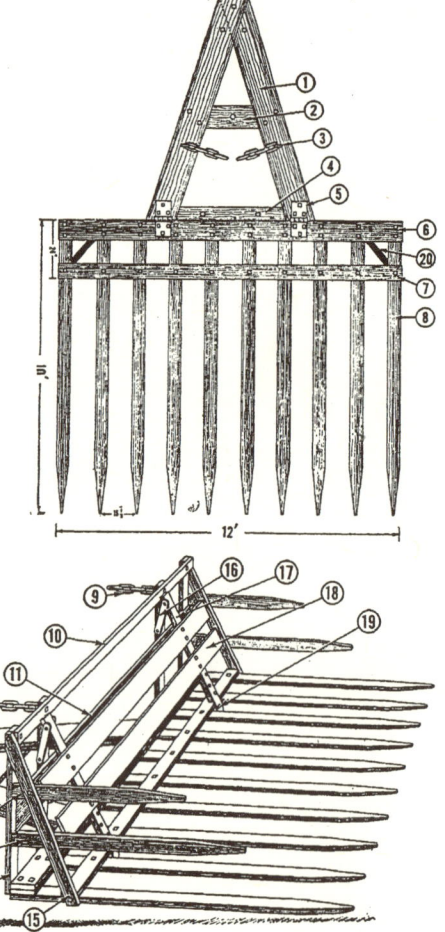

Portable Grain Elevator
Auger Type

The auger type elevator is an easy type to build. It can be made self-cleaning and, therefore, especially suitable to seed growers. The auger can be made to remove the awns of barley and oats by restricting the flow at the delivery end so that the auger is made to

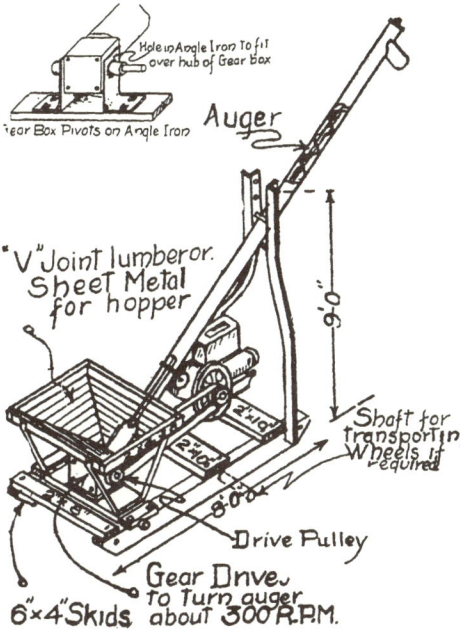

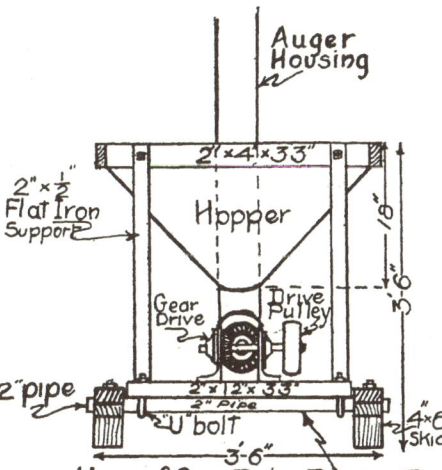

churn the seed against pressure. A bevel gear assembly, either open or enclosed in a housing, is required to drive the auger at the bottom end. The main bearings of the bevel gear assembly or the bearing hubs of the gear box should be made to protrude through heavy angle iron supports. The auger can then be raised at any angle without interfering with the drive shaft. The hopper is supported on 2x½-in. iron straps or brackets. A sheet metal feed gate is required to control the flow of grain into the auger. A four-inch auger, turning 300 r.p.m. at about a 45 degree angle, will deliver about 175 bushels per hour and require about 1½ horse-power. A six-inch auger, operated at the same speed and angle, will deliver about 400 bushels per hour and require 2 to 3 horse-power.

Pick-up Box for Tractor

Here is a sketch of a pick-up box to fit almost any tractor, which will be found very convenient around the farm.

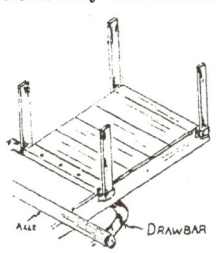

The side pieces are 2 by 6's six feet long, the bottom is one-inch boards except the one over the drawbar, which is a heavy 2 by 6. These boards can be nailed or bolted to the side pieces as preferred. The bottom 2 by 6 is bolted across the top of the drawbar and the front ends of the side 2 by 6's slip under the axle. The width and the distance from the axle to the kingbolt will vary with different tractors. One can easily see how convenient it is when the tractor is brought up at noon and night for refueling.

Stone Boat From Binder Platform

An old discarded binder platform will serve as a stone boat for gathering small stones. It polishes smooth and is easier to draw than a wooden boat. The guards are removed and rods attached as shown. — Albert Leroi, Rose Valley, Sask.

Snow Plow or Ditcher

There are quite a number of walking plows scattered around the country. Even on the open prairie a few superannuated breaking plows are left. This shows how a walking plow can be rigged up to make a snow plow or a shallow ditcher. On the level prairie a very shallow ditch will frequently drain the water away from buildings in wet seasons and avoid the accumulation of mud. A furrow soon fills in, but a wide shallow ditch will continue to function for a long time.

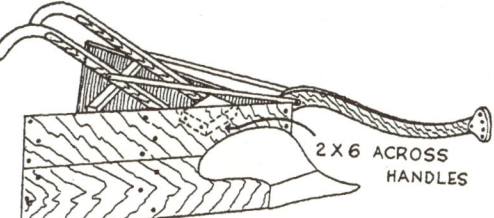

VITAMIN STRONG
VICTORIA
DAIRY BALANCER
Mixed with Farm Grown Grains

IMPROVES MILK FLOW
MAINTAINS HERD HEALTH
INCREASES TOTAL BUTTERFAT

Contains protein, calcium, phosphorus, carbohydrates and other milk-making elements.

IT REALLY MADE A MILKER OUT OF ME

VICTORIA
POULTRY BALANCER

has all the necessary vitamins, proteins and minerals for either high egg production or growing pullets.

VICTORIA HOG BALANCER

ensures health, growth, perfect condition — hustles hogs to market in record time.

Ask your Victoria dealer or McCabe elevator agent

VICTORIA
FEEDS - BALANCERS - MINERALS

Manufactured Fresh Daily by

McCABE GRAIN COMPANY LTD.
WINNIPEG - REGINA - EDMONTON

Wind Blast Seed Grader

Many light and small seeds cannot be graded satisfactorily by means of sieves or other mechanical means of measurements. Such seeds, however, can often be graded according to weight by means of a wind blast from a fan as used in the common fanning mill and various other types of seed cleaning machinery. To make more use of the wind grading method of separation, a wind blast machine can easily be constructed. The chief parts are:

1. A simple wood or iron frame, sides and wind chamber enclosed with wire mosquito screen at one end.
2. A hopper with slide control feed gate.
3. A vibrating pan below the feed opening of the hopper to feed the seed in an even thin stream into the wind blast.
4. A crank shaft or an eccentric on shaft with a short pitman arm to provide a vibrating motion to the feed pan. The length of stroke should be about ½-inch and the number of vibrations should be about 400 per minute.
5. Two pillow block bearings for the crank or drive shaft.
6. A fan and housing to provide the wind blast. These may be obtained from an old fanning mill or a small harvester combine or threshing machine.
7. Two adjustable divider boards and "V" shape partitions for separating the seed into three grades as it falls.
8. Three pans or boxes to catch the graded seeds.
9. Two bearings for the fan shaft. These can be two ordinary plain bushings fitted loosely in a hole in a bar of strap iron so that they will be self-aligning. The bar is fastened across the air inlet on each side of the fan housing.
10. Two "V" belt step pulleys each having three to four different size pulleys. These can be purchased at small cost. One is fastened on the fan shaft. The other is fastened on the main drive

or crank shaft. A rubber "V" belt or a piece of ½-inch round leather belting, such as is used for washing machine drives, is used to run the fan from the drive shaft. An adjustable idler pulley may be required to take up the slack in the belt.

11. A plain or "V" belt pulley is used for the main drive pulley, which is fastened to the drive shaft. Speed of this drive pulley should be about 400 r.p.m.

Operation

Seed is placed in the hopper. The feed gate is opened so that the seed is fed in a wide thin stream into the horizontal air stream from the pan. The fan blast is controlled by means of regulating the speed of the fan with the aid of the step cone pulleys. Further control can be effected by varying the amount of air admitted to the fans by the conventional type of slides on each side of the fan housing. As soon as the machine is set in operation, the divider boards are adjusted to separate the seed as it falls through the air blast into three grades so that each pan below will contain grades of seed as required. If it is desired to use sacks instead of pans or boxes to catch the seed, the machine can be set up on a stand high enough for the purpose and the sacks hung below the machine.

Litter Carrier

Here is a sketch of a litter carrier to carry straw from stack to barn. When loaded, pick up the handles and pull the load. The skids will slide easily on the

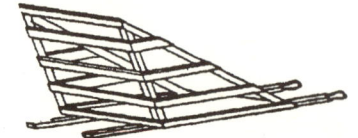

ground or snow. I used light 12-foot poles for handles, spaced two feet apart and the straw frame is six feet long and four feet wide, with the back about three feet high. Laths spaced four to six inches apart do for bottom and sides. The strength and rigidity of such a frame will be greatly increased if the joints are coated with water-proof glue before nailing.—I.W.D.

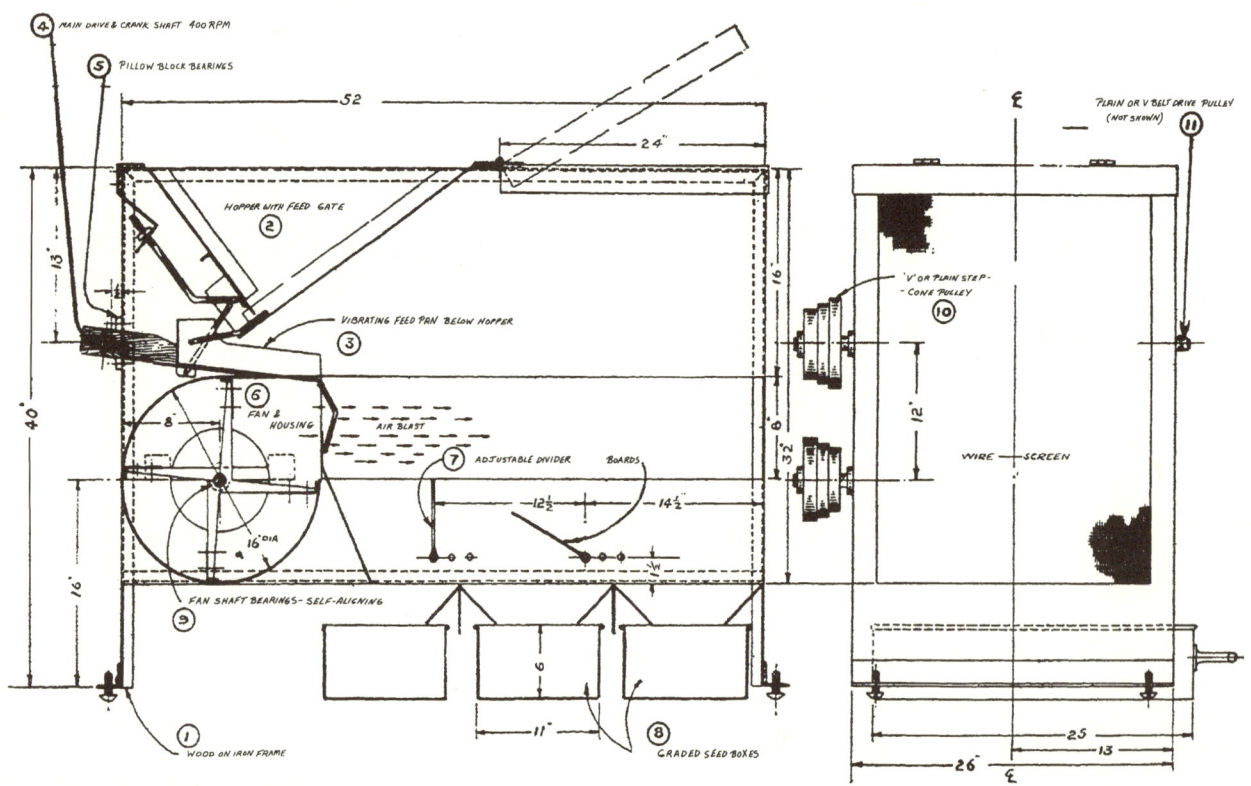

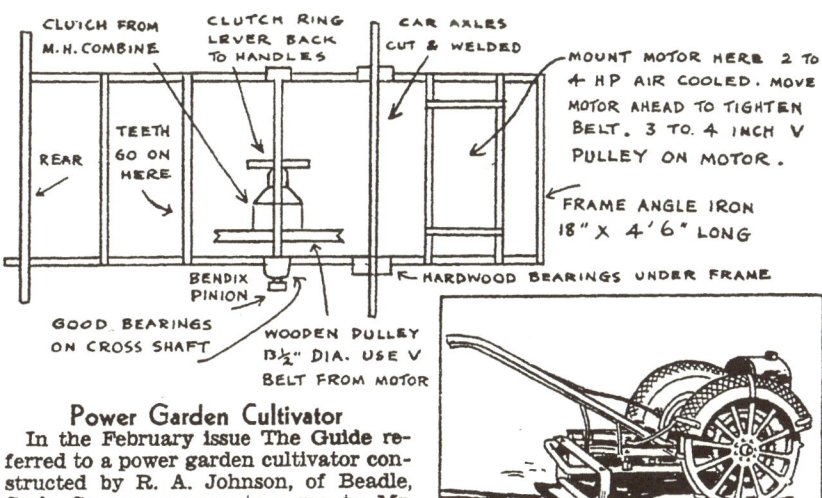

Power Garden Cultivator

In the February issue The Guide referred to a power garden cultivator constructed by R. A. Johnson, of Beadle, Sask. So many requests came to Mr. Johnson and The Guide for a detailed description of the cultivator that further information was supplied by Mr. Johnson. He says:

"It is not hard to build, with the help of a local blacksmith. Some welding and lathe work is needed. The chief difficulty at present is to get a suitable engine. We are using a Wisconsin air-cooled 2 to 4 h.p., depending on speed.

"The drive is taken by a V-belt from a 3 to 4-inch pulley on the engine to a 13½-inch wooden pulley mounted on a counter shaft, along with the clutch from an old style Massey-Harris combine with the low grain tank. The lever from the clutch goes back over the handles to reach the operator.

"The counter shaft should be about an inch in diameter in good bearings. On the right-hand end, looking from the rear, is spot welded a Bendix starter pinion which is meshed with two ring gears from a model T Ford. These are mounted on the right-hand wheel, either by bolting to the spokes, or better still a steel plate is bolted to the hub where the brake drum goes.

"I am using model T wheels with lugs welded to the rims, and Ford axles cut and welded. Wooden bearings are all right on the main axle. No differential is needed; good long handles made from piping give the leverage to turn it easily. The frame is made of angle iron and the teeth from a spring tooth cultivator can be used; also blades made from car springs. I prefer the blades for getting all the weeds.

The teeth or blades should be set so they will cut outside the wheels. Made to cultivate a strip 22 to 30 inches is wide enough to handle and to cultivate. The mounting of the ring gear and starter pinion must be accurate and true. The only thing that wears is the small Bendix pinion and it is easily replaced.

"This cultivator is by no means a toy but is quite practical. We have had ours for four years and would not want to garden without it.

Repairing Gas Engine

It is easy to tell whether an old engine is worth repairing, no matter how old and disreputable it looks. Work a little heavy cylinder oil around the piston and see if the piston holds reasonably well against compression. If so, the cylinder walls are probably not scored or cracked and an easily installed set of new rings will make it like new. If the oiled piston does not hold compression, remove the cylinder head, move the piston to the crank end of its stroke, and feel carefully all over the cylinder walls for deep scores and cracks. If the walls are smooth, the lack of compression is probably due to the rings being stuck, and new rings should make it OK.

If the flywheels wobble when they turn, the crankshaft likely is sprung. A good mechanic can probably press this out. Better have him check it before buying. Loose bearings usually can be rebabbited at small cost. Bad valves or valve seats usually can also be ground or repaired at small cost.

Grain Swath Lifter

In certain seasons when swathing is practised, it may be necessary to lift or move a swath which has started to grow or is not drying properly. Some farmers have used the side delivery rake for this purpose but usually found

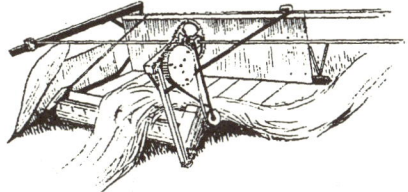

it unsatisfactory. The side delivery rake will move and turn the swath but the lower side of the swath is not raised up onto the stubble. Because of this, should the weather be adverse after turning, the swath is left in a worse condition than before, as it is open to the weather and is resting on the ground.

Some farmers have attached combine

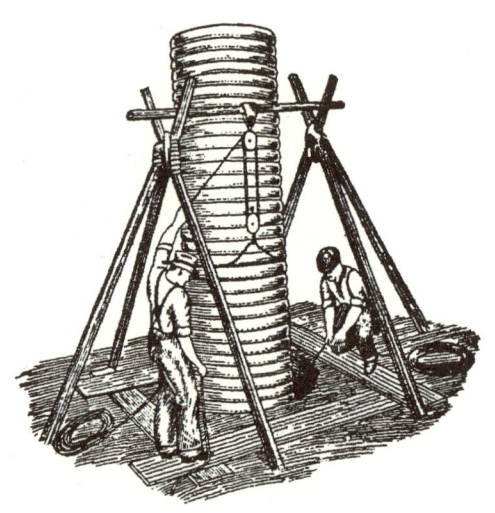

WESTEEL
Corrugated Galvanized Well Curbing

The best possible curbing for wells of any depth or diameter.

The high quality of steel used is a guarantee against rust and corrosion.

Deeply corrugated sheets, perfectly circled, securely rivetted, withstand any amount of pressure, protecting against cave in.

The snug, close-fitting.

Slip-Joint

well lapped, prevents seepage and keeps water clean.

Sanitary, vermin-proof, rust-proof, frost-proof—gives a lifetime of satisfactory service.

Write for prices, etc.

Diam.	Gauge
8"	22
10"	22
12"	22
15"	22
18"	22
24"	22
30"	22
36"	20
42"	20
48"	20
60"	20
72"	20

Order by Diameter

WESTEEL PRODUCTS LIMITED
WINNIPEG
REGINA   SASKATOON
CALGARY   EDMONTON   VANCOUVER
MONTREAL   TORONTO

---

pick ups to swathers. The swath is simply picked up onto the swather table and deposited in a new location on upright stubble. The swath is right up on the top of the stubble which allows the combine to pick it up more readily. If the weather is poor after moving the swath, it will remain in good condition for a long period of time. The swath is lighter than when first cut and the stubble is thus able to hold it up more easily.

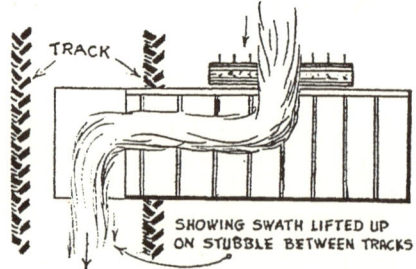

Any farmer who practises swathing usually has the pick up and the swather. Hence if it becomes necessary he can easily put the pick up on the swather and go to work. A simple way to do this is to detach the reel slats and arms. Attach a split V pulley to the reel pipe and another V pulley to the pick up. Secure a belt of the proper length. Use a cross belt.

Note that for a good job of picking up the speed of the pick up should be slightly greater than the forward speed of travel. Hence in selecting V pulleys get suitable sizes so the pick up and the forward speeds are correlated. If the pick up is travelling too slow the swath will push and if travelling too fast, the swath ribbon will break.

When the pick up is attached, make sure that you get it properly located on the table so that the picked up swath will drop on stubble which has not been disturbed by tractor wheels, etc. This is shown in the accompanying drawing.

Gasoline Gauge

This is a diagram of a homemade gasoline gauge that anyone can make and install. Drill a hole in your gas cap, put a small wire through it, fasten a cork on the end of the wire, and bend

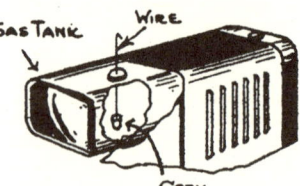

the wire on top so it cannot slip through the hole. When tank is empty, the wire will go to bottom. This simple but effective gauge will work fine on tractor and stationary engine tanks, where the filling cap is over the lowest part of the tank; but not on many cars, where the filler cap is not directly over the tank.

SECTION 3.
Trailers, Wagons, Sleighs

Modern Two-Wheel Auto Trailer
(With frame extension suitable for house trailer)

This trailer may be constructed from a standard front axle, springs, and converted auto frame. It may be built to take the same wheels as your car—car wheels and trailer wheels are then inter-changeable.

Axle Assembly

The standard fixtures of a front axle are left complete. In order to provide rigidity, the wheels are set in the proper position and held by bolting a piece of heavy strap iron to the end of the steering rod and to the axle.

This is claimed to be a superior arrangement to that of welding the spindles to the axle, for, in the latter case, a slight parking lot accident might break the weld or bend the axle; whereas, in the former, the steering rod only would bend and it could be easily straightened or readjusted.

Wheel Assembly

Hubs are selected to fit the spindles. If, as in this case, the desired wheels have the attaching bolts spaced in a large diameter, it will be necessary to bolt a large drum to the hub and then drill holes in the proper position in the drum and bolt the wheels to the drum. The holes must, of course, be centered with calipers in exactly the right position or the wheel will act as an eccentric.

If the desired wheels have the attaching bolts arranged in a small diameter, it might be possible to bolt the wheels directly to the hub.

Frame Assembly

The frame is a converted auto frame arranged and bolted in a V-shape. By selecting similar frame material in a slightly smaller size, and bolting it inside this frame, its length may be conveniently extended and the wheels moved further back. The vehicle would then be suitable for a house trailer.

In making trailer frames always be sure they are V-shaped. Square or oblong frames result in excessive road sway and excessive strain at corners of the frame.

Springs

The springs are bolted to the frame in the usual fashion which permits swinging at both ends.

Hitch Assembly

The hitch needs little explanation for the illustration is self-explanatory. The piece of heavy strap iron shown is bolted directly to the car frame and braced with light metal strips to the bumper. The hitch-bolt is of extra hard steel. The spring on the hitch-bolt is to keep both sections of the hitch on contact and prevent rattling. A suitable gadget is bolted inside the frame to keep the hitch-bolt in a vertical position.

Box Assembly

Plans for the box are unnecessary as each builder will have his own ideas. Three or four 2x4's should be bolted to

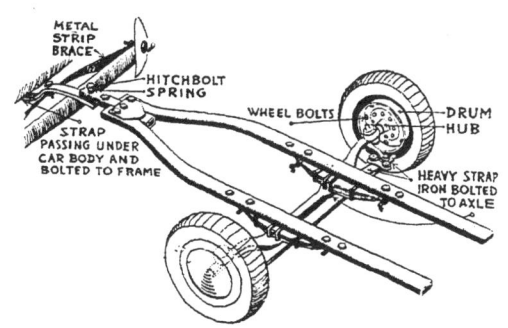

the frame parallel to the axle, and the boards forming the floor of the box should run at right angles to the axle.

A Tractor Trailer for Heavy Farm and Road Work

The tractor trailer illustrated in the accompanying drawing is designed to carry a ton and one-half load and a minimum of steel is used in its construction.

A front axle from a 1½ or two-ton truck can be utilized. If a truck front axle is not available a rear axle can be used with some variations in construction. In this case blocking must be used between the axle and the frame to give wheel clearance and the body frame may have to be narrower to utilize the spring saddles when bolting the stringers to the axle. If the front trucks from a discarded thresher with steel wheels is used the speed of hauling should be correspondingly reduced.

Where a truck front axle is used, the axle should be inverted and bent cold to give the wheels a camber of about one degree. This bend should make the wheels about one-half inch closer together at the bottom than at the top. Six pieces of three-inch angle iron are welded to the axle for mounting the body.

In cases where automobile or truck axles are used, 32x6 heavy duty high pressure truck tires are recommended for a two-ton load. Other sizes of tires may be used with corresponding loads.

The side members of the frame may be made from two pieces of 8-inch channel iron 14 feet long or from two pieces of 4x8-inch clear fir. The cross pieces may be made of 2x8-inch pieces of lumber and the unit bound together with two half-inch rods at each cross bar. The frame should be made as wide as possible to carry the load close to the wheels.

The frame may be braced with pieces of two by four which are attached to the frame with light bolts.

The axle may be attached to the frame with four half-inch "U" bolts which are notched into the top of the frame members so the platform can be placed on a level surface. Keep in mind that the trailer must be strong to be useful.

The pole should be made from a sturdy piece of 4x4, (hardwood preferred) and rigidly fastened to the axle so the draw goes directly to the wheels. The pole piece can be bolted to each cross member at D1 and D2 or attached by a narrow iron plate fastened below the cross pieces. Two pieces of 5/16x-

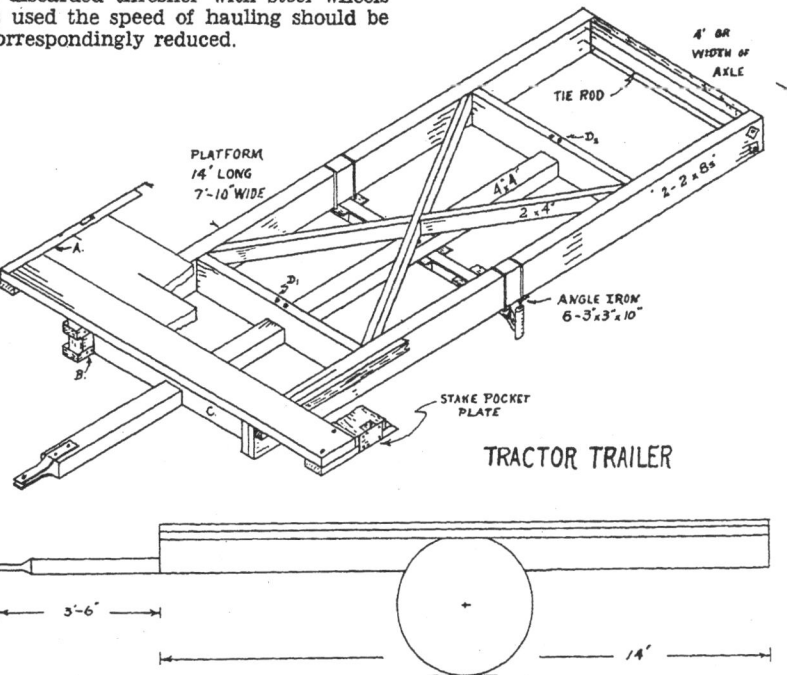

TRACTOR TRAILER

3½x14-inch plate may be used for the hitch. A piece of three-inch pipe or a piece of channel iron may also be used as a pole. Two-inch plank, seven feet ten inches long may be used to build the platform.

If angle iron is available it may be placed along the edge of the platform so that stake pockets can be installed. Where this is not available stake pocket iron plates can be placed on the side of the platform and attached with lag screws. If desired, the platform can be made slightly narrower and a 2x4 can be bolted along each side to make a plate for the stake pockets.

Notice that the front cross member of the frame (C) has been placed four inches back to allow for stake pocket bands on the front of the trailer. The rear end may be constructed in a similar fashion.

The platform may be spiked to the frame members but a better fastening can be obtained by bolting a piece of 2x2 along the outside edge of the frame beams. Bolts from the platform can be run through this rib.

Where brakes are required for highway hauling the master brake cylinder can be attached to the front frame cross member and operated from a lever conveniently located for the tractor driver.—W. Kalbfleisch.

Fuel and Water Tender for Tractor

This tender is constructed mainly of ordinary dimension lumber, plus an old automobile front wheel assembly and a few simple parts made from flat iron. The dimensions of the tender and sizes of lumber are shown in the sketch. The special features are:

1. Ease of loading.
2. Convenient transportation.
3. Choice of pump or tap to supply fuel or water to the tractor.

When the tender is tipped up so that the back rests on the ground, full barrels can be loaded without much lifting, or empty barrels are easily removed. In this same position, fuel or water can be pumped direct to the tractor with the tractor standing close to the barrels. The tender is tilted to a horizontal position for travelling and the leg rests

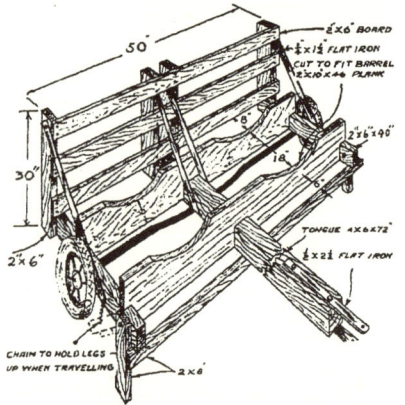

are held off the ground by means of a hook and chain. The tender is also placed in the horizontal position with the leg rests on the ground when the fuel or water is taken from the barrel through a tap.

In order to make tilting of the full barrels as easy as possible, the wheels are fastened close to the back of the machine. The wheels should be two inches off the ground when the tender

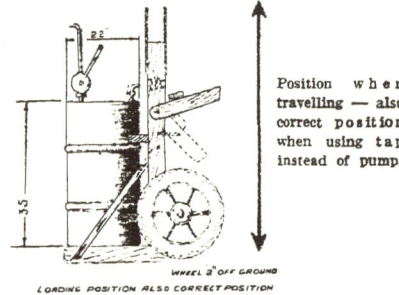

is resting on the back, as shown in the sketch. To view the tender in a horizontal position, turn the illustration shown here on its side.

Livestock Trailer

Prof. L. G. Heimpel, of Macdonald College, Que., designed this livestock trailer and the illustrations are from drawings by his department. The measurements are all given and little descriptive matter is necessary. The sills are 2x3½ oak, but clean, straight grained fir 4x4s could be used. Balance in a two-wheeled trailer is important and it is ensured by faithfully following the drawings in making the measurements. The rear end is hinged at the bottom. Side and front views are given.

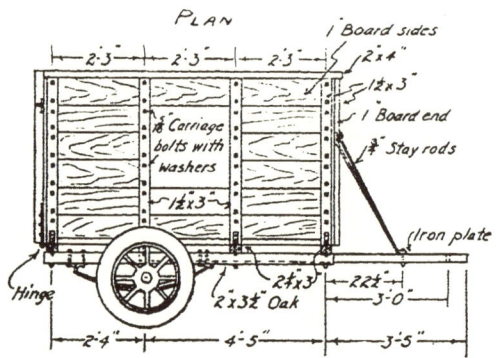
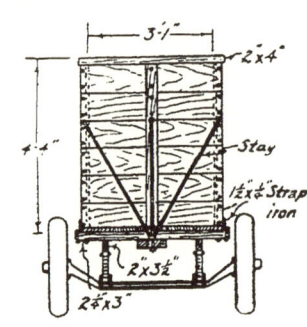

Moving Brooder & Farrowing Houses

Here is a homemade trailer for moving brooder and farrowing houses. The parts are adjustable so the trailer can be used on houses of various sizes. To

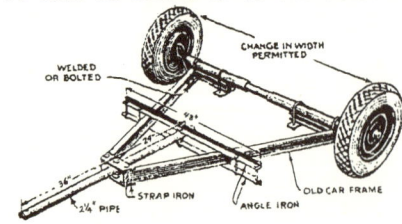

load a house on to the trailer you'll need a jack and some large blocks. Put the jack in the middle of the house, raise one end and set blocks under corners. Raise other end but instead of putting blocks under the corners, put them near the centre so that house will be almost balanced. Run the trailer under the house, tongue first and use jack to lift so blocks can be removed. To unload the house, just reverse the procedure.

Trailer or Buggy

Here is an old auto chassis rebuilt into a trailer or buggy. The side view shows it with a box 84 inches long and 10 inches high. The hind axle is an old buggy axle turned down to fit Ford front wheel bearings. Buggy springs are also used on the rear axle. There are two reaches connecting the rear and front axles, made of flat iron, ¼ by 1½ inches, and given a half turn at both ends. A detail of it is given in the upper left corner of the drawing. The front view shows that the box is 44 inches wide and that the front assembly of the running gear of the car is used, including the spring. Therefore two sets of front wheel bearings and wheels are used as the rear assembly is made from

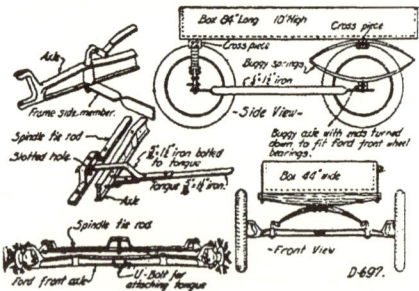

front wheels and bearings. The front axle is shown at the lower left. A U-bolt inserted in the axle takes the tongue. A piece of flat iron, suitably bent, is bolted to the tongue and passes over the axle to the spindle tie rod. A piece of flat iron, bent saddle fashion, takes this iron, which has a slotted hole to engage the bolt and allow for play. Both the tongue and the iron fastened to it are of 5/16 iron, 1½ inches wide.

The same assembly could be made the basis of a horse-drawn buggy. It would need a different box, with suitable seating arrangements, and the tongue would be made to use with horses. Otherwise the general construction would be the same. A little ingenuity would make it suitable for one horse and a pair of shafts. Instead of one iron bolted to the tongue and connecting with the spindle tie rod, two could be made, one for each shaft.

Wagon From Auto Parts

This very serviceable wagon and four-wheel trailer from the front axle of an old car and the rear end of a Model T truck. It will carry five tons, yet can be pulled around the farm yard easily by one man.

Take the worm and ring gears out of the rear axle housing, insert a piece of old 2½-inch well pipe about seven feet long with half a coupling left on the end, and make fast with housing cap. Fit a 4x6, four feet long, over rear housing so as to rest on spring shackles. Fasten to shackles with 9/16-inch U-bolts which extend one inch above the 4x4, and fasten stake holders made from Model T frame irons under the clip for the U-bolts.

For the front truck, use the front axle of an old heavy car and weld the spindle straight and solid. Take the front truck of an ordinary wagon, saw off the old skeins, and fit on top of the car axle. Clip it to the axle with U-bolts where the car springs were fas-

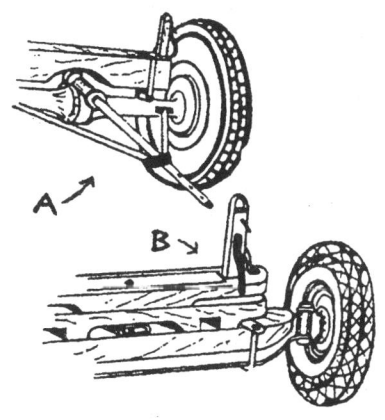

tened. Take a piece of two-inch pipe five feet long, insert flat pieces of steel in one end for reinforcement, flatten and round that end, and have hole drilled for kingpin. The flat steel reinforcement should be welded in place after the front end of the two-inch pipe has been flattened and rounded. Insert in front truck for reach, extending back to telescope into rear reach pipe. Bore holes for reach pin to adjust wagon length.

The front axle should be from a heavy car with tires of a size easily available. If they are 32 by six-inch tires, they will be interchangeable with the rear ones. With large tires front and rear, the wagon will carry loads easily out of spongy spots and soft and muddy fields where an ordinary wagon cannot go.—I.W.D.

BUT...

FOR BETTER SERVICE IT'S STILL B-A

Today — as always — the personnel at your local B-A Branch offers you prompt and efficient service, as well as the highest quality in gasolenes and lubricants.

This year the need for continued top production on Canadian farms is still urgent. To help meet this demand, your tractor and other power-driven machinery should be maintained in top-notch operating condition. By using dependable B-A products, you will have fewer breakdowns and get maximum production at minimum cost.

The manager of your local B-A Branch is glad to discuss your lubrication problems with you. You'll find his knowledge of fuels and lubricants always helpful; what's more, he can supply you with B-A gasolenes, distillate tractor fuel, tractor oils, lubricants and greases especially refined for your particular equipment.

It pays to buy top-quality products. This time buy the best — buy B-A. You'll find, like hundreds of thousands of others across Canada, that

You Buy the Best When You Buy B-A

THE BRITISH AMERICAN OIL COMPANY LIMITED

Steering Device for Trailer

The accompanying drawing shows a drawbar and steering device for either a trailer hitch for a four-wheeled trailer made from an old automobile or for a horse-hitch for a similar vehicle. Steering is accomplished through an extension of the main drawbar to the rear of the axle which is coupled to a shortened drag link (A) made from a part of the original steering mechanism of the Model T Ford car. The drawbar itself is made of extra heavy wrought iron and it is important that this part of the hitch be made of the weight of material specified to prevent it from bending under the heavy load it will have to bear. A one-piece pillow block bearing makes the best type of anchorage for such a hitch and lock nuts should be used on the main draw bolt passing through this bearing. It is important that the trailer tongue is free to move up and down though it must always be possible to keep it tight so that steering of the trailer will be controlled. The large bearing surface of the flat plate on either side of the front end of the hitch where it is coupled to the rear end of the trailer tongue provides the necessary bearing surface to prevent looseness at this joint.

The greatest difficulty usually met with in the use of four-wheeled trailers is the tendency for these vehicles to weave badly when following a car. This is due usually to the destruction or maladjustment of the caster action of the front axle. Any garage man will know how this should be adjusted, but in any car with transverse springs, such as has been employed in practically all old models of Ford cars, it is important that the radius rods be kept in good condition and properly coupled to some reliable anchorage on the underside of the trailer chassis. Too much caster action in the front axle is likely to result in a wheel wobble closely associated with a "shimmy" while too little caster action causes weaving of the vehicle or the destruction of the natural steering qualities of the vehicle. Overloading of trailers or wagons is likely to disturb the set of the springs and destroy the caster action. For this reason farm wagons made from old automobiles had best be rebuilt so as to remove the springs altogether.—L. G. Heimpel.

For Handling Bundle Corn

One of the greatest labor-saving devices in handling bundle corn is the

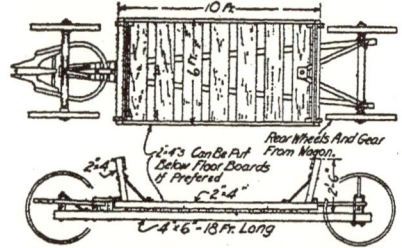

(Short Underslung Silage Wagon)

low-down rack. The short type is the most easily constructed and the most

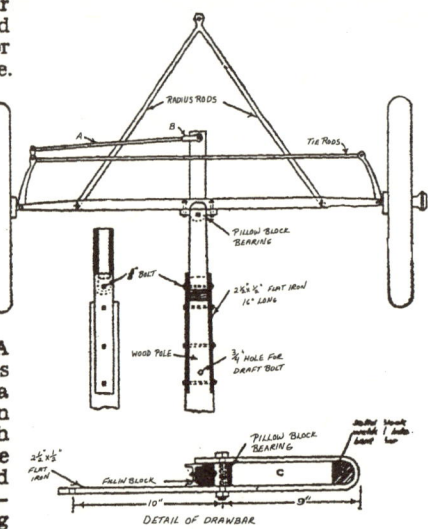

common, and will hold plenty of a load where the distance is not too great. The sills are 4x6's 18 feet long swung underneath the rear axle by means of heavy clamps, the rear reach being bolted on top of the back cross board of the platform. The sills are brought together and bevelled at the front and bolted together.

The regular kingpin on which the front axle pivots is removed and replaced with a longer one, which is

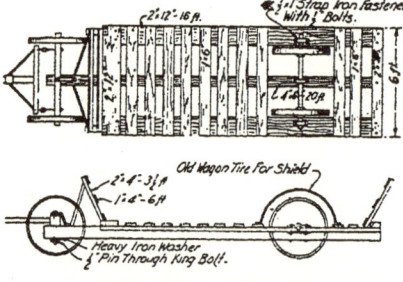

(Long Silage Wagon.)

either threaded at lower end and nut, and lock nut put on below the 4x6's, or else is bored with a three-eighths inch hole and heavy washer and machine bolt put on below.

The long type is not quite so common or so easily made, but has a much greater capacity. It can be built in much the same way as the other type by swinging the timbers under the rear axle of an ordinary wagon, or, as shown in the cut, is made from two mower wheels mounted on the old mower axle or piece of gas pipe of the proper size for an axle. This is clamped to the top of the timbers by clamps made of old wagon tires fastened with half-inch bolts. Guards to go over the rear wheels are made of the same kind of material. The front wheels shown are from a low truck wagon, or the front wheels of an ordinary wagon can be used if preferred. The same arrangement of kingbolt is used as shown for the short type.

Trailer from Ford Chassis

The bolsters of this trailer are made from 2x4s bolted together in such a way that they clamp tightly on pieces of piping which function as bolster pins.

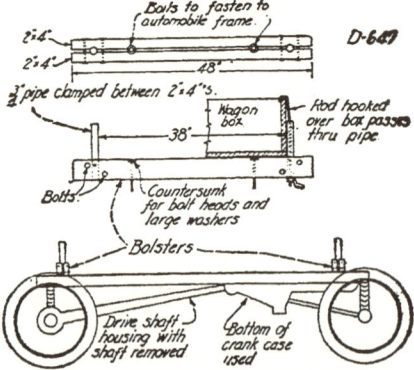

A rod comes up through the pipe and hooks over the top of the wagon box. The heads of the bolts which fasten the bolsters to the frame are countersunk. Notice that the bolts which hold the 2x4s together are staggered, the outside ones being near the top and the inside ones near the bottom to help hold the bolster pins plumb.—I.W.D.

Trailer Hitch

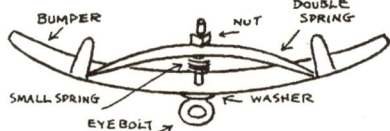

This is the best trailer hitch that I have used. Take two old car springs, one longer than the other by 8 inches and wide enough to fit snug into the groove in the bumper of your car. Make an eye-bolt about 6 inches long and drill holes through the middle of the bumper and the old car springs. In between the bumper and the springs put a coil spring. An eye-bolt goes through the holes as shown, with a washer between the eye and the bumper. Any one using this kind of a hitch will hardly know they have a trailer behind the car.—Ken Nelles, R.R.2., Wetaskiwin, Alta.

Wagon Frame from Auto Chassis

Although every other farmer has made some kind of a wagon from an old car, there is always something new and better. The wagon illustrated is an all-steel construction job which has been designed to trail well at high speed and to back up with a heavy load easily. This feature is obtained by having the complete front trucks turn like a wagon and not like the wheels of an automobile.

Two front axles and wheels from wrecked cars, some scrap angle iron, seven feet of old two-inch pipe, and some pieces of strap iron are the main parts needed for the wagon. By varying the construction according to the scrap iron laying around the yard, the wagon

"Serving Agriculture with Chemistry"

OUT OF A TEST TUBE come these Agricultural Chemicals—INSECTICIDES, SEED DISINFECTANTS, WEED KILLERS and FEED SUPPLEMENTS —to increase farming profits.

INSECTICIDES:

"Deenate 50-W."—This 50% D.D.T. powder serves a three-fold purpose on our prairie farms:

(1) One pound of "Deenate 50-W" mixed with one gallon of water should provide sufficient spray for the average barn.
(2) One pound of "Deenate 50-W" will provide 20 gallons of spray which can be applied direct to the backs of animals.
(3) "Deenate 50-W" may be used to control the Colorado potato beetle on potato tops.

"Atox" (C-I-L Derris Dust)—This non-poisonous dust controls a wide range of insects on horticultural crops.

"Licide"—For lice control on cattle, horses pigs, sheep and poultry.

"Warbicide"—For control of warble fly on cattle.

SEED DISINFECTANTS:

"Ceresan"—Reduces the smut losses of wheat, oats, barley, rye and flax.

"Semesan Bel"—Controls seed-borne disease organisms on potatoes.

"Semesan Jr."—For corn.

"Semesan."—For vegetables and flower seeds.

FEED SUPPLEMENTS:

"Sol-Min"—Minerals for cattle, horses and sheep.

"Pig-Min"—Minerals for hogs, sows and young pigs.

WEED KILLERS:

"Herbate 2, 4-D"—May be used to control mustard in wheat, oats and barley for approximately 75c per acre. At higher concentrations "Herbate 2, 4-D" may be used to control Canada thistle, common ragweed, sow thistle, field bindweed, etc. For further information write Canadian Industries Ltd., Agricultural Chemicals Division, Winnipeg, Man.

"Rodant"—The new Antu formulated rat poison. Two ounces should provide sufficient poison for 250 rats.

CANADIAN INDUSTRIES LIMITED
Agricultural Chemicals Division

WINNIPEG   **REGINA**   **CALGARY**   **EDMONTON**

can easily be built for twenty-five dollars even though much of the iron is purchased.—W. Kalbfleish.

Construction of the Rear Wheels

When obtaining car axles and wheels for the wagon, you should endeavor to obtain wheels which will take the common size of tires so the tires can be

heating the angle in a forge and holding it in a heavy vise while giving the iron a twist.

Construction of the Front Trucks

The half circle hound can be formed by giving the angle iron a series of short bends when red hot. The top hound is a straight piece of angle iron

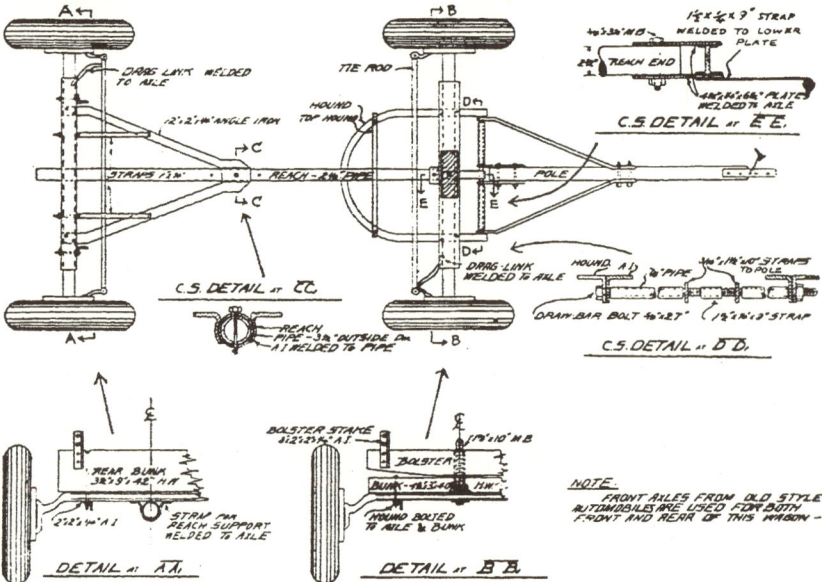

readily purchased second hand. The axles of the car can be laid down in their natural position for a low wagon or can be turned up to give good clearance. When left low as in a car, however, the wagon can not be made to turn sharp unless a narrow box is used.

The rear hounds are attached to the axle by two bolts placed in two of the holes formerly used by the spring clips. The other two holes are used to bolt the rear bunk to the axle.

A flat plate welded to the top of the axle, or a plate clamped on to the axle makes an excellent flat bed for the bunk. In order to obtain sharp turning, the rear bunk is made to be just about as high as the wheels. At the front end the rear hounds are welded or bolted to a short piece of pipe which will easily slip over the reach pipe. This pipe has a large hole in it for the reach pin, so the rear trucks will be free to move over rough ground. Angle iron can be bent by sawing a "V" out of the flange and then joining the iron by electric welding. The bend can also be made by

placed above the reach pipe and spaced by two short pieces of pipe about two and a quarter inches long.

The plates used to connect the front end of the pipe reach can be formed by putting a "U" shaped piece of heavy flat iron around the axle. This eliminates a welding job and is possible because the central strap to the cross rod for the pole has been found to be unnecessary. A heavy bolt (at least ⅝-inch) should be used in the front end of the reach because of wear on the bolt due to turning.

The king bolt must be placed in the lower bunk with its head down as illustrated, and set in place before the lower bunk is bolted to the axle. Since the lower bunk is not heavy, the king bolt can not do much work and therefore I advise you to put cleats on the wagon box alongside of the rear bunk stakes and omit them on the front.

Once you have a sturdy rubber-tired wagon to go behind horses or a rubber-tired tractor, you will consider it the handiest piece of equipment on the farm.

SECTION 4.

Derricks, Hoists, Presses

Braced Boom Hay Derrick

THIS stacker, from plans put out by the Lethbridge Experimental Farm, takes considerable construction and building one should not be left till haying time approaches. It possesses a number of advantages not found in other types. It is easily skidded from one location to another and since guy wires

are not necessary, it is ready to be put to work at once. Either slings or hay fork can be used. Care should be taken not to overload the slings unless the stacker has been sturdily constructed.

The following suggestions will be helpful in building the stacker: Metal parts found in salvage heaps, and discarded telephone poles can be used in the construction, but good quality poles should be obtained for the mast and boom. The mast should be at least 26 feet long and is placed with the large end up. The braces running from the mast to the boom should be attached to the mast at the surface of the platform and not above it. This construction is necessary because the portion of the mast between the platform and the point of attachment of the cross braces must support all the weight exerted on the stacker. It follows that the greater the length of this section the more likelihood there will be of breaks occurring.

The boom is a 35-foot pole with a diameter of at least five inches at the small end. The base may be constructed of poles or squared timbers. It should not be any smaller than the dimensions shown on the attached plan as a smaller base might lack weight and so allow the stacker to upset. For the hoisting cable a 5/16 or a ⅜-inch flexible wire cable is desirable but a ⅝-inch rope may be used. Heavy duty pulleys should be used throughout as it has been found that cheap, light weight pulleys will not last for more than one season.

Butchering Hoist

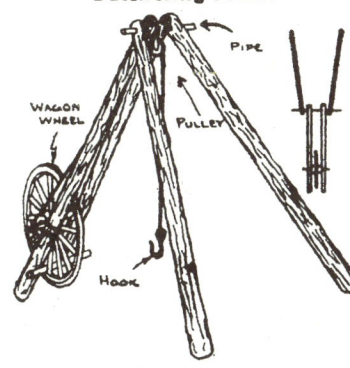

To make this one-man butchering hoist get four straight poles or 2x6's 14 to 16 feet long. Lay in relative position on the ground and bore holes 6 inches from the top end. Then fasten them together with an 18-inch bolt, with a pulley between the two inside poles. Now raise them into a tripod with the two centre poles apart at the bottom just wide enough to allow a wagon wheel to go between them when the wheel is just clear of the ground. Put a short piece of piping or bar through the wagon wheel hub and lash it firmly to the centre poles. Run a rope with a hook from the pulley at the top then down and around the wheel hub and fasten to a spoke. A piece of 2x4 or other stick will hold the wheel from turning. This hoist can also be used for lifting an engine out of a tractor or an automobile, or in changing a tractor wheel. One man can exert all the force a ⅞-inch rope will stand.—I.W.D.

Samson Hoist

Equipment for lifting barrels of fuel, beams, heavy boxes, tractor engines and heavy weights becomes increasingly important. A handy portable lifting device is the Samson hoist shown here. It is mostly built of wood.

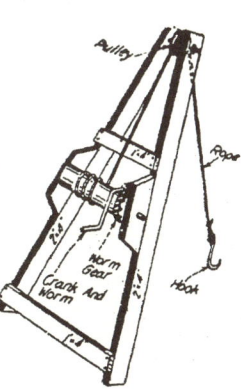

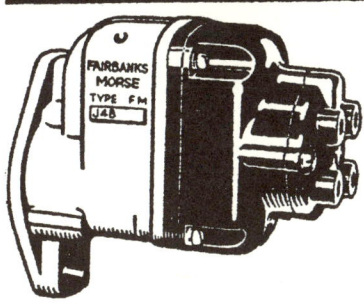

SPECIALIZED SERVICE FOR MAGNETOS

SKILLED WORKMANSHIP
SPECIAL TOOLS
GENUINE REPLACEMENT PARTS
REASONABLE CHARGES
PROMPT ATTENTION

Repair estimates made without obligation

SALES and SERVICE —
- MAGNETOS
- GENERATORS
- STARTERS
- MOTORS
- SPARK PLUGS

SPECIALIZED EXCHANGE SERVICE

For

AUTOMOTIVE ELECTRICAL REPAIRS

SEE OR WRITE

WILSON
AUTO ELECTRIC

Head Office:
242 MAIN STREET

Branch Office:
891 MAIN STREET

WINNIPEG - - - MANITOBA

Roll-in Stacker

This roll-in hay stacking assembly is used in North Dakota. Three men are required to operate it, one to do the stacking, one to drive the team and arrange the slings and the other to operate the sweep rake.

The material needed for building this equipment is as follows: 80 to 100 feet of 1-inch rope, 3 pieces of ½-inch rope 35 to 40 feet long, 3 iron stakes made of ¾-inch iron bent to form a ring at one end and pointed at the other. One piece of 2x2-inch hardwood 6 feet long, 3 rings 3 inches in diameter made of ½-inch iron, and two single block pulleys and a telephone pole 25 to 30 feet long. The pole is held in place with guy wires.

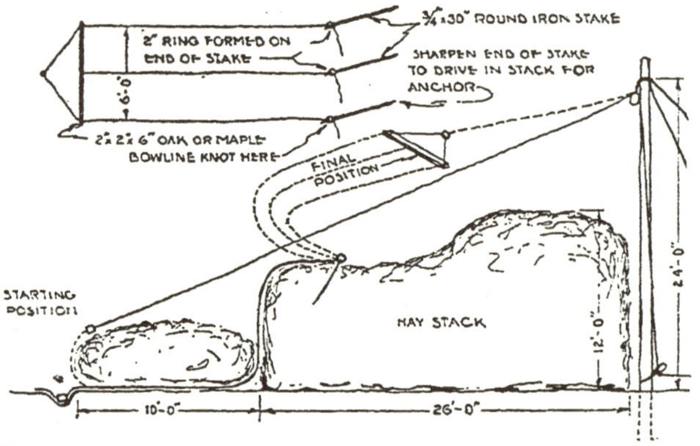

Swinging Boom Hay Stacker

Trees, sturdy fence posts, or dead man logs are usually used to anchor the guy ropes. When a 70-foot guy rope is used on a 40-foot mast, the anchor posts can be placed about 40 feet from the base of the mast. When the mast is to be hoisted, the base is laid in a hole in the ground which has a sloped trench at one side. The top end of the mast is then lifted up onto the top of a load of hay. To raise the mast from this position a rope is attached to the mast about 18 feet from the ground and a tractor or a team is used to pull the mast to a vertical position, guided by the three guy ropes.

The boom can be hooked onto the eye bolt and raised by the boom hoisting rope. If the alternative method of attaching the boom to the pole (Marked S) is used, it is necessary to place telephone pole step cleats on the mast in order to fasten the chain around the boom. A cant-hook may be used to turn the mast around so the boom is conveniently located for different parts of the stack. This information and the sketch are supplied by the Central Experimental Farm.

A. Steel Pin—1-in. x 20-in. Drill a ⅛-in. hole in top of mast 10-in. deep.
B. Band—Mast bound with wire.
C. Pulley—Six-in. pulley block attached to ⅝-in. x 8-in. eye bolt.
D. Boom Rope—1-in., about 80 ft. long, tied to top of boom, and snubbed to bracket at point P. This rope should be of good quality and securely fastened to a heavy bracket or tied around the mast below to sturdy pins.
E. Mast—10-in. base, 5½-in. at top, 36 to 40 ft. long, spruce, etc.
F. Guy Wires—Three ¼-in. cables or 1-in. hay ropes, 60 to 100-ft. long. See Items N and O.
G. Clevis Bolt—⅝-in. x 6-in. bolt.
H. Clevis—$\frac{5}{16}$-in. x 1¼-in. strap iron clevis about 6-in. deep.
I. Hoist Rope—1-in. in diameter.
J. Hay Fork—Or hay sling and trip rope.
K. Boom—20-ft. to 24-ft. boom, 6-in. at base, 5-in. at top, spruce, etc. Lower end bevelled and corners cut to allow boom to swing.
L. Boom Hook—½-in. x 1½-in. flat iron, rounded at one end for hook.
M. Eye bolt—1-in. x 12-in. eye bolt with 1¼-in. to 1½-in. eye hole.
N. Details of Mast Top—(Items N. & O.)
O. Guy Plates—Three $\frac{5}{16}$-in. x 1½-in. x 8-in. plates. A small clevis may be used on the end of each plate to attach rope or cable to the plate.
P. Bracket—For snubbing boom rope marked D. A ½-in. x 8-in. pin on opposite sides of the post may also be used to hold the rope when it is tied around the post.
R. Hoisting Rope Pulley—Six-inch pulley block attached to mast by chain.
S. Alternative Method of Connecting Boom to Mast—(Items S. T. U. V. W. X. Y and Z.)
T. Boom Chain Pin—Place pin 3½-ft. above boom if this method of mounting the boom is used.
U. Boom Support Chain—The boom supported by a chain instead of by a hook and Eye bolt (L. and M.).
V. Mast—(See Item E.)
W. Boom Guide Plates—2½-in. x 2-in. x 26-in. strap iron or light angle iron.
X. Boom—The same unit as "K."
Y. Chain U Bolt—⅝-in. round iron U bolt.
Z. Top View of Boom—Showing guide plates.

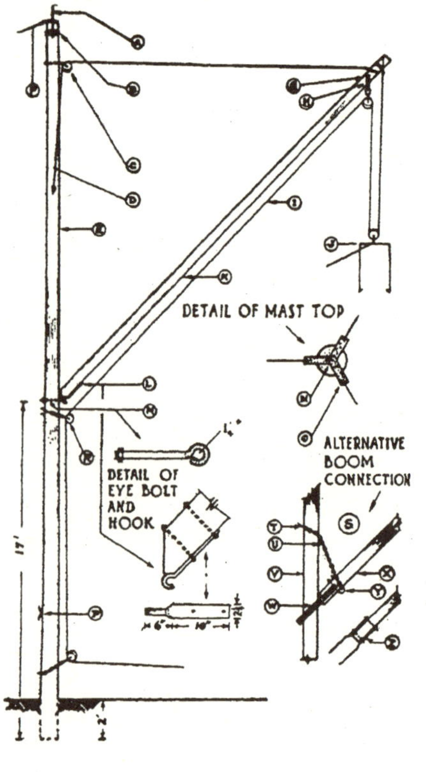

Overhead Hay Stacker

THIS overhead haystacker was designed by F. F. Parkinson, School of Agriculture, Olds, and the drawings are by L. E. Pearson, Institute of Technology, Calgary. A side view and a bird's-eye view are shown, and the other drawings are numbered to refer to the parts shown in the main drawing.

Item 1 shows the runners, which are fir, preferably creosoted, 6"x8"x20' long. Notice that the section of an old tire casing is attached to keep the cable from catching under the back part of the runner. Also notice in the main side view drawing that a piece of 3"x6" by 8'6" long is put in place to hold the hoist up when travelling.

Item 2 is the front cross piece to which tackle is attached by means of a U bolt shown in Item 26. It is 6"x8" fir 7'8" long.

Item 3 is a cross piece 3"x6" by 7' long, which holds the iron rods supporting the springs.

Item 4 is the rear spreader 3"x6"x7'.

Item 5 of which two are required are 6"x6" by 20' long. They are the main arms of the hoist.

Items 6 and 7, the main cross pieces of the hoist, are 4"x6"x12' of fir.

Item 8. The teeth are staggered, the short ones shown in Item 8, are 2"x6" by 12' long and tapered down to 2¼". Six are required. Item 9, shows the long teeth 2'x6" by 13'6". Of these five are required.

The upright teeth shown in Item 10, are of 2"x4" and are 6'6" long. Twelve are needed. They are tapered down to 1¼" at the point.

Item 11 shows the uprights brought together in a V shape at the top and taking the upper end of the tackle. Two are required, a right and a left, and they are 4"x6" fir, 14'5" long.

Item 12 is the spreader across the uprights near the bottom and is 2"x4" fir.

Item 13 shows metal spreader at the top of Item 11. It is made of ¾"x2" iron with 11/16" holes.

Item 14 is a 1¼" steel bar 15½" long with a ⅜" hole ¾" from the end to take cotters. It takes the upper end of the tackle.

Item 15 forms the axle of the main hoist. It is a piece of 1½" standard black pipe 8'6" long with holes ¾" from each end to take ⅜" cotters. It is fastened to the runners by Item 16, of which two are required. An iron plate ½" by 6"x16" is bolted on to each runner and a strap ¼"x2" is welded on to these plates and the pipe as shown.

Item 17 shows the similar mechanism by which the uprights (Item 11) are attached to the main runners, while Item 18 shows the attachment of the axle for the wheels (Item 22) to the runners.

Item 19 is a piece of 1½" pipe 7'2" long for supporting the uprights.

Item 20 is the 1½" standard pipe 7' long which serves as an axle for the main wheels, while Item 21 is a 2"

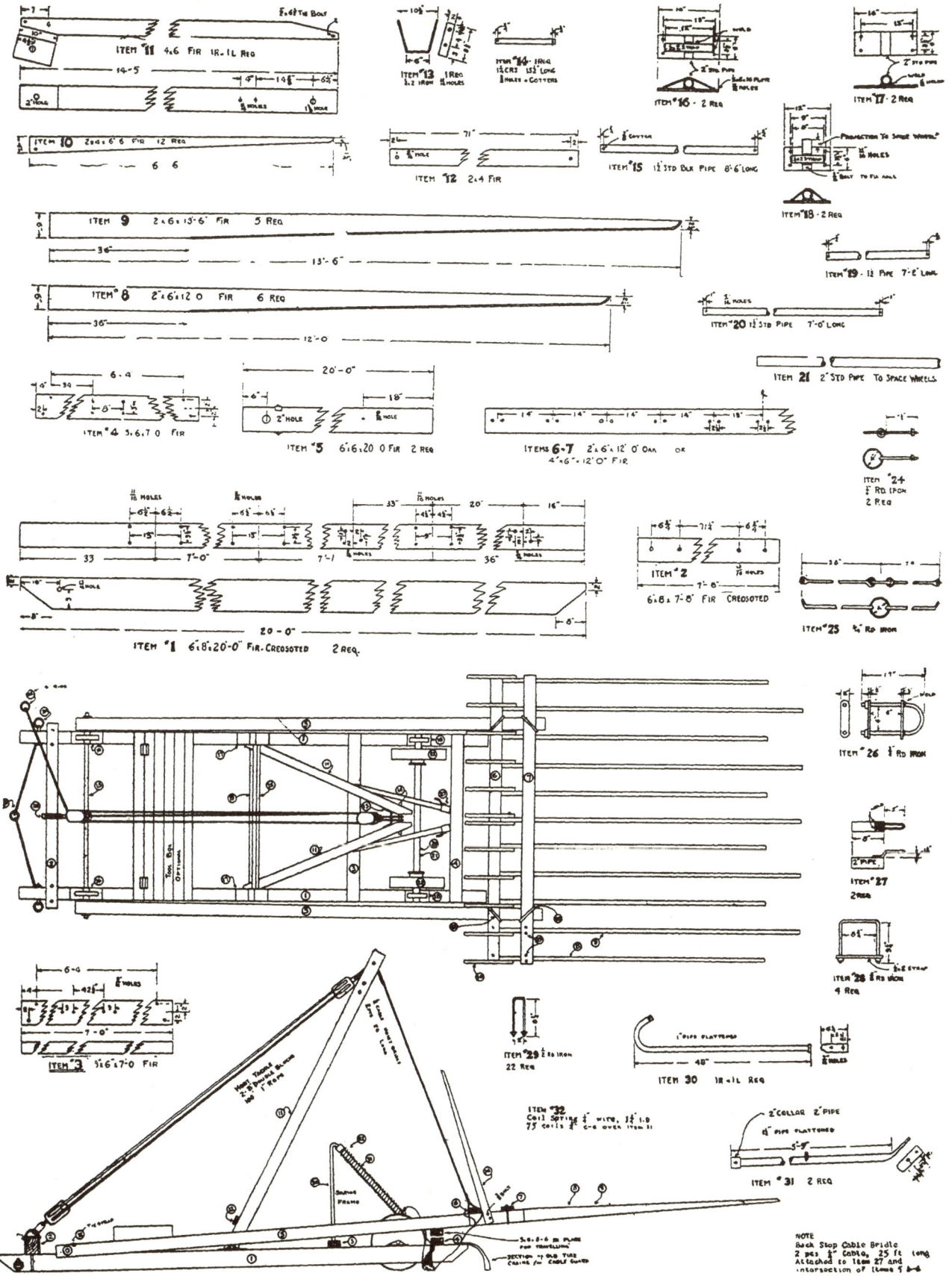

standard pipe to space the wheels. Item 22 shows the wheels.

Item 23 is a 4" ring attached to the end of the hoisting cable.

Item 24, of which two are required, are made of ¾" round iron through which the front end of the stacker is staked to the ground, while Item 25 takes the hitch for hauling.

Item 26 shows the lower U bolt attachment for the tackle.

Item 27 shows a piece of 2" pipe with a lug, which slips over the 1½" pipe carrying the coiled spring.

Item 28, of which four are required, fasten the main fork to the hoisting beams.

Item 29, the U bolts for fastening the teeth of the work to the cross timber (Item 7).

Item 30, of which two are needed, the right and the left, 48" long of 1" pipe flattened to hold the upper end of Item 31.

Item 31 is a piece of 1½" pipe flattened at the bottom which carries the spring.

Item 32 is a coiled spring of ¼" wire, 1¾" inside diameter, 75 coils at ¾" centres, carried by Item 31.

The back stop cable bridle is of two pieces of ½" cable, 25 feet long attached to Item 27. The spring takes up the jolt when the hoisted load is discharged.

Beaver Slide Haystacker

Some time ago The Guide published an article, The Haystacker Saved the Day, telling how an aged farmer in the Swan River Valley, Manitoba, with the help of his daughter, saved a second crop of alfalfa by pressing into service a stacker he had used years ago in putting up prairie hay. We wrote for a description of the construction of the haystacker and received in reply full information from Fay A. Stewart, the lady who helped out in the emergency, together with drawings and bills of material used.

The base of this haystacker is a pair of timbers 6 or 8 by 12, and 24 feet long. Logs would do where they are available. They are held apart in front by a cross piece 4x4 and 12 feet long. There is no cross piece between them at the back or stack end as it would interfere with the stack. They are braced with two diagonal pieces of strong 2x6 stuff, bolted to the runners at each end and to each other in the middle as shown in the diagram.

The next step in the construction is the frame work at the rear or stack end. The upright posts are two pieces of 4x4, 14 feet high. Three or four crosspieces are bolted on as shown. For these, 2x6 scantling, 12 feet long, are used. They are braced diagonally with two 2x6 scantling 16 feet long. The pieces are thoroughly bolted to the uprights and bolted or nailed together where they intersect as this adds to the rigidity of the frame.

Now take the side view, shown in the diagram. A pair of centre posts is provided, with a cross piece across the top, to support the slide up which the hay is drawn. These posts are seven feet high. It would be well to brace these also. Near the front there is also a pair of short posts, 18 inches high, of 4x4 stuff, with a crosspiece. This crosspiece carries the lower ends of the floor boards and the upper edge of the folding apron, which is described below. There is also a brace from each end post to the crosspiece which is located on the frame as shown. Note by the photograph that these two braces meet at the upper end in the middle of the crosspiece. This gives greater rigidity to the frame.

The main timbers of the sloping slide, which carry the load, are 32 feet long and should be 6x6 at least. Between

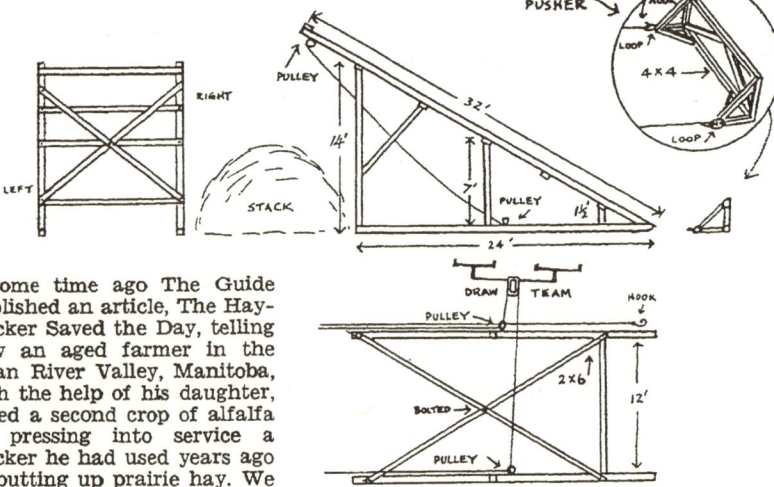

them are the five crosspieces to carry the floor. These are 12 feet long and are bolted on under the timbers so that the timbers form sides for the slide. At the top of the sloping timbers a 2x6 crosspiece is firmly bolted across the top. It projects out on the side on which the horses pull. To it are attached the two top pulleys, the one on the projection keeping the rope on that side clear of the framework.

The floor is of inch stuff nailed on top of the crosspieces between the timbers.

The load falls through the space between the end of the floor and the 2x6 crosspiece at the top end of the timbers, which carries the pulleys. The lower end of the floor ends in the middle line of the crosspiece supported by the short, 18-inch posts.

From the lower edge of the flooring to the ground is a hinged apron, made of the same material as the flooring. This can be folded back when the stacker is being moved or is not in use. Two or preferably three heavy strap hinges are used. When the load of hay is brought in with the sweep it is drawn up on this apron, which extends a foot or two beyond the lower end of the main frame.

The construction of the pusher for elevating the hay, is shown in the circle. It is 14 feet long and is made of a 4x4 with triangular end frames of 2-foot bases and uprights, thoroughly braced. Two ¾-inch ropes are used. The long one, 80 feet long, is on the side away from the horses and it is fastened to the pusher with a wire loop. The shorter one, 68 feet long, carries a hook at the bottom, so that it can be detached from the pusher. When the pusher is empty, the rope is unhooked from it and it is swung to the side out of the way of the sweep.

The ropes are carried from the pusher up through the pulleys attached to the top crosspiece. The long one is on top of the frame and the short one, on the pulling side, is clear of the frame. This allows the ropes to come down to the lower pair of pulleys clear of the frame work.

The sweep of hay is swept up on to the apron. The pusher is swung around behind it. The go-ahead signal is given. The team moves away sideways from the stacker. The load is pulled up the slide and dumped on the stack.

Three Types of Hay Stackers

THE single pole hay stacker shown at right is from drawings made by R. W. Peake and H. Chester of the Lethbridge Experimental Station. The sketches are so complete that but little text is necessary to explain them. The main pole is preferably a single pole but two poles may be spliced together as shown to make a minimum height of 45 feet. The top of the pole should be not less than five inches in diameter. When a splice is used it should be six feet long, with the ends matched as uniformly as possible.

The 20-ft. cross-arm has an average diameter of six inches and is suspended 10 feet from the top. As shown in the diagram the cross-arm is in the position assumed when delivering the bundle on the stack; that is, at right angles to the base. Because the cross-arm must make a half turn to take hay from the wagon or on the ground, it is suspended by two chains that wind around the top of the pole. As the sling load of hay is raised clear of the stack,

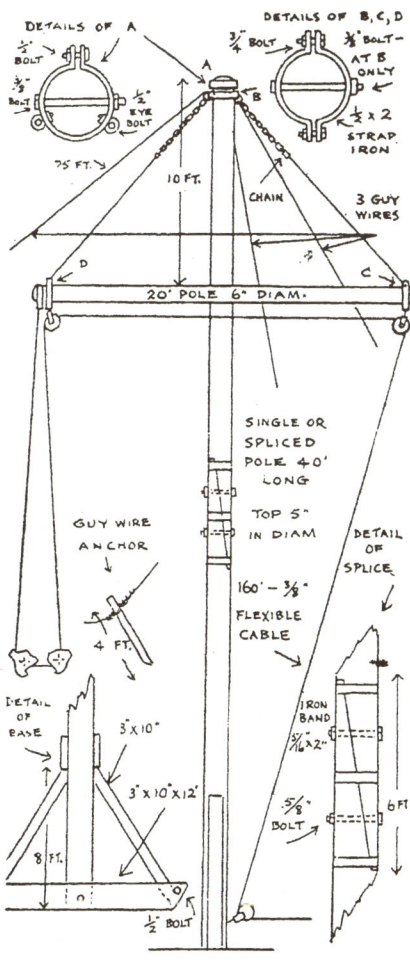

the chains unwind, the cross-arm returns to its normal position and the sling load is carried immediately over the stack. Part of the chains may be replaced by lengths of cable.

One of the three guy wires passes over the stack. Three or four strands of No. 12 gauge galvanized wire are used or ordinary 2-strand fencing wire is satisfactory. The base is made of planks as shown or of stout poles. They are kept in position by stakes. Three 8-inch heavy duty cable pulleys are used and two binding pulleys for the slings. A ⅜-inch flexible cable 160 feet long will pick up a sling load from the ground.

The top illustration on page 72 shows a 2-pole stacker. It is from sketches

SIMPLE FARM ARITHMETIC

THE COCKSHUTT *Self-Propelled* COMBINE

A COMPLETE LINE of Modern Farm Equipment

Tractors.
Self-Propelled Combines.
Power Take-off Combines.
Swathers, Pick-ups.
Tractor and Horse-Drawn Grain and Corn Binders.
Mowers, Rakes, Hayloaders "Tiller Combines."
Tractor and Horse-Drawn Moldboard and Disc Plows.
Walking and Riding Plows.
Grain and Fertilizer Drills.
Broadcast Seeders.
Corn and Cotton Planters.
Disc and Drag Harrows.
Field and Garden Cultivators.
Scufflers, Weeders, Horse Hoes.
Pulverizers, Rollers.
Manure Spreaders.
Lime and Fertilizer Distributors.
Pulpers, Straw Cutters, Milkers.
Specialized Equipment for Tobacco, Sugar Cane, Sugar Beet, Peanuts, etc.

★ BOOSTS Yield Per Acre
★ CUTS Cost Per Bushel

With grain prices pegged, the only way you can increase your profits is by boosting production and cutting costs. The Cockshutt Self-Propelled Harvester Combine eliminates the cost of binder twine and the expense of stooking and threshing. It does all your harvesting jobs in one simple operation. Yet one man operates it ... simply and speedily.

Make your farm a completely Cockshutt-equipped farm ... it is the modern way to efficient, profitable operation.

SEE YOUR AUTHORIZED COCKSHUTT DEALER

COCKSHUTT
PLOW COMPANY LIMITED

TRURO   MONTREAL   **BRANTFORD**   WINNIPEG   REGINA   SASKATOON
SMITH FALLS   CALGARY   EDMONTON

furnished by the Swift Current Experimental Station. In operation, one of the guy wires is slack. When the load is being elevated the guy wire coming over the stack is taut and its opposite slack; with the stacker leaning over the wagon. When the sling load is elevated, the guy-wire over the stack is pulled. This swings the stacker with the sling load over the stack until the opposite guy-wire is taut.

There are two lifting systems shown in the diagram. The A system has a single double-block at the bottom into which both ends of the slings hook. The B system has two double blocks at the bottom. Each hooks into a ring at the ends of the slings. The latter system is preferable as it keeps the load well packed while lifting and will lift the load higher than the A system. It gives the team a 4-to-1 advantage. Half-inch cable is usually used. A good stacker will take a large 10x20 foot basket rack load of over two ton at one lift.

The guy cables are anchored at from 50 to 60 feet from the sill of the stacker. The cable which supports the poles when they are over the stack must be quite strong to take the shock when the poles are swung over. A ½-inch steel cable is sufficient; often an old cable of a larger size is used. The other guy-cable may be lighter.

The load is driven in as close to the stacker as possible and the poles swung over it. In this position the poles are just past centre so that there is no danger of their swinging into the stack before desired. When the load is lifted and the poles swung over the stack, the load is tripped with the sling rope when it has reached the desired position. It is not necessary for the top of the poles to be over as far as the centre of the stack as the swing of the load permits considerable range in placing the load. This stacker is easy to work and will lift a heavy load for making stacks as high as 40 feet.

At the bottom is shown a cable stacker. The A-shaped frames are usually made from poles. The purchased material consists of a reversible cable carrier, a double harpoon fork or grapple fork, or slings, 150 feet of ½-inch galvanized steel cable on which the carrier travels and which is used for the end guys, 130 feet of manila carrier rope and necessary bolts, etc.

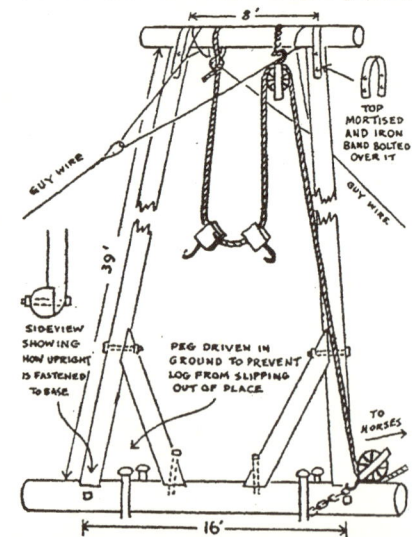

the hoist. Hook the two pulleys, one in each end of the sling. Hitch the horses to the end of the long lifting rope, and drive them ahead. When all the slack in the rope is taken up and the weight comes on it the top of the framework will be pulled to the load as far as the guy-wires permit. As the slings leave the hay rack they will swing away from the load, the momentum will make the hoist fall toward the stack till stopped by the guy-wire. Pull the trip rope on the sling and down the hay falls right where you want it.

Drive the horses back, pull down the slings and you are ready for the next load.

Using a hoist like this it is possible to build a high stack beside the barn. This means one can put more hay in a small stackyard where it is handier for feeding.

Roll-up Stacker

Remove one side of the rack and put the sling shown across the bottom of the rack and load the hay on top of it. At the stack set a couple of saw horses under that side to keep the rack from tipping too

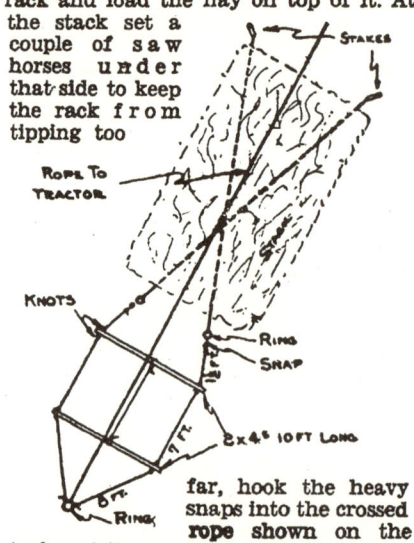

far, hook the heavy snaps into the crossed rope shown on the stack, put the tractor or guide rope over the load and tie it in to the ring on the

Hay Unloading Derrick

All over the country there must be many who are looking for some device to do the job of unloading hay, for different reasons; some to save work, some to save the leaves of alfalfa hay that are broken off in handling.

This hay hoist does the job very satisfactorily and can be made at very little cost. The poles are made into a quadrangle with the top narrower than the bottom. The corners are well braced and the whole thing is bolted firmly together as shown in the illustration.

The guy wires, made of a few strands of fence wire twisted together, are fastened on, the rope tied and the pulley fastened before the framework is raised.

When the framework is raised, lengthen or shorten one guy wire so that the framework will be standing up straight when the ropes are lifting the slingful of hay off the load.

Lengthen or shorten the other guy-wire enough to allow the top of the frame to fall to a position exactly over the centre of the place where you want the stack to be. The action is this:

Drive the load of hay to position by

opposite side of the sling, and roll it off the side of the load and right up on the stack. Get up on the stack after every load and straighten it a little and see that it is shaped properly and is kept full and well tramped in the middle. A team can be used instead of tractor if desired.

The 2x4x10-foot pieces can be bored near each end to put a ¾-inch rope through, but the middle rope should be tied around the 2x4's so as not to weaken them. The 2x4's should be about 7 feet apart. The crossed ropes can be pulled from under the load when it is in place.

SECTION 5.

Power Saws

Power Drag Saw

This design of a power drag saw is by Prof. L. G. Heimpel, of Macdonald College, Que. It is run by a 2½ h.p. gasoline engine which is belted to two old mower wheels from which the lugs have been removed. One of them runs free and there is a belt shifter to shift the belt from one to the other. An end and side view of the belt shifter is shown. It is connected to the lever by a rope.

The mower wheel which drives the saw, is cranked to a swing to prevent excess swaying. From this swing an arm connects directly with the drag saw. The arm runs through a guide, which slides up and down in a frame so that the saw can be raised and lowered. The second man pushes the log or pole along the grooved roller and holds it in place while the block is being sawed off.

The skids are of 3x12 planking, carefully framed together. The saw blade is thick and strong and is specially made for such machines as this.

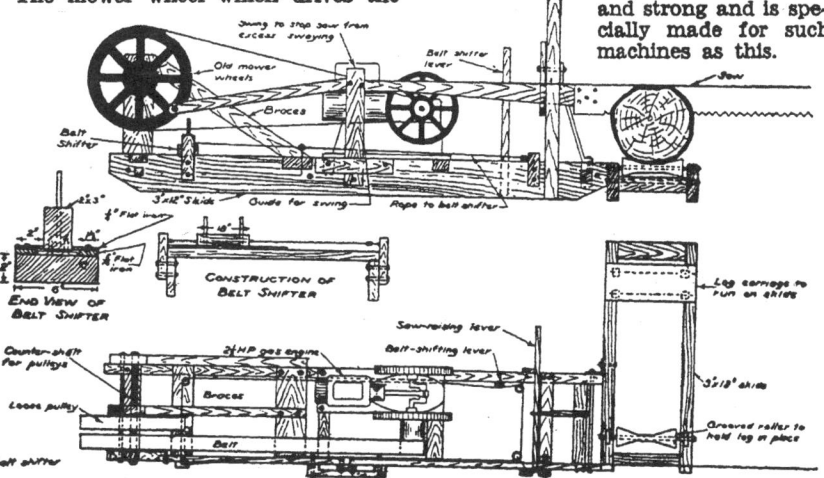

Power Saw For Big Logs

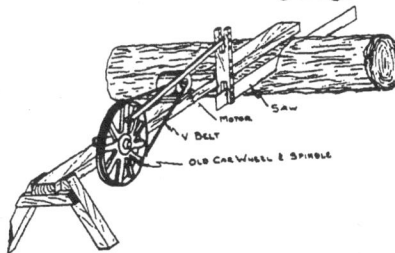

Here is a diagram of a power saw for large logs. It is powered by a washing machine engine or a ¼ to ½ h.p. electric motor where electric power is available, which is bolted to a 2x8 about 10 feet long, with saw horse legs at one end. The motor pulley is belted to an old car wheel whose spindle is bolted to the timber as shown. A rocker arm with each end slotted and bored for three changes of length of stroke is also clamped to the 2x8, and the top connected to the car wheel by a wood pitman, and the saw to the lower end of rocker arm. If thought desirable sharp pointed nails can be put on the under side of the 2x8 to hold it in place on the log.

Homemade Saw Frame

The sketch shows a steel pole saw frame with a swinging table which works very well. The frame is made of 2x2 inch steel angle iron taken from an old seed drill. The table is made of 1½x1½-inch angle iron taken from an old binder. The hooks are 2x½-inch flat steel. The hooks should be made according to the size of blade used. For a 24-inch blade the hooks should be about 11 inches. The base is an old car frame about five feet long. The mandrel is made from an old drive shaft taken from an old Chevrolet car. The frame stands 6 ft. 6 in. high and is about 3 ft. 6 in. wide.

DO A BETTER CUTTING JOB

with a VICTOR

When it comes to metal cutting jobs around the farm, Victor Unbreakable Flexible Hack Saw Blades are ideal. They cut tough metals faster and cleaner. Because of their flexibility, they make sawing easy in hard-to-get-at places.

For cutting extra hard materials, the Victor high-speed steel Moly* blades are recommended. These blades are highly resistant to wear and stay sharp longer.

Either blade properly strained in the Victor improved hack saw frame provides a handy cutting tool to have around for making emergency repairs to farm equipment.

*T. M. Reg. — Blades bearing the name "Moly" are made only by Victor Saw Works, Inc. and affiliated companies.

VICTOR SAW WORKS, INC.
Middletown · New York · U.S.A.
Makers of hand and power hack saw blades, frames, and metal cutting band saw blades.

Canadian Distributor
DAVID CLARK
365—43rd Ave., Lachine, Quebec, Canada

Plan of Swing Saw

When heavy logs are being sawn, a swing saw can be used to advantage, because the logs are in a stationary position while they are being cut. In using the swing saw, one end of the log is placed on the saw table and the other end of the log is placed on a trestle

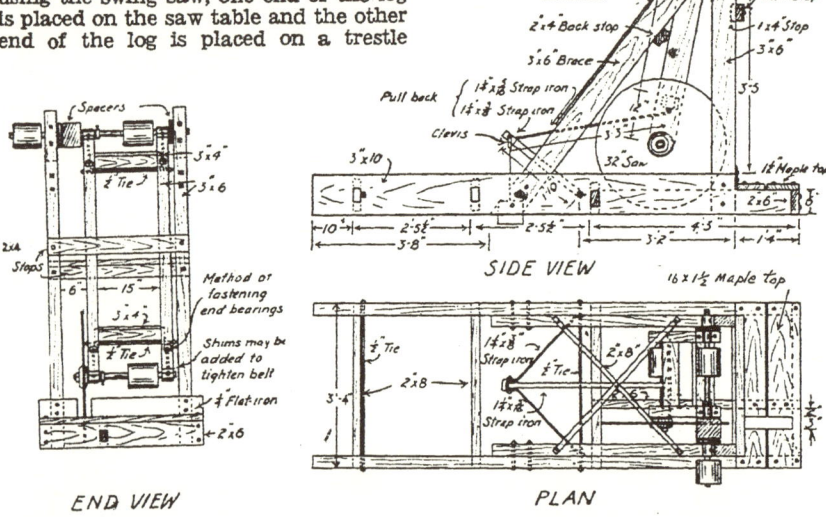

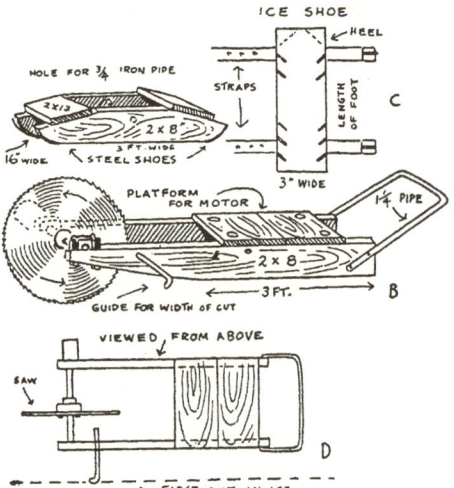

which is the same height as the saw table.

Where a portable sawing device is desired, the swing saw unit is usually mounted on a low wagon gear. For stationary work the saw frame is blocked up so that the saw table is about 30 inches from the ground.

In constructing the sawing unit, all parts of the frame should be well braced and rigidly constructed. The pull back device consisting of a tie strap, a 1¼-inch by ⅜-inch strap iron hinge, and a weight, is designed to hold the saw back firmly and to reduce pull as the saw is drawn through the log. A suitable weight can be selected for the pull back device and it must be securely attached to the tie strap and hinge.—W. Kalbfleisch.

Homemade Tilting Buzz Saw Frame

Prof. L. G. Heimpel, of Macdonald College, Que., designed this tilting saw

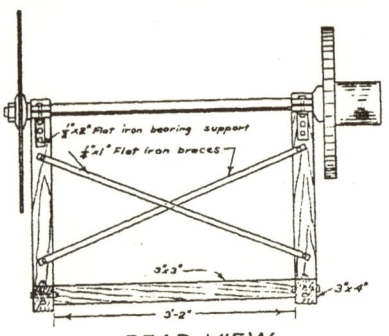

table. These views of it are given from his drawings. They make it self-explanatory. The sills are 3x4, or could be made of 4x4, and are securely framed. The uprights are of the same dimensions. A saw guard is provided and this should not be overlooked, as a whirling saw can be an instrument of destruction and proper precautions should always be taken with dangerous machinery. A 32-inch saw is recommended.

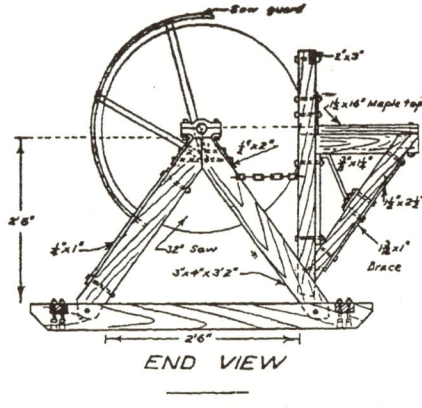

An Ice Saw

I have used this ice saw for 11 years and it has given entire satisfaction. I use a 3-h.p. engine and a 30-inch saw, and can cut 12 inches deep. I used an old brake drum out of a model T Ford car for bolting the saw to, using two ⅜-inch bolts which go through the saw and collar. The holes in the platform on which the motor is set has oblong holes so that the motor can be slid backward for tightening the belt. This assembly is mounted on the sleigh, shown at the top left. The weight is carried by a piece of ¾-inch iron pipe which passes through the holes in the runners of the sled and in the side pieces of the part carrying the saw. This allows you to tilt the saw up and down, according to the depth you want to cut. You go backwards while pulling the saw. Ice shoes can be made from flat iron, with slits cut in at an angle of ¾-inch and the points turned down for spikes.

I cut the ice both ways and then cut about every five cuts with a cross-cut saw and split the balance out with a bar. The blade of the bar is two or three inches wide and tapered from half-an-inch thick at the top to a thin edge. It is the wedging that splits the ice.—Lee Bussard, Wetaskiwin, Alta.

SECTION 6.
Gates, Fences, Clotheslines

Wherever fences are required, the need to get from one side to the other is almost always an inevitable certainty. The means, however, may vary with requirements. The old fashioned stiles are useful, convenient and inexpensive where travel is only occasional. Gaps are more desirable where travel is more frequent and pails of water, etc., must be carried and at the same time livestock is to be excluded as from the grounds around the farm home. Gates that are used frequently must be well constructed and well hung. Back pasture gates, which are opened only a few times a year, need not be expensive but should be strong and safe. Gates which provide easy access to cars and trucks, without the driver having to leave his vehicle to open or close them, invite the greatest appreciation. The following sketches are intended to provide suggestions. Materials available, as well as the need, are intended to determine the construction and means employed.

Barb Wire Stretcher

This stretcher is made from a U-frame of ½-inch pipe or angle iron and a gear from a discarded washing machine; the large wheel being riveted to an inch pipe collar. Then an inch pipe, 8 or 10 inches long is screwed into the

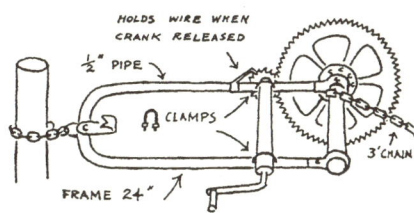

collar and a 3-foot light log chain is attached to this pipe near the gear side. The hook at the end next the post swings both ways. In the forward position it holds the wire while taking a fresh grip on the wire with the grip on the 3-foot chain in case one operation is not enough. In the backward position it holds the wire in case you want to splice wire between posts. The ratchet holds the wire at the desired tension when the crank is released.—A. A. Dickman, Langham, Sask.

Simple Pole Gate

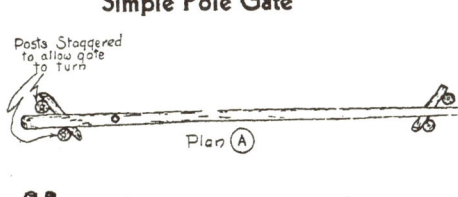

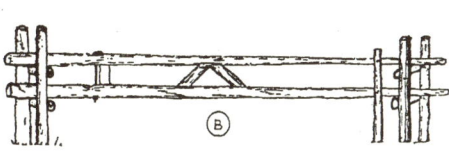

The simplest of all gates. It has neither hinges nor latch. The upper view shows the arrangement of the posts.

Cantilever Gate

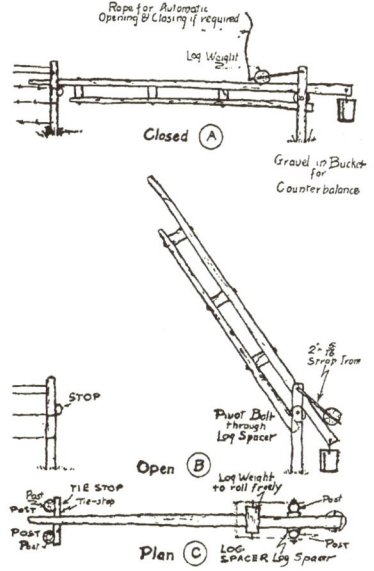

A simple automatic opening and closing gate. The rope attached to the weight is carried through pulleys on high posts and the end drops down where it can be reached from a vehicle. Three views of the gate are shown.

Gate From 2 x 4's

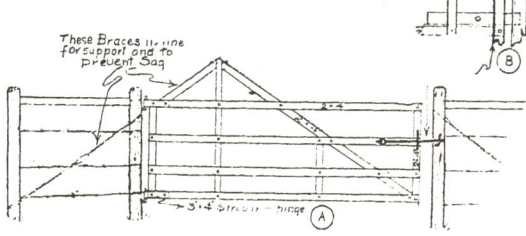

A simple gate. Two well known designs of latches are shown.

Chain Gates

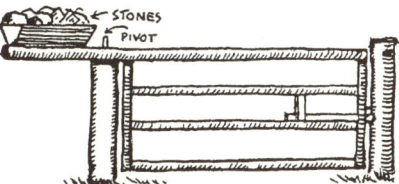

Two designs of gates which can be lowered to allow a vehicle to pass over. A and B show one closed and open. Short pieces of chain drop down which assist in turning stock. It is tightened by a lever, which comes to rest on a lever stop below dead centre. C shows a double chain gate with the chains passing through holes in the post and tightened by a crank and roller. Plow springs are used where shown.

Sagless Gate

This gate will not sag and will last. I have seen one in Ontario in good condition after 50 years of service, and occasionally have seen one in western Canada. It is generally made of poles, as shown in the diagram and the stones are just enough to balance it. Where poles and stones are available the cash money cost is nil.—D.C.R.

Improved Wire Gate

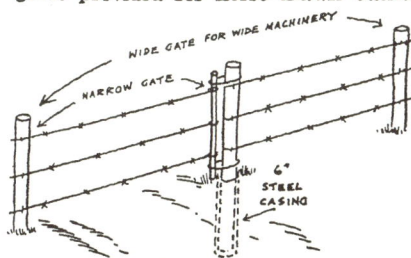

Here is a handy wire gate that is not hard to open. We use them all the time. Fasten the cross rail to the upright 2x6's and let it project past the post, dropping it behind the bent strap iron. This gate can be made any length or height and with as many wires as the farmer wishes. — T.C.S., High River, Alta.

Widening the Gateway

Now that nearly every farmer has wide tractor machinery the narrow gates provided for horse-drawn outfits are not wide enough. To remedy this set one of the gate posts into a six-inch iron pipe. When drawing one of the wider outfits simply take the post out of the casing and you will have twice the former width. It will be necessary, of course, to rearrange the bracings of the posts.—A.S.W., Alta.

Emergency Gate

To avoid driving a team on the highway when exchanging work with an adjoining neighbor, an emergency gate made in the dividing line fence comes in mighty handy and saves much time in passing through binders and combines. Cut each barb wire, fasten a hook to one cut end and a few links of chain on the other end, and this will make a neat and serviceable gate. If for any reason the wire becomes loose, tension may be taken up by dropping a link from the chain. If this is to be used frequently it would be well to brace both posts. A strong spring in each wire would help to keep up the tension.—I.W.D.

Dogs Preferred

Here is a sketch of a doorway in a picket fence which will let the dog through but excludes the feathered denizens of the barnyard. It works. It is simply a piece of board hung in an

GOODYEAR Sure-Grips PROVED BEST on every job!

- In spring mud or winter snow—on tractor jobs the year round—you get more pull with Goodyear O-P-E-N C-E-N-T-E-R Sure-Grip Tractor Tires.

Sure-Grips are *tried* and *proved* by years of farm experience, by farmer preference, by unbiased scientific tests.

Play safe. Insist upon Goodyear Sure-Grips. No other tractor tire gives you all of their time-saving, money-making advantages. No other tire can equal Sure-Grips for ability to pull.

SAVE WITH "SOLUTION 100"
Exclusive Goodyear method of filling tubes 100% with liquid. Pressures remain constant, summer and winter. More weight, less slip, greater draw-bar pull, smoother ride. Saves up to 10% on fuel.

GOODYEAR IMPLEMENT TIRES are the finest that can be made for the job. Any implement is easier to pull or push when on rubber tires.

KLINGTITE THRESHER BELTING
For better power transmission, less slippage regardless of weather... wet, dry, cold or hot... use Goodyear Klingtite Thresher Belting.

GOODYEAR
THE GREATEST NAME IN RUBBER

opening which has been framed in the picket fence. It is hung on an iron pin held by staples and will swing both ways. The dog just pushes it up and slips underneath but the fowls are fooled. — John P. Napier, Kilgobbin Farm, Royal Oak, Saanich, B.C.

Simple Wire Stretcher

The handle is made from two pieces of ⅛x1 inch strap iron, bent and riveted together as shown. The holes are two inches apart. The chain which runs back to the post is five or more feet long and the other two, 18 inches. The chains

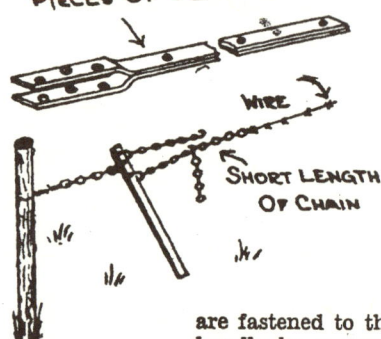

are fastened to the handle by a small bolt or rivet passing through the end link. To the end of each short chain fasten on a singletree hook by means of a split or lap link. Another piece of chain about five feet long completes the assembly. The latter has an old mower guard on the end to hold the barb wire.

Mending Broken Wire

Make a loop on both ends of the broken wires. Then take a piece of soft baling wire or No. 9 gauge and fasten it to it to loop (3), pass it through loop (1) and back through (3). Take a hammer and catch the end of the patching wire

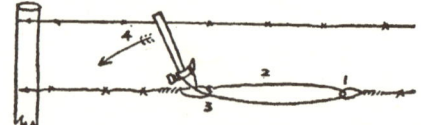

with the claws and then twist the hammer around so that the wire comes over the neck of the hammer and down through loop (3). Now turn the handle away from the break as shown by the arrow and keep on turning until the wire is as tight as wanted. Then turn the handle back and wrap the end of the wire around the splice. Allow 14 or 16 inches between loops to take up the slack of the wires.

Peter Pepper says:

"Score a Bull's-eye – BUY BRANDED SHOES FOR THE WHOLE FAMILY"

Be Satisfied

DURABILT AND KING PIN

MINER RUBBERS *Too!*

DURABILT and KING PIN
WORK BOOTS for MEN and BOYS
CRITTENDEN / ADMIRAL } *Men's Fine Shoes*
MINERVA - STYLE *plus* COMFORT – *A Woman's Fine Shoe*
. Togs . FOR THE *SMART* YOUNG MISS	PETER PAN FOR HAPPY CHILDREN

FIT YOUR FEET *before* YOU BUY

ASK FOR These *BRANDED* Lines At Your HOME-TOWN Store

BRANDED FOR *YOUR* PROTECTION!

Footwear — By Congdon-Marsh Ltd., Winnipeg, Man.

Performance is what Counts!

Today's Quality name in FARM MACHINERY is MINNEAPOLIS-MOLINE --- based upon proven performance: low-cost operation; low-cost upkeep; fewer hold-ups; highest "box score" in Time and Labor-Saving and efficiency

"THE HARVESTORS"

MM HARVESTORS Get and Save the Crops

HARVESTORS, in six sizes, point to a faster, surer way of harvesting all grain, seed, bean and rice crops at lower cost per bushel, per acre and per dollar invested. MM introduced the *original* HARVESTOR in 1934 in the 12 foot size . . . the FIRST lightweight, big-capacity combine for all crops . . . it weighed nearly a ton less than previous combines of that size. The HARVESTOR proved its time-saving and extragrain-saving features in its first year, and in 1935 became the leading seller of all 12 foot combines. Every year from 1936 to 1941 the original MM HARVESTORS were the most popular with farmers and custom operators alike. Yes, sell-outs on MM HARVESTORS were a regular occurrence year after year even before the war—attesting to their popularity. Production of Harvestors, greatly curtailed during the war years due to shortage of materials, has now been expanded to the limit of materials available.

There is now a HARVESTOR for every size farm. The 5 foot HARVESTOR 69 is the ideal unit for a speedy harvest of all crops on the family-size farm. For larger acreages there are the tractor-drawn 9 foot and 12 foot HARVESTORS and the Self-Propelled 12 foot and 14 foot HARVESTORS. A special Self-Propelled 13 foot Rice Harvestor is also available.

One of the main reasons for the outstanding success of all MM Harvestors is their threshing mechanism consisting of MM *rasp bar* cylinder and one-piece, all-welded steel concave and grate, which provides a positive threshing action without breaking straw and weeds to bits and without cracking the grain. Because of rigid construction and positive adjustment of spacing between the cylinder and concave, it successfully threshes even the smallest, most-delicate seeds.

The grain pan is so designed that it assures a steady flow of grain toward the cleaning shoe, and prevents bunching of the grain when combining even on rolling land.

There are many other things about MM Harvestors which enable them to harvest all crops faster and surer.

See your MM Dealer for the MM Harvestor of your choice. They're worth waiting for.

MINNEAPOLIS-MOLINE POWER IMPLEMENT COMPANY
OF CANADA, LIMITED
REGINA, SASK.   WINNIPEG, MAN.

MM DEALERS NEAR EVERYWHERE

Stone Puller from Plow Beam

Mr. E. J. Stansfield, Atwater, Sask., has furnished The Guide with a snapshot of a stone puller taken on his farm from which this drawing has been made. It is an old idea to him, but may be new to many. It is made from an old P and O plow beam and the assembly is left just as it was on the plow after the share, mouldboard and frog are removed. The clevis is arranged so that the rock puller remains vertical. The puller is strong enough to take all you can give it, but a man can handle it. It can be attached to any tractor. Mr. Stansfield never tried it with horses, but he knows some men who have used it successfully with them.

Extended Wagon Box
Facilitates Scooping

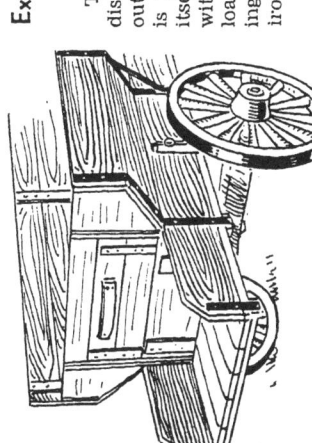

The design of this wagon box has two distinct advantages. By sloping the box outward, about 30 per cent bigger load is possible and by extending the bed itself past the endgate, no grain is lost with those first few scoopfuls when unloading. Angle iron forms the reinforcing at the ends of the endgates, strap iron binds the rest of the box together.

THE WESTERN FARMER
and
UNITED GRAIN GROWERS
LIMITED

Co-operatively owned and operated since 1906

Child-proof Gate

Our little girl got the habit of going out through the garden gate and getting into danger. I stopped that by putting an ordinary snap on the gate and a ring in the post. By the time she is big enough to undo the snap she will be more able to take care of herself.—Geo. Ray.

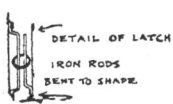

Simple Gate Latch

Diagram of a simple gate latch. It will take care of considerable changes in alignment, is proof against rubbing by cattle and most horses, and is easily repaired. All that is needed are two

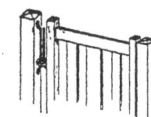

heavy pieces of strap iron bent as shown, and a neckyoke ring. The cut shows only one bolt in the strap iron fastened to the gate. Two bolts or screws would be necessary here to keep this piece from turning.—I.W.D.

Wire-Gate Tightener

This is another way to make a wire-gate fastener. The bit of plank is 4½

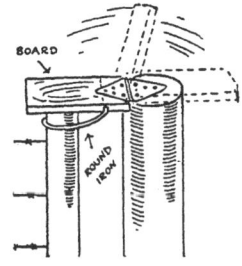

inches wide and 20 inches long. Fasten it to the gate post by means of a hinge. It gives a good purchase in tightening the gate.—Vernon E. Hotz, St. Boswell's, Sask.

For the Wire Gate

Instead of wire loops to fasten the free end of a wire gate sections of an

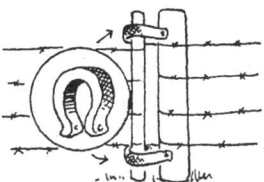

old auto casing can be used, according to directions in Popular Mechanics. These are said to be flexible, easier on the hands, and quite resistant to wear.

Lifting Posts

If the fence post is stubborn and refuses to leave the hole, tie a chain onto the post and use the car jack to lift it out of the hole. Tying the chain as tightly as possible will prevent it from slipping up the post. A piece of wood at the base of the jack will assist it from sinking into the ground.—Adam Szczepanowski.

Self Closing Gate

A farm gate that is self closing is made like an ordinary gate except that the top bar is four inches shorter than the bottom one. Also the upper hinge pin is longer than the lower one by the same amount. When the gate is swung open the latch end will be elevated so that it will swing shut of its own weight no matter how much or how little it is opened.— Bernard Schick, Carmel, Sask.

"Cattle Preferred" Gate

In cases where it is desirable to keep hogs from going into the field with horses or cattle, try using a gate built along the lines shown in the sketch. As there is considerable strain on such a gate it should be made with planking. Horses and cattle will step over it, but hogs go to one end or the other. In looking for a place to come out they

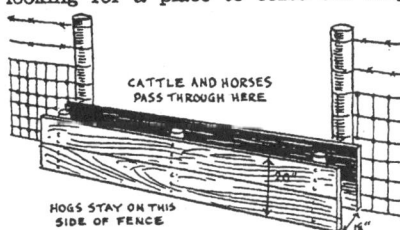

simply walk through to the other end and arrive on the same side of the fence. Hog psycho-analysts assure us the I.Q. of a hog is so low that he will never know that he is being tricked.

Texas Gate or Auto Crossing

This is the most convenient means by which automobiles and trucks can pass into fenced yards, fields or pastures without the necessity of getting out to open and close gates.

The opening in the fence should be about 10 feet wide to provide ample room for wide trucks, as well as for cars. A pit 10x9 feet by 5 feet 8 inches is dug so that 4½ feet of the pit will be on each side of the fence line. Two sills 10x8 feet by 8 inches and four beams 9 feet by 8x8 inches are required. Each sill is set about 16 inches below the ground line, along the front and rear

---

THERE'S BRIGHTER, LONGER LIFE IN GENERAL DRY BATTERIES

Still the most reliable battery for your needs. The Super Bilt A & B Pack for radio.

The 6 Volt Multiple for electric fence and ignition.

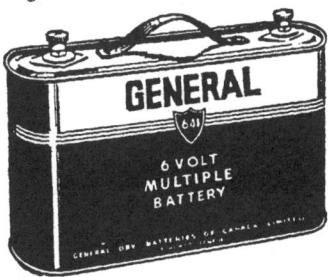

And all other GENERALS for signalling systems, hearing aids, flashlights, etc.

GENERAL DRY BATTERIES
OF CANADA LTD.
Toronto, Ontario

edges of the pit as shown in the sketch. The four beams are then set on the two sills and equally spaced. Either poles of about 4 inches diameter or 2½ to 3-inch pipe are then laid across the beams. The poles or pipe are then spaced 6 inches apart with pieces of 2x6 lumber. The 6-inch spacing is essential to prevent injury to the feet of all livestock should they attempt to cross. The poles can be nailed to the beams. When pipe is used, the ends can be held down with a piece of 2x6-inch lumber. A piece of bevelled plank is used on each side of the crossing for the approach.

outside upright of the swinging gate. A weight balances the sliding gate.

People-Preferred Gaps

Designs for Stiles

Two Designs of Turnstiles

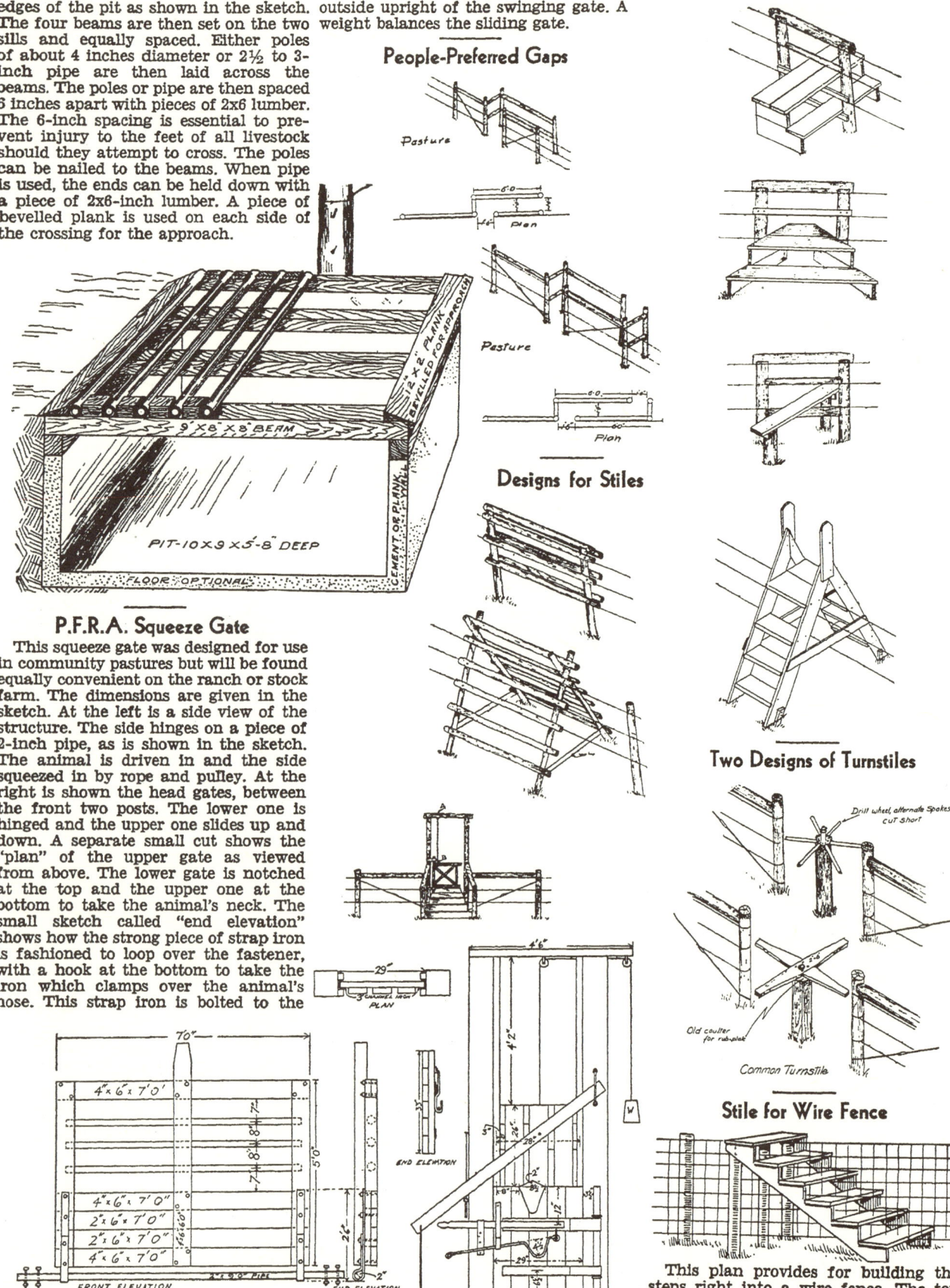

P.F.R.A. Squeeze Gate

This squeeze gate was designed for use in community pastures but will be found equally convenient on the ranch or stock farm. The dimensions are given in the sketch. At the left is a side view of the structure. The side hinges on a piece of 2-inch pipe, as is shown in the sketch. The animal is driven in and the side squeezed in by rope and pulley. At the right is shown the head gates, between the front two posts. The lower one is hinged and the upper one slides up and down. A separate small cut shows the "plan" of the upper gate as viewed from above. The lower gate is notched at the top and the upper one at the bottom to take the animal's neck. The small sketch called "end elevation" shows how the strong piece of strap iron is fashioned to loop over the fastener, with a hook at the bottom to take the iron which clamps over the animal's nose. This strap iron is bolted to the

Stile for Wire Fence

This plan provides for building the steps right into a wire fence. The top step rests on a fence post and a post is set under each end of the top step. The stringers are put in place and the remaining steps are slipped through between the wires. If any of the upright wires are in the way they may be cut to let the step in. The horizontal wires should not, of course, be cut.—Mrs. Dan Harris, Edgeworth, Sask.

Helps Climb Gate

Here is a plan to keep small livestock and hogs from following me through an open gate. The top section is in two parts, hinged at the middle on a bolt, so it can be raised while

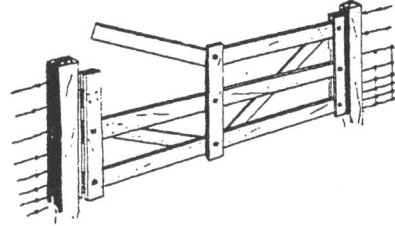

stepping over and then dropped back into place. When down, the movable part rests on a block in the end upright of the gate at the latch. When the gate has five boards the top two could be fastened together and hinged in the same way.—I.W.D.

Loophole in Fence

Popular Mechanics must have received this idea from a prairie farmer who knows what it feels like to get his overalls caught on barbed wire. It is a double frame, which slips in between two strands of the wire and through which

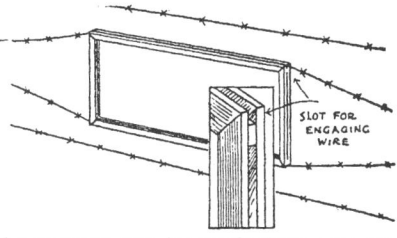

the passage can be made from one enclosure to the next without engaging the barbs. One precaution seems necessary. It should not be placed opposite any tempting mouthful of green feed or the old cow, reaching for it, and finding she was also protected, might put on enough strain to wreck the fence.

Guards for Electric Fence

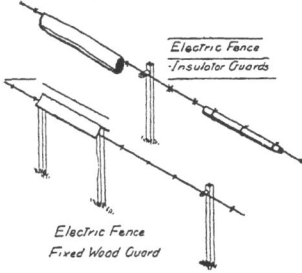

A piece of old rubber inner tube, a piece of split rubber hose, or a wood guard is useful to protect a person from electric shock when stepping over an electric fence. These can be placed where most frequent crossings are necessary. Gates of any kind are not then required and the circuit of electric current in the entire fence is less likely to be impaired or completely broken. When part of the electrically charged wire is disconnected to provide access, it is sometimes left on the ground and short-circuited.

A Latch and a Hinge

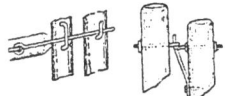

This shows the design for a latch and a hinge made out of common round iron. Where the hinge iron passes through the post good large washers should be used as they have to carry the weight of the gate.

Barnyard Gate Fastener

Here is a gate fastener which will prevent livestock from opening a gate. The locking device and latch can be made of either wood or iron. A pivoted block in the top of the locking device can be moved to permit the latch to drop into a slot and the pivoted block is allowed to swing back to prevent the latch from being lifted.

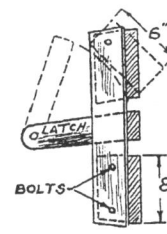

Old Horseshoe for Gate Hanger

Oft-time farmers fence in fields temporarily, or erect light fences that have equally light gates. An easy way to hang a light gate, and still make it secure, is to use four old horseshoes, mounting two on either gate post. If the gate is made of light wood, it can easily be hung on the horseshoes, and to open, merely lift it off its hangers. The farmer who used this idea told the writer it was one of the finest tricks he had ever employed.—Grover Brinkman

Mending Break in Fence

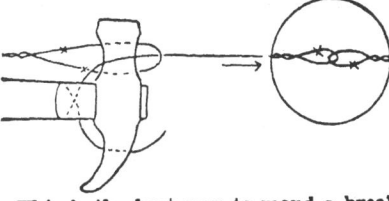

This is the best way to mend a break in a wire fence, much better than using a wire stretcher. Connect on a new piece of wire to the broken strand and remove the barbs for a short piece. Fasten this to the claw hammer as shown and wind up until tight. Then pull against the loop strongly and wind the wire in the hammer around the strand.

THERE'S A BURNS VIGOR FEED FOR EVERY FARM ANIMAL

Burns VIGOR PROTEIN & MINERAL SUPPLEMENTS

Drive Gate Latch

This gate latch will hold the gate securely. A half-inch rod goes through

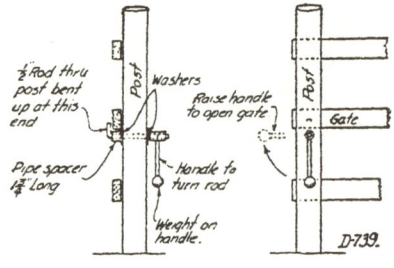

~Drive Gate Latch~

the post and is bent up to hold a bar of the gate. A bit of pipe the thickness of the gate bar is used as a spacer. A handle with a weight holds the latch in the closed position by gravity. Details are all given in the cut.—I. W. D.

Wire Tightener

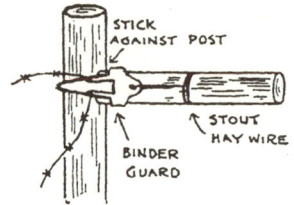

Here is a simple wire tightener which proves to be very handy around the farm. It consists simply of a bit of stick 2½ inches in diameter and two feet long. An old binder guard is wired to the stick, about four inches from it. It costs nothing and often comes in very handy.

Fence Stretcher

A lever, four clevises that can be borrowed from the doubletrees, four bits of chain, two of them with hooks, and a clamp to be tightened on the fence with bolts, make up this stretcher. As you pull on one 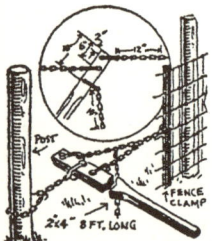 chain and take up some of the stretch of the fence you get a chance to catch the chain to the fence a few links further on with the other hook. The lever gives lots of power.

Gate Latch from Old Hinge

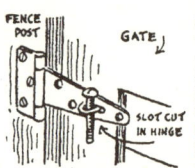 Old hinges make good gate latches when used in the manner shown herewith. After being slotted with a cold chisel the hinge is mounted on either the gate or the post and a good heavy staple provided to hold the latch bolt. Fasten the bolt to the gate or the post with a bit of string and all danger of dropping it is avoided.

Post Puller

This post puller is made by taking a piece of heavy plank a foot square and a bit of hardwood about three feet long.

The stick is fastened to the plank with a strong hinge. The stick is notched at the top and a bolt put through to keep it from splitting. Fasten the chain to the post just above the ground and pass it over the upright in the notch. Then

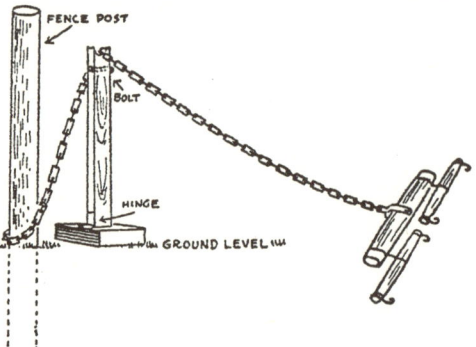

drive ahead.—Stewart A. Glauser, Delisle, Sask.

Handpower Post Lifter

This jack will lift a post without the trouble of using horses. The two uprights have holes bored in them at intervals so that they can be adjusted for any height. Through 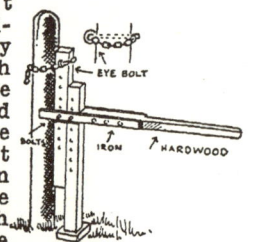 the one next the post to be lifted there is an eye bolt. A short piece of chain goes round the upright, through the eye of the bolt and around the post. It is then a simple matter of lifting by pumping on the handle, getting a new catch with the chain every time the handle is lifted.

A Gate Anchor

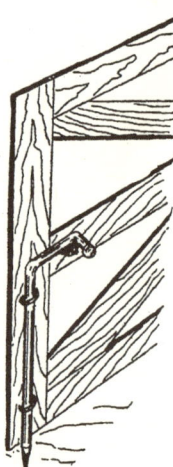

Gates that will not stay open when required are frequently a nuisance. A simple drawbolt type of anchor can be fastened to a gate as shown. A piece of ⅝ round iron for the drawbolt is made sharp at the end, which is to contact the ground or floor. The opposite end is bent to form a suitable handle. Two eyebolts or "U" bolts serve as suitable guides for the drawbolt. An ordinary carriage bolt serves to hold the drawbolt off the ground when not in use.

Just what the Farmer ordered

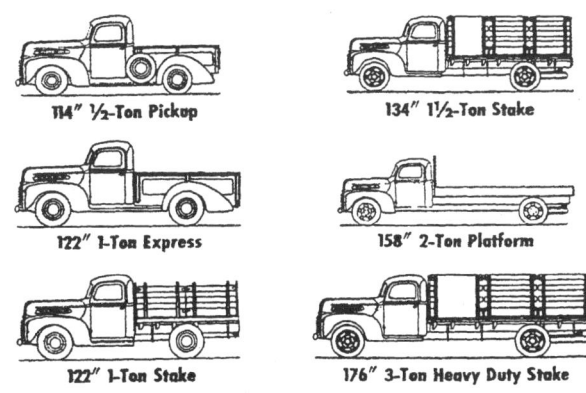

A FEW REPRESENTATIVE BODY STYLES

- 114" ½-Ton Pickup
- 134" 1½-Ton Stake
- 122" 1-Ton Express
- 158" 2-Ton Platform
- 122" 1-Ton Stake
- 176" 3-Ton Heavy Duty Stake

Built into this amazing Mercury Truck is every feature a farmer could ask for.

Back of its massive, gleaming grille is a big, powerful Mercury Special High-Torque V-8 Truck Engine, with Aluminum Cylinder Heads for extra power. It's your best bet today for low-cost hauling, and for long-lived dependability.

The Mercury Truck chassis is truck-engineered, truck-built throughout... with dual sun visors, horns and windshield wipers, cigar lighter, ash tray and rear ventilating window in all models.

And best of all, Mercury trucks, from ½-ton to 3-ton, are priced right down near the lowest. See your Mercury and Lincoln dealer.

MERCURY AND LINCOLN DIVISION
FORD MOTOR COMPANY OF CANADA, LIMITED

THERE'S A MERCURY AND LINCOLN DEALER NEAR YOU

Horseback Girl's Gate

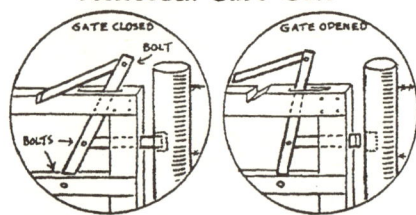

The only iron on this gate latch is three bolts and a piece of strap iron for the upright. The gate may be closed and opened from the saddle. It is much handier if the gate is made to swing both ways.—Horseback Girl, Northmark, Alta.

Clothes Line Pole

Common clothes line poles have a way of slipping and causing the line to drop, dragging the clothes in the mud. If the pole is split at one end and a spacer nailed about a foot from the end and then a screw eye placed in the middle of the spacer the weight of the clothes on the line will cause it to lock and prevent slipping.—PAUL TREMBLAY, St. Paul, Alberta.

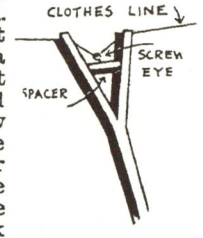

For Removing Fence Posts

Please find enclosed a sketch of a post jack of my own construction which will help many farmers in the problem of pulling out posts which must be replaced by new ones or removed to another location. With this jack two men can pull out from 75 to 100 posts per hour. It is very simple to make and inexpensive. All it takes is a piece 2x4 by 8½ feet long, two pieces 2x4 by 2 feet long, a piece of 1x4 for bracing, an old mower guard, a bolt ⅝-inches by 6½ inches, two 6-inch clevises, two ½-inch bolts 3 inches long, one 4-inch spike and a few 2½-inch nails.

The frame and lever are assembled as shown in the drawing. A hole for the large bolt is bored 18 inches from the working end of the lever. One man does the jacking by setting the jack the proper distance from the post and poking the end of the mower guard into the post about 24 inches above the ground and prying up the post while the other man pushes the post against the guard. With a little practice you will master the use of this jack.

Foot-opened Gate

When you are coming through a gate with your arms full it is handy to have a gate that you can open with your foot. This gate opens with the foot from either side. The fastening on to the gate slides under the part of the latch which is fastened between the twin gate posts. A piece of rubber from an old inner tube pulls the latch down when the gate is shut. The cross piece projects out on each side to put your foot on.

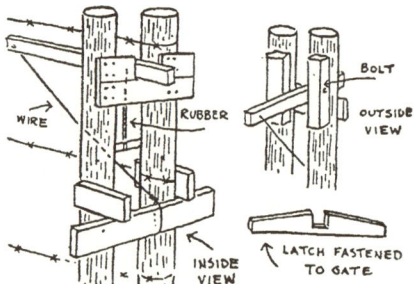

The gate should be on hinges that will allow it to swing both ways. The cross piece would be best notched to take the two short pieces that project from the posts so that it would not get out of place.—Evan Price, Hemaruka, Alta.

SECTION 7
Electricity and Electric Wiring

Alarm Clock Time Switch

An alarm clock can be used for making a simple time switch for turning lights on or off, or for automatically switching electric current on or off for other mechanical purposes. A mercury switch, such as is used with various electric controls is fastened to a metal clip, which in turn is soldered on to the alarm winding key. The clip is made out of a piece of tin, copper or brass and is bent to fit the mercury switch. The two lead wires from the mercury switch are connected to the electric light circuit in the same way as an ordinary toggle switch. An adjustable stop is fastened on the back of the clock to limit the travel of the alarm key sufficiently to allow the mercury in the glass tube to flow to or from the enclosed contact points. In operation the clock is wound up as usual. The alarm is also set in the conventional way. When it is desired to have lights turned on, or a motor started at a set time, the mercury switch is set as shown in the sketch. When the lights or electric power are to be turned off, the mercury switch is taken out of the clip and reversed and then placed back in the clip.

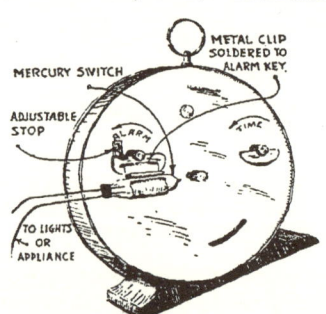

Alarm Clock Starts Radio

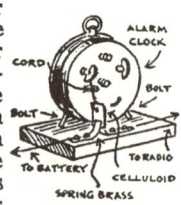

If you would sooner be awakened in the morning by your favorite morning program than by the raucous call of an alarm clock you can hitch the clock to the radio. The clock is mounted in place as illustrated, the bell removed and the alarm set as usual. When the alarm goes off the winding of the string around the winder lifts a piece of celluloid allowing the metal points to form contact. Obviously the regular switch must be left turned on so the clock can do its stuff.

Wiring Model T. Engine

A subscriber states that he wishes to make a circular saw engine out of a Model T engine, using all four cylinders and also using clamp to the battery. He states: "What I wish to find out is how to wire the coil timer and the spark plugs." Prof. G. L. Shanks answers: "What your enquirer wants is a complete wiring diagram for an old Model T Ford. The attached diagram indicates

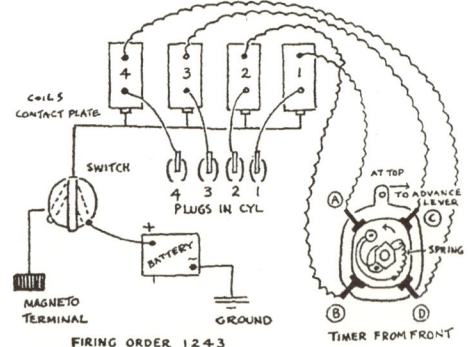

all the necessary connections and names the parts."

Another subscriber writes: "By using the Model T spark coils is it possible to build up the voltage of the 6-watt battery or generator? I have a 6-volt outfit but find in many cases it hasn't the power needed. Could you tell me how high I could jack the voltage, if any, and also the watt power?"

To this question Prof. Shanks replies: "Your subscriber is apparently on the track of a perpetual motion idea. It is true that a Ford spark coil will step up voltage from 6 volts to approximately 5,000 volts, but it is done at the expense of the amperage. In the end the volts times the amperes or watts, is less than in the beginning and there is a loss in watts rather than a gain."

How To Correct Faulty Timing

Faulty timing or ignition is the great cause for gas engines being discarded. This is especially true of the older type of low tension ignition which does not use a spark plug. The high tension or spark plug ignition is so much better and more reliable that now few if any low tension engines are built. In general it will not pay to replace worn parts of low tension ignition, whether battery or magneto, but they should be changed over to battery spark plug ignition, using the method shown clearly in the diagram. Take the igniter block to a machine shop and have the larger hole drilled and tapped for a standard size spark plug, while the smaller hole is plugged by tapping or welding in a short stud or bolt. The timing device is simply a drop welded on or a short pin tapped into the half-time or cam gear in such a location that it will touch the insulated spring just before the piston reaches compression dead centre. The coil used is a model T Ford or other vibrating jump spark coil. Four dry cells or a hot shot battery furnishes the current, or a small auto storage battery can be used if preferred.

If the engine has spark plug ignition furnished from an oscillating or rocking magneto and the magneto gives trouble better discard the magneto and arrange the battery ignition as shown, as it gives much less trouble from getting out of time. Over ten years ago I bought for five dollars a three h.p. engine on which the magneto had gone bad, and installed the timing pin and spring with model T coil, and it is still giving satisfactory service.—I.W.D.

Mercury Cut-off Switch for Pump Engine

A simple wood float, fitted with a

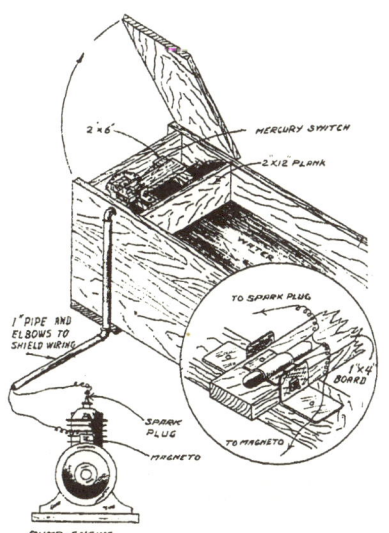

mercury switch, is ideal for stopping a pump engine and preventing overflow of water and waste of gasoline. The mercury switch is used in many electric controls, such as temperature regulator control switches used for automatic coal stokers. They can be obtained from any well stocked electrical store and are inexpensive. The sketch shows the mercury switch attached to the hinged arm of a wood float, in a manner similar to that of a spirit level. When the water is below the full level of the tank, the float arm is tilted downward and the mercury in the tube flows towards two lead-in wires in one end of the tube and thus contacts both of them. One of the lead wires from the mercury tube is connected to a high tension cable from the magneto. The other lead wire is connected to a high tension cable attached to the engine spark plug. When the tank is full, the float arm becomes tilted upwards so that the mercury in the switch flows away from the two contact wires. The flow of electric current from the magneto to the spark plug is thus interrupted and the pump engine stops.

Electrifying Farm Fence

"I wish to electrify a fence surrounding my feed yard. A six-volt battery will supply the current. What type of coil should I use and how should the whole outfit be wired?"—R.L.B., Swanson, Sask.

To electrify a fence using a six-volt

For **SUSTAINED** *Power* use **MARLENE** HEAVY DUTY **MOTOR OIL**

- Increases load carrying capacity.
- Prevents accumulation of hot engine sludge.
- Eliminates sticky rings.
- Prevents the formation of varnish.
- Resists oxidation.
- Reduces scuffing of pistons and cylinders.

MARLENE HEAVY DUTY **MOTOR OIL**

ANOTHER **MARSHALL-WELLS** PRODUCT

storage battery you require a six-volt ignition coil such as is used in most modern cars. In addition an interrupter is desirable as this will conserve the battery and under the worst conditions of shock will produce less serious results. The complete distributor of a wrecked car mounted in a stand with a small wind propeller is a simple interrupter. A motor driven interrupter made from

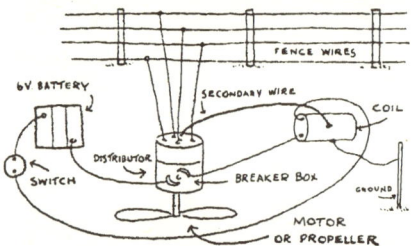

a six-volt fan motor could also be used and driven from the battery. This latter plan, though harder on the battery would be the most satisfactory as it would work in calm weather. The wiring of this outfit is identical with the wiring of the car from which the parts were taken except that the wires which lead to the spark plugs are now connected to the various fence wires and the ground of the coil system is connected to a rod driven into the ground. The diagram would be typical for most car wiring systems.—Prof. G. L. Shanks.

Electric Persuader

This home-made electric livestock prodder is used for touching up stubborn cows, hogs, or other animals when they object to going up loading shutes or into trucks or cars. It consists of a Model T Ford coil clamped to an old billiard cue and actuated by a three-cell flash-light battery and a simple push button switch. Insulated wires leading from the coil out to the end, deliver a hot spark where it will do the most good. This is much more humane than a club or a pitchfork and does not bruise the flesh or damage the hide.—I.W.D.

Mounting Rear Tractor Light

Here is a handy mount for the rear light on the tractor. It is made of two pieces of light ½-inch by ½-inch angle iron each about three or four inches

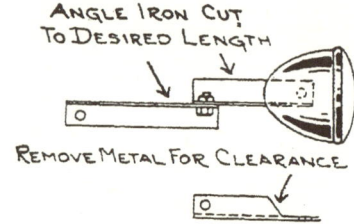

long. The two irons are held together by a bolt as shown, and are fastened to the fender or tractor frame by a second bolt. The light may be adjusted for height by loosening the first bolt, and from left to right by loosening the second bolt.—I.W.D.

Handy Battery Charger

A handy portable arrangement for quickly recharging auto or radio batteries or for furnishing light at camp consists of a small washing machine engine mounted on a two-inch plank with a V-belt driving an auto generator clamped to the same board with the bolt holes slotted for belt adjustment.

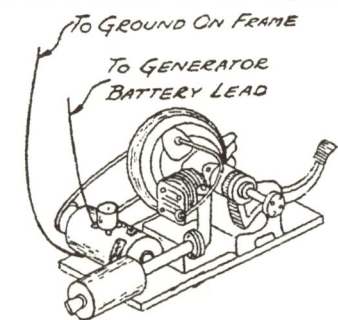

The heavy wires from the generator are fitted with heavy battery clamps,
one being fastened to the car generator terminal or to the ammeter terminal and the other grounded on the engine or frame. Such an outfit will deliver from 10 to 15 amperes and will recharge a battery in a few hours.—I.W.D.

Jiffy Connector

I found that a paper clip serves excellently when temporary test connections are to be made. While the diagram shows two phone tips held together with a paper clip, almost any connection could be made in a similar manner. Flexible wires, as well as solid ones can be joined together without the trouble of twisting them.—Wm. J. Dutka, Emerson.

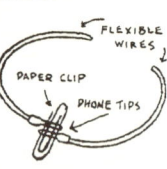

Repairing Electric Element

Join the wires together loosely, using as little wire as possible. Switch off the

current when the join becomes white hot, and put on a little boracic acid. Then shake the join until it arcs and you have a weld.—Grant Macleod.

Lighting System for Tractor

All parts with perhaps the exception of the V-belt can be taken off an old car. The principal trick is to take the generator and attach it to the tractor in such a way that it can be driven by the tractor belt pulley. Most tractors have a 10-inch belt pulley which runs at about 900 r.p.m. The proper speed for a generator is about 2,200. Therefore, one must see to it that the correct size of pulley is used on the generator to develop the proper speed. In this particular case the generator pulley should be about four inches. Your tractor dealer can give the rated belt pulley r.p.m. for your tractor if you do not have a speed indicator.

Even the old dead battery can be used, and it is not necessary to buy a new one. However, if the battery is dead

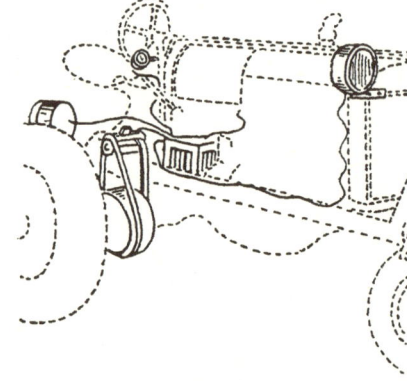

there will be no lights while the motor is not running. But a battery of some sort must be used to take up the excess current when necessary. Otherwise, it might damage the generator.

It is not a difficult trick to do the wiring if you follow the wiring diagram. The wire should be well insulated against shorting and also against dampness during wet weather. Old rubber garden hose and the use of friction tape should solve the problem.—Kansas Circular.

Tower for Wind Charger

To make a tower for a battery charger, take boiler flues from an old steam engine and get them acetylene welded, making four pipes each about 16 feet long. A piece of 4x4 about 6 feet long and three old wheel rims are needed, one from a drill wheel, one from a plow wheel and a third about 12 inches in diameter. Four long bolts are needed, two about 10 inches long for the top of the flues and two about 14 inches long to make the 4x4 secure in the small rim at the top. Twelve U-bolts are also needed, four to fasten each rim in its place between the upright flues. This makes a tower 18 feet high and one that will last a long time. It can be made secure to the ground by means of iron stakes or cement blocks.—Arnold Smith, Peebles, Sask

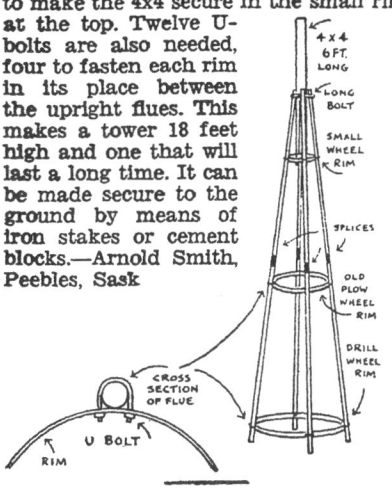

Switch for Van

Here is a very easy switch to make for a van with a double contact bulb. I used the T top of a Ford coil. It snaps on two round headed screws. The coil top swivels on a little wooden plug from a twine ball. It is notched so that when you turn the switch on it does not slip.

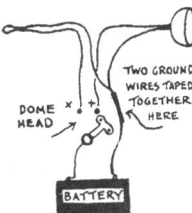

Alarm Clock Time Switch

An alarm clock can be used as a time switch for any electric circuit on the farm or in the shop. It will turn off the poultry house lights, the battery charger, the milk cooler, turn on a small stove and other similar jobs.

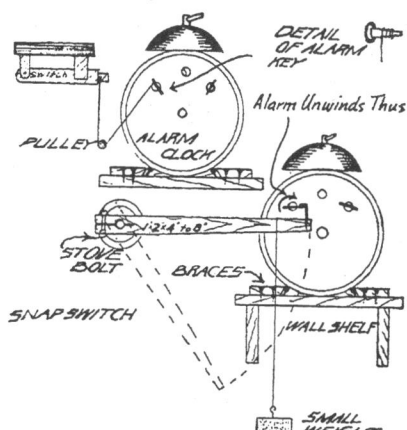

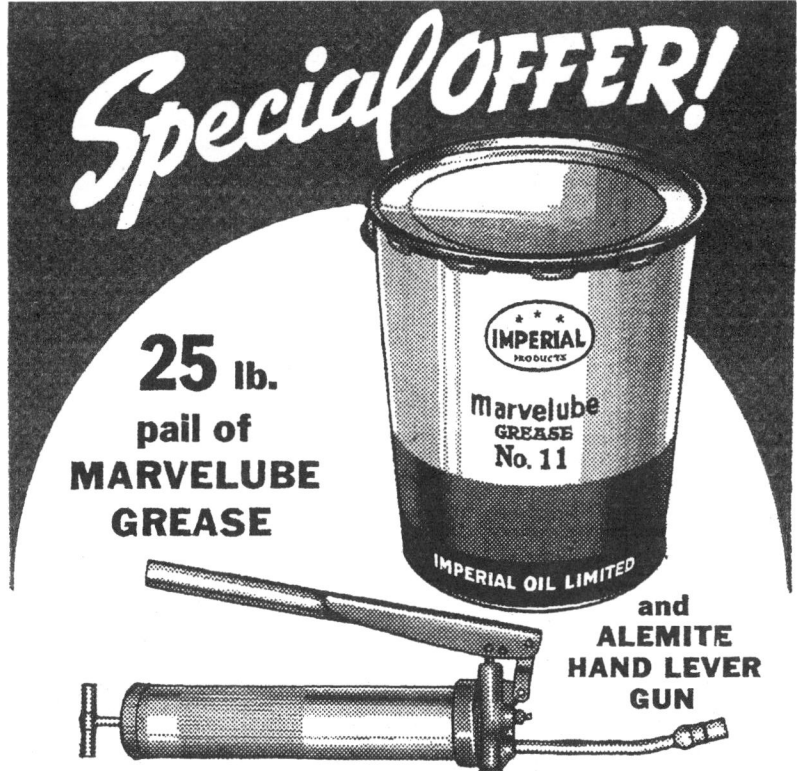

Special Offer!

25 lb. pail of MARVELUBE GREASE

and **ALEMITE HAND LEVER GUN**

Combination Offer at a **SPECIAL LOW PRICE**

Correct lubrication prolongs life of farm machines, increases speed and efficiency

Here's a grease gun for you that will save you money and speed up your work by making your farm machines operate more smoothly. It is an Alemite hand lever gun designed and manufactured by the originators of high pressure lubrication. It was developed and used for army vehicles during the war and can handle all types of light or heavy bodied and fibrous lubricants. It is strong, sturdy and efficient. It is offered along with a 25 lb. pail of Marvelube Grease at a *special low price*.

Several other attractive grease equipment deals. See your Imperial Oil Agent.

IMPERIAL OIL LIMITED

Where electricity is not used, the clock and wooden control arm can be set up to control the furnace dampers. For this arrangement connect the front damper of the furnace to the rear damper on the stove pipe by a light chain over a set of rollers. Adjust the chain so the front damper is three-quarters open when the rear damper is just closed. Balance the system with a small weight so the front damper stays open. In the evening connect a strand to the alarm clock control to hold the weight on the chain up (front damper of furnace closed, rear open). The alarm clock (remove the bell) will drop the weight and reverse the dampers at six a.m. in the morning.—W. Kalbfleisch.

Six-Volt Soldering Iron

I have a diagram here showing how to make a six-volt soldering iron. The wire from the battery clips on one post of the battery and a wire from the work to be soldered to the other post thus making a circuit. I have found this iron to work very successfully.—John Nickolcon, Cairns, Alta.

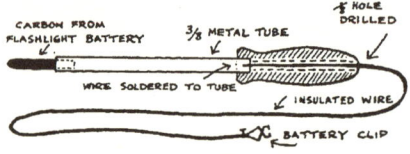

Avoiding Bad Connections

You can avoid bad connections at battery terminals by soldering the terminal clamps to the posts and joining the leads a foot or so away. The joins in the leads will have to be wrapped with tape to protect them and prevent short-circuiting.—Grant Macleod.

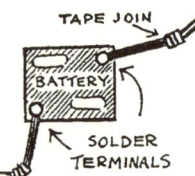

Rear Light for Tractor

A rear light for the farm tractor can easily be made by cutting a flashlight down as shown. A hole is made through the flashlight cap where an auto tail light connection and bulb is soldered on.

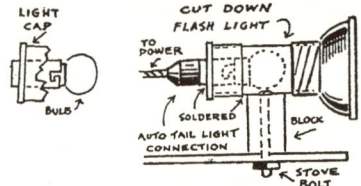

The light is mounted to the rear of the tractor by means of a wood block and stove bolt, which is first pushed through from the inside of the light case. Such a light gives considerable illumination at a minimum cost and with very little power.—A. S. Wurz, Jr., Rockyford, Alta.

Getting Most Out Of Batteries

Where a multiple dry battery is used for starting or for ignition in an internal combustion engine during cold weather, the battery should be brought into the house at night so that it may be at room temperature when put into use in the morning. Flashlight batteries, in cold weather, will give much better service if kept at ordinary room temperature when not in use. Cold does not permanently injure a dry battery, but it does interfere with its functioning.

Generator Makes Grinder

This simple grinding wheel mounting is an old car generator with an extended, threaded shaft, strap iron brace strips and the belt running on the commutator. Copper tubes, bent to shape and soldered in place, carry oil to the bearings.

For a price usually less than a dollar, you can get a burned out generator with

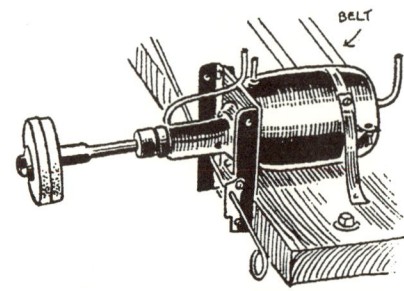

good bearings which, after months of service, will be tighter and smoother running than the bearings in a grinding wheel stand with plain bearings, but built especially for the purpose and costing a great deal more.—Dale Van Horn.

Timing the Engine

One of the common causes of poor engine performance and decreased engine efficiency is incorrect timing of the ignition. I have personally known of many cases where the efficiency of the engine has been increased as much as 40 per cent by correct timing. This is true of small engines, automobiles and tractors. Ignition usually occurs at dead centre for starting and in advance to fire before dead centre for running. Timing will usually be correct on new engines and it is only when the unit has been removed, repaired or replaced or where wear has taken place that the ignition timing will be incorrect.

On practically every engine ignition marks will be found to aid the operator in checking timing. The marks are found in a variety of places and if the operator is not acquainted with the procedure it would be best to read that part of the instruction book that deals with timing. On some machines the timing can be advanced or retarded while the engine is operating on a steady load.

In order that the operator may get the extra miles to the gallon it is necessary to have the breaker point gap and the spark plug gaps correctly spaced as well as seeing that all wiring is in good condition.—O. H. Lovelace, Saskatoon, Sask.

Alarm Clock Time Switch

The diagram shows a very effective but easily constructed time saver for any poultryman using electric lights.

Secure an alarm clock and fasten it down to a board or timber to which the light and pull switch are attached. Make the distance from clock to switch at least four inches. Place a piece of lath or light board on a nail or stick as a fulcrum, place one end of the lath not more than half way under the alarm wind and on the other end a weight sufficiently heavy to operate the pull switch when the weight drops. When the alarm unwinds, the weight falls and jerks on the light. If desired, a light spring may be used to fasten weight to pull chain.—I.W.D.

Care of Battery

I would suggest owners of two-volt radios using a six-volt storage car battery for operation by cabling to a single cell at a time; to change over to a new cell at least every three days and repeat this process until the three cells are all discharged at the same rate and about at the same time. Using a single cell until it is completely discharged is a bad practice, as after the third or last cell is discharged the first cell reaches a deplorable condition; excessive hard sulphate forms on the plates which offers a high internal resistance. Unless a low charging rate is used, the cell will heat, possibly causing the plates to buckle and throw out the paste from the grids. Thus each cell would have to be charged separately, or the charging current maintained low or the battery may be ruined; but by having all the cells at a more or less equal discharge; recharging then will be faster and the life of the battery prolonged.

Electric Wiring Tips

In wiring your home for electricity or in making any changes in the wiring

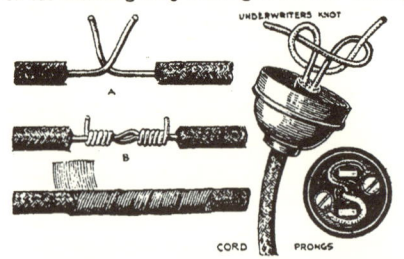

it is important that you do a good job of splicing the wires. Poor connections may arc and cause a fire. See drawings A, B and C for splicing details. In making a splice, remove the insulation for a length of about three inches from the end of each wire and scrape the wire bright with a knife or piece of sandpaper. Then bend the ends at right angles to the wires, hook them together and twist each tightly around the other wire with pliers so that a firm contact is made. To prevent corrosion and to obtain a good electrical contact the joint should be soldered. It should then be wrapped securely with rubber tape applied while the joint is still hot from soldering so the rubber will be vulcanized. Friction tape should cover the rubber and extend at least one-half inch beyond the points where insulation was removed and should be compressed firmly.

The other figure shows how to tie a knot that will relieve strain on individual wires attached to an electric socket or also how wires should be placed around the bayonet prongs of a plug to give a connection maximum strength.

SECTION 8.
Constructing Simple Gadgets

A Sagless Door

Here is the way to make doors so they will not sag. The joints are matched as shown, and this construction is extremely rigid. It is not difficult to match

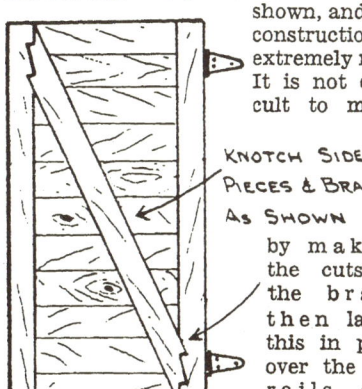

by making the cuts on the brace, then laying this in place over the side rails and marking with an awl or sharp pencil, and sawing to the marks with a thin saw.—I.W.D.

Firewood Holder

Often when chopping kindling, pieces will fly high in the air and come down on one's head. To prevent this, cut a strip of old belting or tire about 4 inches wide by 24 inches in length and nail it in the form of a loop to the side of your chopping block, leaving a space about 5 inches in diameter above the edge. If you do happen to hit the gadget you will harm neither it nor the axe.—Robt. J. Roder, Reist, Alta.

Re-babbitting Bearings

The average job of re-babbitting bearings on farm equipment is one that can be done in the farm shop without any trouble. First, remove the old babbitt from the bearing. If the bearing is one that will stand heating you can melt out the old babbitt, otherwise, it will probably be necessary to use a chisel and light hammer. When this has been finished, wrap one thickness of thin paper around the shaft being sure there are no wrinkles in the paper. Then fasten it with paste. This will help to keep the new bearing from binding on the shaft. Next cut one cardboard washer to fit on to the shaft on each end of the bearing. Use a wooden peg to plug the oil hole. Block the shaft up so it will be centered in the bearing and cover both ends of the shaft with putty or clay. Shape the clay on one end to form a funnel for pouring the metal. Leave a small air hole in the top of the clay in the other end. Skim the surface of the melted babbitt before pouring and when you start be sure to pour continuously until the bearing has been completely filled.—F. Mix.

Fastening Picture in Frame

If driven in the corners of a small picture frame and then covered with a triangular piece of gummed paper old razor blades of the type shown are more effective than the brads generally used.—Marvin Levi Wall, Great Deer, Sask.

Straightening 2x4's

I have found that curved 2x4's can be salvaged and straightened in a very practical way with the aid of C clamps and can then be used for double plates and studs around doors and windows while framing buildings. The scantlings are laid face to face with an inward curve of one toward the outward curve of the other. First spike the two ends together and then the clamp is screwed

on in the middle until the wedges are brought together. They are spiked in this position and are ready to put in place in the building.—Paul Tremblay, St. Paul, Alta.

TO HELP VETERANS *and* FARMERS

In the sincere belief that farmers and returned men living in rural communities are entitled to a fair share of today's housing materials, it is the policy of this firm to do everything possible to see that materials are provided for their needs in the greatest volume consistent with supplies available.

You are invited to call at the nearest Beaver yard and consult us regarding your building problems. Here you will receive preferred attention.

We will help you to plan your building, aid and advise you in the selection of the most suitable and economical materials, and exert every effort possible to procure the needed materials with the utmost dispatch.

- We have the experience—
  40 years in business.
- We have the facilities—
  Yards throughout the West.
- We have the interest of our customers at heart.

Whether you need a single piece for repairs, or a new house or barn—it will pay you to call at our nearest yard.

Serving the West Since 1907.

IF IT'S PROFITS YOU'RE AFTER—

SEND FOR ANY OF THESE FREE BOOKLETS

Pigs, poultry, cattle—milk or eggs ... whatever you raise, the final pay-off is in market grades and prices. You can learn how to reduce costs and increase profits by following the detailed, scientific methods for the correct care and feeding of livestock given in these five booklets.

SEND FOR YOUR FREE COPY TODAY
Fill in the coupon below.

PURITY FLOUR MILLS LTD.
Winnipeg, Manitoba.

Please send me your free folders on:
HOGS....CATTLE....CHICKENS
TURKEYS....EGGS.

Name.................................

Address..........................318

To Protect the Wall

Oftentimes walls of newly painted buildings are scratched and damaged by the sharp corners of a ladder. This may be prevented by nailing pieces of old tires or of inner tubes over the ends of the ladder. To the bottom ends of the ladder nail pieces of ordinary one-inch strap iron so that they will extend down about an inch. This will prevent the ladder from slipping when placed at a long angle.

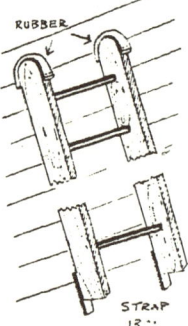

Old Drill Pressure Springs

We have found old drill pressure springs to be quite useful in preventing rubber hose from kinking. Especially is this true with the one-inch hose of our gasoline pump. The springs are inserted inside of the hose, the hose itself being filled with springs from one end to the other. A heavy wire ring is inserted after the final spring to prevent the springs from coming out again. The end spring where the hose is fastened on to the pump is made small enough to slip inside the pump opening.

The springs used in this manner prevent the hose from making sharp bends which in the case of a pump handling gasoline or tractor fuels rapidly deteriorates the hose and causes fuel leakage and carburetor grief through small quantities of the deteriorated rubber being pumped into the fuel tank with the fuel.—Roy E. Stokes, Coronation, Alberta.

Punch from Tine

A dandy pin punch for use on small work can be made of a five-inch piece of a pitchfork tine, properly ground on the small end and tempered.—Bob Larson.

Imitation Frosting for Windows

White lead paint can be used to make imitation frosting on bathroom or other windows where you want to obstruct the view from the outside. Reduce lead paste to the consistency of buttermilk by adding turpentine. Clean and dry the glass, put a coat of the paint on the inside of the glass and stipple it with a wad of cheesecloth. It's not a bad idea to do a little practising on odd pieces of glass before starting on your windows.

To Remove Sprockets

Usually straight keys can be pulled by setting a vice-grip pliers tightly on the protruding part and driving against the pliers with a punch through the sprocket. Occasionally a drift punch can be used in the keyway from the back of the hub. In extreme cases it may be necessary to drill out the key.

Hold a heavy bar or sledge against the back of the hub as close to the shaft as possible. Use a large punch or short iron bar that is slightly smaller than the shaft. Hold it squarely against the end of the shaft and strike with a heavy hammer. If a screw type gear puller is available it may be set in place and the pressure gradually applied with the screw, while at the same time a heavy punch set against the back of the hub is struck a sharp blow with a heavy hammer.

Block for Skid Chains

The diagram shows a safer and much more convenient plan worked out for putting on skid chains. It is especially handy when the snow or mud is deep, as it does away with the danger of the jack settling in the mud or toppling over just when you are ready to put on your chains.

Make a block about four inches wide or a piece of 2x6, and 18 to 24 inches long, with the ends bevelled and cleats

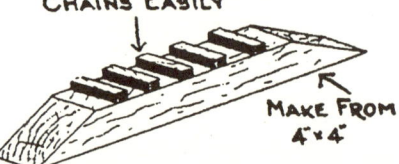

nailed across the top as shown. The block is laid back of the wheel with the chain stretched over the block with the cross chains between the cleats. The car is then backed onto the block, and the chain can then be easily pulled back and forth and fastened and adjusted to the desired tightness.—I.W.D.

Strengthening Wooden Wedge

A wooden wedge that will not shatter or split is made by shoeing it with band iron held on by rivets as shown. Then on top of the wedge nail another piece of the same kind of iron. Use a wooden maul. The split is started with an iron wedge and the wooden one finishes the job. I often split rough logs eight feet long ready for the cordwood saw. The one I am using now has been in use

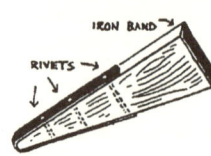

for 10 years. I call it my separator. It is made of hard maple. Don't make one less than four inches wide.—Alex. Woods, Sicamous, B.C.

Wood Fur Stretchers

"Because of the shortages," says a correspondent, "I have been unable to buy any steel fur stretchers, so have made an adjustable one of wood which works to perfection. I made this as shown in two sizes, small for muskrats and large for skunks; and a few of each

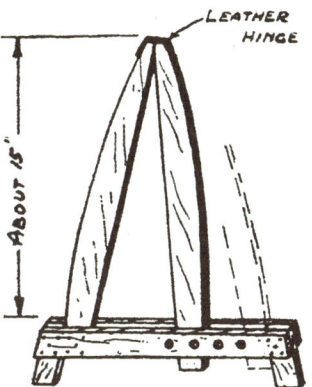

size takes care of all my needs. The curved pieces are made of about 1½-inch stock; the side pieces at the bottom are one inch thick, and the supporting legs about two inches thick. If a short board is nailed to the leg and ends of the side pieces at each end, it will stand upright. The size can be varied by moving the right hand curved piece in or out and fastening by a nail or peg through the holes. A heavy strap or piece of tire makes a good hinge."—I.W.D.

Basket Repair

A galvanized steel bushel basket or measure may break at the top rim, and if not repaired at once must be thrown away. To mend such a break, pull the broken edges together as well as you

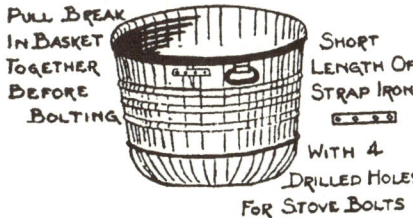

can, cut a piece of light strap iron and bend it to fit the curve, drill about four holes, and fasten with rivets or stove bolts. Such a repair will last for years. To prevent scratched hands, the broken edges should be filed smooth and preferably be coated smoothly with solder.—I.W.D.

MODERNIZE
THE FARM HOME

● We make Farm Improvement Loans at 5 per cent per annum for a variety of purposes, including not only the purchase of implements and equipment, but also additions and repairs to buildings, painting and interior decoration of farm dwellings, and the installation of heating and plumbing systems.

These loans may also be obtained for the purchase of household and dairy appliances of particular interest to the farm housewife, such as:

Washing machines	Refrigerators
Water heaters	Cream separators
Stoves	Churns

ASK OUR MANAGER
at your nearest branch about terms of repayment and other details.

697

THE CANADIAN BANK OF COMMERCE

Straightening Bolt Thread

If you want to straighten a bolt with the bend in the threaded part without damaging the thread, cut a nut in two with the hack saw. Be sure the thread of this nut exactly matches the thread on the bolt. Place one-half of the nut under the end of the bolt and the other half on top of the bend. Smart blows with the hammer will straighten out the bend without affecting the thread in the least.—Frank O. Johnson, Bagley, Sask.

Barn Ventilator

A subscriber cut the bottom from an old ten-gallon cream can and put it in an opening in the side wall of the stable with the large end inside. A 1/6 h.p. motor attached to a fan from an old car was mounted at the middle of the can as shown, so that the air was drawn out whenever the current was turned on. The motor was mounted on a piece of 2x6-inch held down by ¼-inch bolts.

The subscriber's idea seems quite good as a 1/6 h.p. motor would move enough

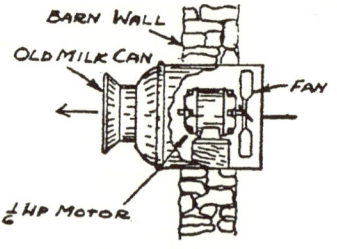

air to help materially in ventilating the stable. A light cover hinged to the top of the can would be lifted enough to let the air out, but would drop down and close the opening when the cold air started to move in.—I.W.D.

Playground Device

A small consolidated school without funds for special playground equipment was provided with one good device by a handyman. It consists of six old hay rake wheels with two out of every three spokes cut out. The wheels were mounted on a centre pipe, were welded two feet apart, and three smaller pipes were welded to the rims, evenly spaced. The result is a myriad of rungs through which youngsters love to climb. Several games have already been devised. The sketch shows the assembly.—Dale Van Horn.

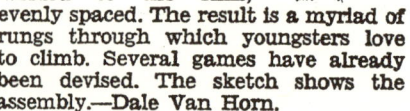

Merry-go-round for the Kiddies

Now that the children are playing outdoors again here is a simple merry-go-round for them. Trim a fence post to fit easily into the hub of an old waggon wheel. Put it in the ground and tramp the earth firmly. Grease the axle and put the wheel on. The children love it and mother and dad like to take the cushions out and sit on it too. We use ours very much.—Mrs. Emma Schmelzer, Winnifred, Alberta.

Fitting Handles

Ordinarily new handles for axes and other tools are made of hardwood and take a lot of whittling and scraping to make a good fit. One way to fit them easily and quickly is to put the end of the handle in the fire so as to char it evenly but not too much all around. You can then scrape the charred part off in a jiffy and the job is done. This method should work very well for handles which have little tendency to work loose; but for axe and hammer handles the charring will have to be done very carefully so the handle would not be too small in places. Charring would also have the advantage of showing plainly where material would have to be scraped off.—I.W.D.

Sheet Metal Cutter

This sketch of a sheet metal cutter made out of odds and ends around the shop was found a great time and labor saver. For the cutting edges two discarded Oliver Raydex plow shares were used. By using the edge that fits up against the moldboard as the cutting edges, they have about the same bevel as a pair of tin snips. Use a half-inch steel bolt to fasten them together at one end. Then fasten a 24-inch handle to one share and clamp the other share in a vise, with a short length of chain to fasten to the bench to help hold against the heavy pull. With this device you can cut sheet metal up to one-eighth inch with ease. After being used for several months the cutting edges were still sharp.

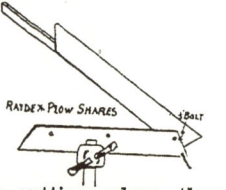

Bonding of Babbitt

The bonding of babbitt to a connecting rod is a considerable problem, due apparently to the fact that all rods and babbitts are not of the same composition. In general the following procedure should produce good bonds with ordinary babbitt:

Melt or otherwise remove all the old babbitt from the rod, keeping the temperature as low as practical. This might be satisfactorily accomplished by dipping the connecting rod in a pot of molten scrap babbitt.

Brighten all the surface to which the babbitt is to adhere. This may be done with a steel scratch brush, steel wool or emery cloth.

Apply flux to the surface to be tinned. Zinc chloride solution is usually a satisfactory flux.

Dip the fluxed rod in a pot of molten pure tin. Lift out and examine. Wipe the tinned surface with dry waste to remove excess tin and to observe the completeness of the coating. Where incomplete add more flux and dip again. Small spots untinned may be attacked by use of a soldering iron.

When all the rod is tinned it can then be babbitted and the babbitt will adhere to the tin. Some trouble may result if the rod is not hot before pouring.—G. H. Shanks.

Protecting Paint Brushes

A reader gives this sketch of a method for protecting paint brushes. He suspended them in oil or water. But good brushes should not be allowed to soak

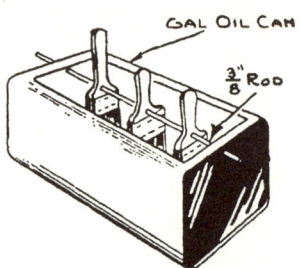

indefinitely in water; as it swells the bristles and injures their resiliency and ability to hold and transfer paint. They may be washed in turpentine or thinner and supported in raw linseed; or washed thoroughly and wiped as dry as possible and then wrapped in oil paper.—I.W.D.

To protect a cordwood saw while not in use I slit a 28-inch high pressure bicycle tube open around the inside and placed it around the saw. This provides good protection.—M. H. Schab, Calder, Sask.

Keeps Auto Chains Ready

Last winter I wasted several valuable minutes putting on my auto chains in an emergency, because they were badly tangled from the ends twisting through

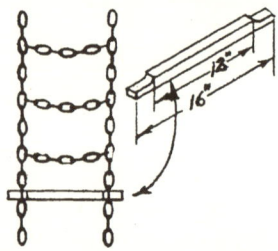

the body of the chain. When I took them off I cut for each chain a stout stick about 16 inches long with a two-inch shoulder at each end to go through the middle ones of the loose links which fit into the fastening hooks, so that the end links stick out several inches beyond the width of the chain. Then I rolled the chain up tightly around the stick and finished by fastening the hook into the spread links. This makes it impossible for the chain to become tangled when the hooks are unfastened and the chain unrolled. When the chains are put on, I drop the sticks into the trunk to be used again.—I. W. D.

Anvil Support

It is customary to go to the woods and get a large block to make a support for an anvil. But in many districts there are no wooded lands. However, a good support can be made at home from scrap lumber and sand. A square box is made slightly larger than the anvil, and of the right height to suit the worker, and this is filled with sand. The anvil is simply placed on the sand, which will absorb the shock of the hammer.—Paul Tremblay, St. Paul, Alta.

Moving a Wall

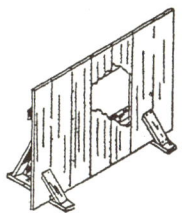

This sketch shows how to move a piece of wall safely, such as the end wall when lengthening a garage. First slide two 2x6's about 8 feet long under the wall, quite near the ends. The base of the wall is toe-nailed to these skids. Then four 2x4 braces three or more feet long are nailed from the wall to the skids as shown. In this way the wall can be slid to where it is needed without any danger of it toppling over. A few pieces of small pipe under the skids might make the moving easier.

Bolt Fastening Kink

I have found this trick invaluable while working on machines or overhauling tractors or automobiles. Often we have to put a bolt in such an awkward place when re-assembling that it is impossible to reach it with the fingers.  A piece of soft wire is wound a complete turn around the bolt, under the head and then bent right over the top of the head. It is then easy to insert the bolt in the hole. After the nut has been turned on just enough to catch the thread, the wire is pulled off. Hay wire or stove pipe wire is used.

PROTECT your buildings from all three...

● You'll find Johns-Manville asbestos materials a lifetime investment in security. They can't burn, rot or deteriorate and they save you money because they never need paint to preserve them!

Johns-Manville makes a wide range of asbestos roofing and siding materials that are ideal for fireproofing barns, implement sheds, poultry houses and homes. For sanitary, rodent-proof interiors in dairy barns, grain bins and feed storage rooms, choose J-M Flexboard. This versatile asbestos-cement wallboard can also be used on exteriors.

For full details on Johns-Manville Building Materials and how they can help you to build or remodel for greater efficiency mail coupon today!

JOHNS-MANVILLE BUILDING MATERIALS

DURABESTOS ROOF SHINGLES ... CEDARGRAIN ASBESTOS SIDING SHINGLES ... ROCK WOOL INSULATION ... FLEXSTONE ASPHALT SHINGLES ... ASBESTOS FLEXBOARD ... ASBESTOS ROLL ROOFING

Send for the new J-M "Farm Idea Book"... a big, colourful 64 page handbook packed with information of real value to farmers.

Canadian Johns-Manville Co. Limited, Dept. FW17, 199 Bay St., Toronto 1, Ontario.
I enclose 10c in coin for which please send me a copy of the J-M "Farm Idea Book." I am specially Interested in Roofing [ ]; Rock Wool Insulation [ ]; Asbestos Siding [ ]; Asbestos Wallboard [ ] (check which).

Name..
Address...
Town.. Prov.............

B-458

Band Cutter

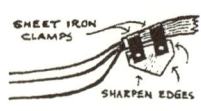

Last fall I had occasion to put a good many stooked sheaves through the combine as there was no threshing machine available. To cut the bands with a knife, then put down the knife and pick up the fork to throw the sheaf into the machine was altogether too slow for one man. I found that by clamping a section of sickle to the handle of the fork in the manner shown that it was possible for one man to cut bands and feed the machine almost as fast as he could feed a regular grain separator. The idea works just as well for feeding a hammermill, cutting box or silage mill. Be sure to first sharpen the section all the way up each edge so that there is no unsharpened spot to come in contact with the twine.—E. J. McFarland, Loma Stock Farm, Vulcan, Alta.

Trace Drop Preventor

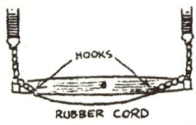

This arrangement keeps the drop links from hooking into the other horse's singletree hooks. A strip of inner tube can be used with hooks made from spikes attached to each end. If necessary it can be wrapped around the singletree once to keep it from hitting the horse's hocks.—Norman Harris, Edgeworth, Sask.

Don't Let It Get Longer

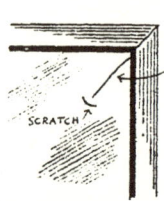

In many cases a crack in a window pane is short to begin with. Unless it is stopped it becomes a broken pane. To stop it from extending, take a glass cutter and make a score on each side of the glass across the path of the crack as shown. The scores needn't be more than half an inch long. The crack will not cross the scores.—D. C. R.

Handy for Melting Babbitt

If you ever have necessity to do a job of babbitting in the field, or even around the shop, and are short a blow torch or forge to melt it in, here is a suggestion for a handy temporary melting pot. I have used it several times myself and have found it very useful. Take a five-pound syrup pail and fill it half full of ashes. Punch a few holes around the tin just at the level of the ashes to facilitate draft. Then pour a quantity of coal oil or distillate over the ashes to form a fuel reservoir. Next touch a match to the ashes and they will burn until the fuel is exhausted, furnishing plenty of heat to melt even the harder types of babbitt.

Ordinary dry dirt can be used in place of the ashes although ashes seem best.

A further improvement to the pail is two small rods inserted an inch or so below the top of the can. The purpose of these rods is to offer support to the babbitt ladle while the babbitt is being melted.—Roy E. Stokes, Coronation, Alberta.

Handy Milk Shelf

Here is a handy milk shelf to keep the milk clean and cats away from it during milking time. Build the shelf against the barn wall. Cut an old tire chain in half and staple one end of each to a studding about 12 to 15 inches

higher than the shelf, and fasten heavy hooks in the outer corners of the shelf, so that the chains can be hooked up until it is level. When not in use, both the shelf and the chains hang down against the wall.

Another Use for Nutcracker

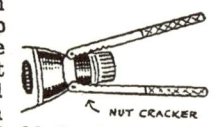

Screw tops of such things as catsup bottles can easily be removed with a nut cracker, for it will get a firm grip on the edges and will hold the tops much as a pipe wrench holds a pipe.—Bob Larson.

Match Striker

The matches we get these days do not light easily on thumb nails, and it is hard to find a suitable striking surface in frosty or wet weather. I save myself a lot of grief by pinning a small square of sandpaper onto the side of my cap. In summer the paper can be fastened under the brim or peak of headgear.—Walter Schowalter, Hayter, Alberta.

Hand Protection

Heat from an open camp fire won't burn your knuckles when roasting "hot dogs" if you punch a hole in the centre of a piece of cardboard about ten inches square and slip it over the stick to serve as a shield. Of course, you must not allow the cardboard to get too close to the fire.

Box For Combination Stone

This box is for a combination oilstone, fine on one side and coarse on the other. The measurements are for holding a stone 6x2x¾ inches. Leather is used for the hinges, with grooves made the thickness of the leather so that the top of the box will close snugly. The depth of the seating is half the thickness of the

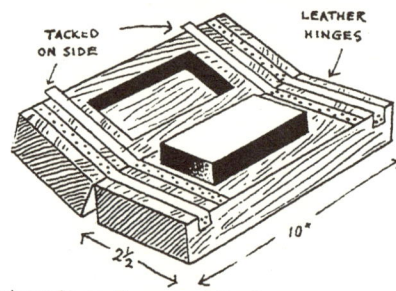

stone in each part of the box. By tacking the leather strips in the grooves alternately as shown the box will open in both directions similar to a billfold. It will open no matter which side is up. The stone is just loose enough so that it will remain in the side of the box which is down, exposing either side as desired.—Harold Tenove, St. Paul, Alta.

Combination Lock and Doorstop

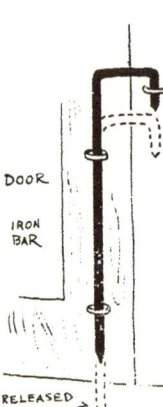

Here is an idea that can be used for holding a garage or barn door open and shut. A rod is bent as shown and is held in position by two eyebolts. When the door is open the sharp point is pushed into the ground. When it is shut the upper hook drops into the eye of another eyebolt or staple fastened in the jamb. — MacM., Laito, Montreal.

Holds Log Solid

You have probably experienced, when cutting a log for firewood, that the log won't stay put on the sawbuck. It

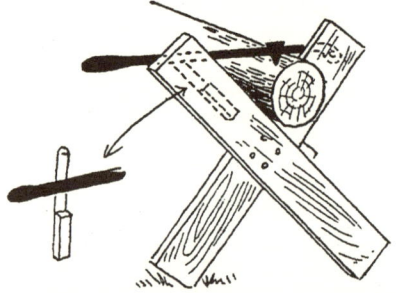

can be held firmly by using this simple device. Use an ordinary farm machine handle or lever and a chisel made from good steel, sharpened to a point at the lower end so that it will penetrate into the log when tapped with a hammer. A guide, shown separately in the illustration, is bolted to the sawbuck to hold the lever in place.—G. C. Glubrecht, Hackett, Alta.

Emergency Reducer in Pipeline

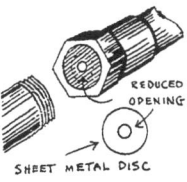

In a pipeline job where a necessary reducer was not available, one of the farm hands quickly made one from a disc cut from sheet metal. The disc was made the exact size of the union packing with the necessary sized hole punched in its centre. It was then placed inside the union and screwed together with the packing, resulting in a reduced pipeline.—A. S. Wurz, Jr., Rockyford, Alta.

Changing Tone of Horn

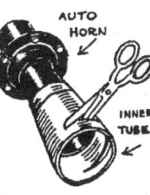

If you don't like the tone of your auto horn, you can change it by the method shown. A little experimenting will be required to get just the right length of tubing.

Saw Horse Extension

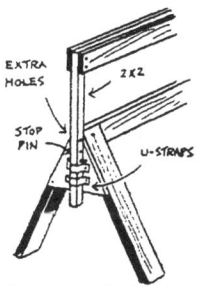

The extension is upward to make a convenient table on which to work. Note that at each end two U-straps are nailed or screwed to the end of the horse and that holes through the upright, with pins which rest on the upper U-strap, regulate the height. The extra height is frequently needed for different kinds of jobs and this is a simple way to provide it.—D.C.R.

Wax Bullets for Rats

If you wish to shoot rats with a .22 rifle, and do not want to run the risk of the bullet going astray, you can remove the lead and replace it with a nose made of crayon or wax. When the lead is removed, be careful to prevent loss of the powder in the shell. It is suggested that you place the shell upright on a flat surface, hold it between the fingers of one hand, and remove the lead by bending it sideways with the other hand. A wax bullet will kill a rat at a distance of 100 feet, and if the bullet goes astray, it will not do much damage.—Bob Larson.

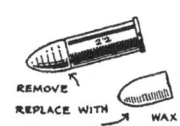

Mud Scraper

If you have an old skate lying around you have an excellent boot scraper al-

ready manufactured. Just turn it upside down and tack it on the outside step and it is there ready for business.

For Small Loose Papers

A block of wood with a small spike driven through it makes a handy gadget to keep small papers, like grocery bills, or cream or check stubs which gather around the home. The nail can be filed to a long point. First run a small gimlet through the block to prevent splitting.—M. Lambert.

Mail Box Signal

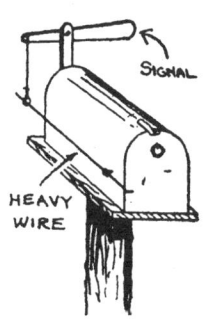

The sketch shows a handy mail box signal which saves us many trips to the mail box. It operates automatically. The wire running parallel with the bottom of the box is drawn forward when the door is opened, and the signal arm being then released drops down by its own weight. The arm is reset when the mail is removed. Quite a help when the housewife is busy and the box is a distance from the house.—I.W.D.

Uses for Horseshoes

Heat the shoe in the fire until it is cherry red, then straighten and flatten it slightly, and bend the ends nearly at right angles. Then take a rectangular gallon oil or varnish can, cut out one side, fill it with concrete, and stick the

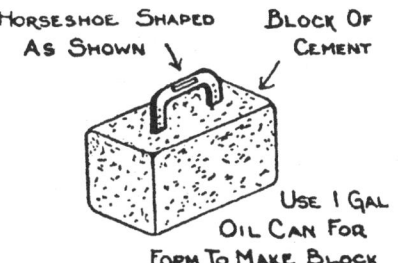

STORM HAT
Fully Automatic
TRACTOR EXHAUST PIPE COVER

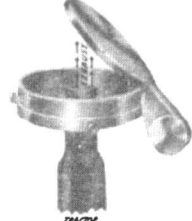

Keeps Rain out of Tractor Exhaust Pipe.

89c

UNIVERSAL — FITS ALL TRACTORS

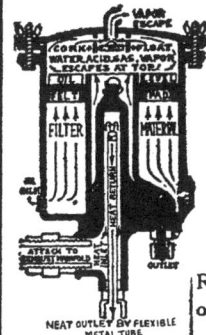

Oil Kept Free From Pollution Is Good Indefinitely!

Reclaimo Rids It of All Impurities

Reclaimo is More Than A Filter!

Impartial authorities agree that motor oil does not wear out . . . that the lubricating quality is not destroyed as long as dilution and dirt are removed. RECLAIMO does both—evaporates dilution and filters dirt. There is a size for every tractor, truck and car . . . Try one at our risk — they are sold on a liberal money-back guarantee.

Write for Free Book

Reclaimo Company
(CANADA)
301A 10th Ave. W.   Calgary, Alta.

or

Diecast PRODUCTS LIMITED
235 Garry Street
Winnipeg   -   Manitoba

ends of the straightened shoe well down into the concrete, working and tamping the concrete closely around it. When well cured, you have a foot scraper that is quite effective in removing mud and will last indefinitely.

Old horseshoes can be nailed to stable posts for harness hooks, set ends down in concrete granary floors for bolting the lower ends of the studding, nailed or bolted to gate posts to serve as hinges, hooks for fastening gate latches, and so on.—I.W.D.

Belt Punch from Hinge

A handy belt punch can be cheaply made by drilling two or three additional holes into a common door hinge. This punch needs no wooden base as the

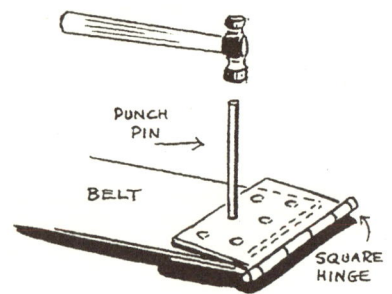

bottom flap is drilled so that the top and bottom holes line up. Place the belt between the flaps and use a common straight solid punch to drives the holes. —A. S. Wurz, jr., Rockyford, Alta.

Wheeled Step Ladder

Outdoor work of washing windows, putting up storm windows, picking fruit, and so on is much easier with this

wheeled stepladder. After you have finished with it in one place, merely pick up the handles and wheel it to the next location, which is much easier than carrying it.—I.W.D.

Handle on Chopping Block

A very handy, yet simple idea is to drive a staple in on each side of the chopping block and attach to them a wire handle. The block can then be carried easily about the yard as needed. —Marvin L. Wall, Great Deer, Sask.

Insecticide Duster

A satisfactory insecticide duster can be made on the farm. Materials needed are a five-pound salt sack, an old broom handle, a tomato can with both ends cut out, and two stove bolts with thumb nuts. By spacing the holes in the can

for the bolts four inches apart and by spacing the holes in the broom handle two inches apart the duster can be adjusted to the proper heighth for the various varieties of garden crops. Stamping it on the ground causes the dust to fall on the plants.

Shock Absorber

When the car is stuck in a deep rut or mud puddle, and it requires a stiff jerk to pull it out, put an automobile tire in the centre of the pulling chain or rope as shown. This will act as a shock absorber, and will lessen the damage to the bumper, and will save strain on the chain or rope.—Bob Larson.

Stump Burner

To burn out stumps surely and safely, use a steel barrel, when they are available, rigged up as a portable stove as shown. Cut a hole in the top for a stove pipe, a door in the side for kindling and feeding the fire, a draft hole four inches in diameter four inches from the bottom. The bottom is removed and the stove set in place over the stump and banked up with a little earth. This makes the stump burn as though it were in a stove in the house.—I.W.D.

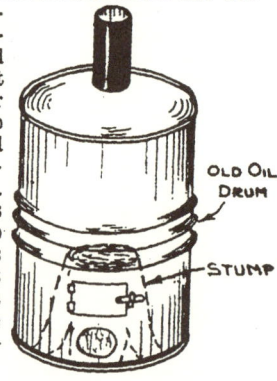

Toe Pull Opens Screen Door

Did you ever have to yell for help when you had your arms full and a screen door blocked your path? A toe pull attached to the bottom of the screen door is a great convenience in such circumstances. This may be a simple little block of wood securely nailed to the outside of the screen door and under which the toe of the shoe may be hooked.

Drying Rack

A very efficient drying rack which takes up little or no space can be made from a short length of galvanized iron of such a length as to encircle the hot

water tank. On to the iron band are riveted three or more small brackets,

made from the same size iron. On to these brackets are fitted pieces of hard wood, size 1"x¼"x9", holes are drilled in the brackets and screws inserted to hold the arms firmly.

Kitchen cloths and other laundry dry quickly from the heat of the tank, at no increase in fuel costs.—Sidney Pott, Victoria, B.C.

More Service From Gear Shift

When a gear shift is worn out so that it slips into neutral you can still get a lot of service out of it. Take a strong screen door coil spring and hook one end over the lever and fasten the other end to the seat at the right hand of the driver. I have used this for quite a while and it works well in high and low gears.—Peter P. Entz, Magrath, Alta.

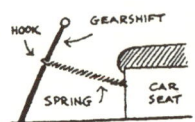

Makes Loading Gas Drum Easy

This skid takes the "back-breache" out of loading a barrel or gas drum on a wagon or truck. It is made of stout material as shown and is just wide enough so that the barrel or drum will lie in the top position without projecting below the side pieces. When the

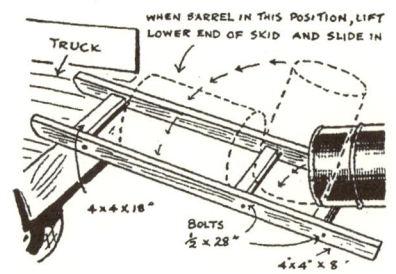

drum reaches the last position on its side the lower end of the skid is raised and the skid and drum are slid forward onto the floor of the wagon or truck. By reversing the steps taken, the drum is lowered to the ground again.—C.L., British Columbia.

To Start Small Brads

This is an easier way to start brads than holding them in the fingers. Use a paper clip, of the size to hold the brads used. It is easier to hold the clip than the brads in the fingers.—Paul Tremblay, St. Paul, Alta.

RIGHT FIT
IS IMPORTANT IN CAR AND TRUCK PARTS TOO

FORD-MADE PARTS ARE RIGHT FOR FORD CARS, TRUCKS AND TRACTORS, MERCURY CARS AND TRUCKS, MONARCH CARS AND LINCOLN CARS

Ford-made Parts give better satisfaction for Ford, Monarch, Mercury and Lincoln cars or trucks because they are made in the same plant in exactly the same way as the parts they replace. Genuine parts are made right to fit right. They make the repair job easier and quicker.

You'll get better satisfaction from any of our dealers or from other garages using Genuine Parts for all Ford Products.

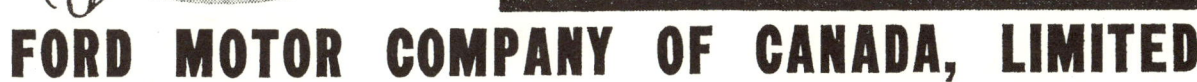

FORD MOTOR COMPANY OF CANADA, LIMITED

FORD AND MONARCH DIVISION   MERCURY AND LINCOLN DIVISION

Pull-Outs for Windmill

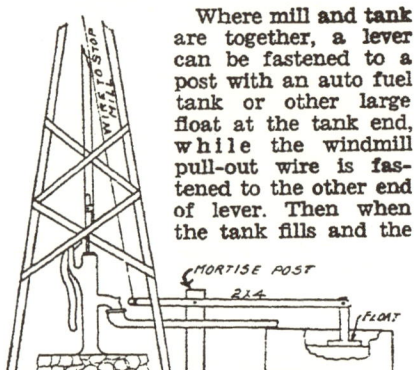

Where mill and tank are together, a lever can be fastened to a post with an auto fuel tank or other large float at the tank end, while the windmill pull-out wire is fastened to the other end of lever. Then when the tank fills and the float rises, the other end will pull down on the pull-out wire and throw the mill out of the wind.

Where the tank is several rods from the mill, an ordinary tank float valve can be put on the supply pipe at the tank. A branch pipe from the tank pipe is taken off near the pump, brought up a foot or so, and then led down into a five-gallon milk can hung below the well platform by being attached to the windmill pull-out wire. When the rising water in the tank lifts the float and closes the shut-off valve, the pump will force the water up through the branch pipe and into the milk can until its weight pulls the mill out of the wind. A small pin hole lets the water leak slowly out of the can and its lightened weight will permit the mill again to come into the wind.

Hedge Trimmer

This knife I cut from an old crosscut saw. After removing the temper by heating in a wood fire built on the ground, the saw was left to cool in the ashes over night. Next morning I marked it off, leaving a little for finish-

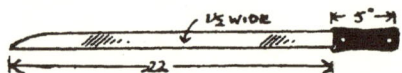

ing the rough edges. I then put it in a vise, cut out the blade with a cold chisel, straightened and sharpened it, drilled two holes for a hardwood handle and then retempered it, taking care not to get the steel too hard, otherwise it would become too brittle. To trim the sides, the hedger keeps the hedge on his right hand side as he walks slowly along, striking downward, keeping his eye slightly in advance of his work so that no time is lost in deciding what must be trimmed and what not.—R. H. Brooks, Half Moon Bay, B.C.

Harness Scrubbing Board

In overhauling harness, first take it apart and make the necessary repairs. Then, to clean it, allow it to soak thoroughly in a washtub of warm water containing a handful of washing soda. When taking it out, a piece at a time, to scrub clean, this scrubbing board is a great help. The water drains back into the tub. Apply the harness oil when the harness is still wet and rub it in

well. As the moisture dries out, the oil penetrates still further. Drop the oiled pieces on top of each other, so that any oil that drops off from the top pieces will drip on to the lower pieces. More than one application of oil can be made if the leather will take it.

Turning Off Distant Windmill

This diagram explains itself. The chain from the windmill comes down around a pulley well up in the tower and is attached to a heavy wire. This wire is carried by line posts to the buildings. A wire stretcher is used to apply the pull. This device can be used on a windmill 80 rods or more away.

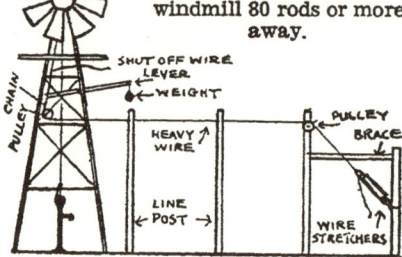

Saving Syringe Needles

Tie about 18 inches of bright red cord to the neck of your vaccine syringe needle. Then if it is knocked out of the operator's hand it will be quite easily found, even though it falls among litter or is trampled underfoot.—Robert J. Roder, Reist, Alta.

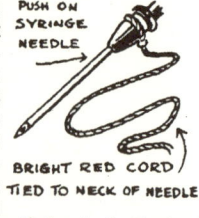

To Heat Soldering Iron

To heat a soldering iron in a coal or wood stove, cap a piece of pipe, place it in the stove and insert the soldering

iron in it. This will overcome burning the tinning off the iron.—Grant McLeod.

Homemade Spray Pump

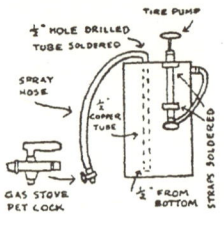

A simple pressure type sprayer for applying chemicals to plants or trees can be made at little cost from a small oil drum, a tire pump and a few fittings. Near the top of the drum mount a valve stem from an old inner tube. Clamp the tire pump to the side of the drum, attaching the hose to the valve stem. A ½-inch copper tube which extends to within about half an inch of the bottom of the drum is next soldered in position and a coupling attached, to which a 10-foot length of garden hose can be coupled. At the outer end of the hose attach a gas stove pet cock.

To Hold the Clothes Pins

A clothes pin holder can be made from a piece of wire bent in the form of a long U. It will be found especially useful in cold weather when gloves or mitts have to be worn when putting the clothes on the line. It is easily slid

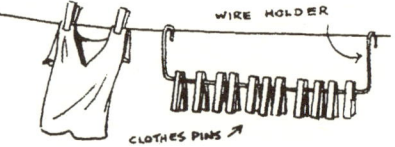

along the line as the clothes are pinned on.—Paul Tremblay, St. Paul, Alta.

For Sewing Jute Bags

I like this simple little device for sewing jute bags better than a million needles. Take a piece of stove pipe wire and bend and twist it, using a nail to make the eye. Use the end which has the bend as the point of the needle. It can also be

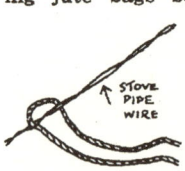

used to sew on pant buttons. Use an awl to make the hole in the bagging.—John McKay, Rembro, Ont.

Fitting Ladder Rungs

When making a ladder, a baby crib or a basket type rack the work will be much more easily done and greatly improved if the rails or rungs are fitted in the way shown here. The size of the end of

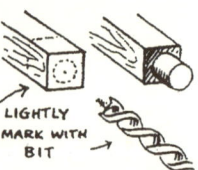

the rung is marked with the same sized bit that is used for boring the holes to take the end of the rung.—Paul Tremblay, St. Paul, Alta.

Machine Jack

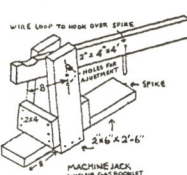

A few pieces of good lumber make a sturdy and serviceable handy machine jack for use when repairing and greasing. Three holes are put in for adjustment of height. From the pin to the end of the lever is eight inches. The base is a piece of 2x6 cut 2 feet 6 inches long. A wire loop, attached as shown on the handle, slips over a spike on the base to hold the machine up.

Buckle Tongue From Split Key

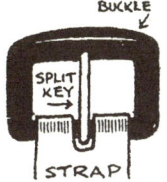

An old buckle can sometimes be repaired by removing the old or broken tongue and putting in its place a split key or cotter pin. Sharp edges can be removed by filing. I have found this to be a very satisfactory repair. — Donald H. Clark, Neepawa, Man.

Repairing Chain

To repair a broken chain, use a nail bent as shown. This comes in handy these days when repair links and even haywire are hard to get. You can always find a nail around the place.—Henry Schuett, Westmark, Woking, Alta.

Handy Farm Level

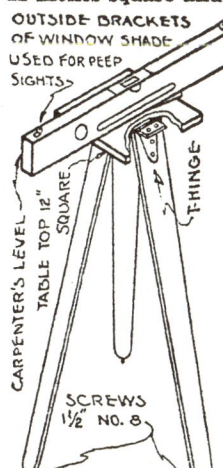

To make this assembly just take a piece of board or better still of plank, 12 inches square and plane the top side until it is smooth and true. Three legs, about 4½ feet long are made from inch stuff, preferably fir. These are attached to the under side of the table with T-hinges as shown. The hinges must have tight knuckles to prevent play. Put a screwnail in the bottom of each leg and file off the flange. This will help get a firmer footing in some locations. The stand is set up with the table as level as it can be conveniently made. An ordinary carpenter's level is used, with outside brackets of a window shade to serve as peep sights. It is almost impossible to get a level sighting along

Yes, MINER laced work rubbers are built to "take it" through all kinds of wear and weather! They offer real protection . . . real comfort; on the roughest, toughest days.

MINER rubbers are firmly welded with the famous Miner Pressure cure, which toughens the rubber against cracking, peeling or leaking. A surface film protects them against barn yard acids and gives a lasting gloss to the rubber.

They are designed with comfort in mind . . . as well as long, sturdy, waterproof wear.

the top of a square and these sights are necessary for accurate work. A thin wedge is used to make the level level.

Proper Use of Cotter Pin

This sketch shows the right and wrong ways to fasten a cotter pin in place. Do not, as shown at the left, have the eye of the pin crosswire of the stud. Place

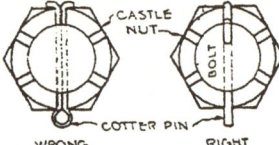

it lengthwise so that the head of the pin goes right into the slot of the castle nut, close to the stud. One half of the lap is brought up over the top of the stud and the other half is bent downward, in the opposite direction.

Rocker-Mixer

This cement mixer does the job easier than it can be done with the usual mixing box. It is operated by rocking back and forth. The dimensions of the various pieces are given in the sketch. One ad-

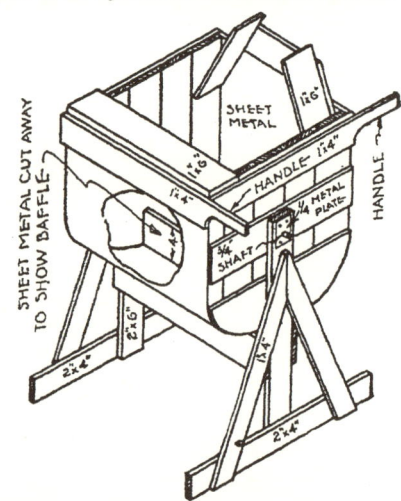

vantage is that the concrete is dumped directly into the wheelbarrow. Note that inside at the bottom there is a baffle board which greatly hastens the mixing.

Tell-Tale Tank Float

It is very convenient to be able to tell the water level in a large supply tank, especially where it is up in the barn mow. The diagram shows how to arrange a simple marker worked by a float in the

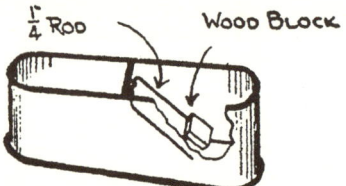

tank. The float was made from a light 2x4 about eight inches long, coated with hot tar and fastened to a one-fourth inch iron rod, bent into a U-shape and hinged to the cross brace on the tank so that one arm moved up and down on the outside of the tank. If desired, an arc could be painted on the tank and marked in any desired units.—I.W.D.

Concrete Mixing Aid

In mixing concrete, all materials, including water should be accurately measured for every batch. A bottomless frame, which holds 1 cubic foot, is just

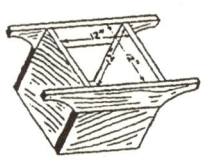

the thing to measure sand and pebbles. A pail marked off on the inside to indicate gallons and half gallons is handy for measuring water.

Flares Gas Line Tubing

When gas line tubings break on the farm engine or tractor, it is necessary to re-flare the coupling joints. This can be done by grasping the tube firmly in one hand and revolving the tapered head of a common stove bolt of correct size inside

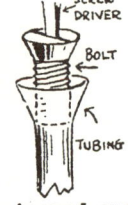

the tubing. A screw driver is used and a very neat even flare will result.— A. S. Wurz, jr., Rockyford, Alta.

Cap Over Heater

Not being satisfied with the performance of my circular heater, I took a piece of metal, curved it as shown, riveted some short legs under it, and fastened it over my heater. Before, the heat would go straight up to the ceiling. Now, it is thrown out to the side before rising, and seems much more effective in heating the room at living height.—I.W.D.

Hauls Firewood

I made a small wheelbarrow for carrying wood into the house, out of scrap lumber. I can load this in the woodshed, push it all the way into the house, and then unload it into the woodbox. I made the wheel out of two thicknesses of inch

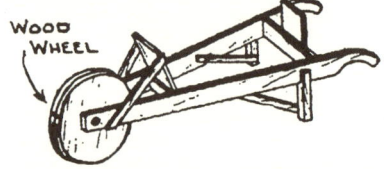

board nailed together at right angles, cut out in a 12-inch circle, and with a small piece of flat iron on each side to act as a bearing. I tacked a piece of old inner tube around the edge of the wheel to keep it from picking up dirt. A few boards tacked on top of the wheelbarrow permits it to be used for other purposes.—I.W.D.

Handy Barrel Tipper

The diagram shows a convenient outfit I made so that one man can easily handle a full barrel of gasoline or oil. I took the two flywheels and crankshaft off a 1¾ horsepower gas engine and bolted the barrel to them as shown by putting a one-inch strap-iron around the barrel and through the wheels and drawing it up with bolt and nut. Before tightening this up I slipped pieces of old tire casing between barrel and wheels to prevent rubbing the barrel and to make it easier to hold the barrel solid. If engine flywheels are not available, two small wheels of any kind can be put on a shaft of the proper length and be used in the same way.

Sack Protects Waterer

The diagram shows my way of preventing the hens from roosting on the water fountains and making the water unfit for drinking purposes. I simply take a gunny sack, slip it over the fountain, and hang it up to the ceiling. I've had no further trouble. — I.W.D.

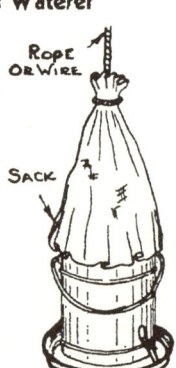

Chain Repair

Often a chain will break at a very inconvenient time. Much delay can be avoided by the use of a simple homemade repair link, made of two pieces of strap metal and

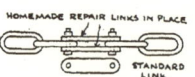

two short bolts. Use a drill to make the holes in metal. A few sets of these ready made will come in handy.

Wind Direction Indicator Bearing

One of the easiest and quickest ways to make a bearing for a wind direction indicator is to use part of an old

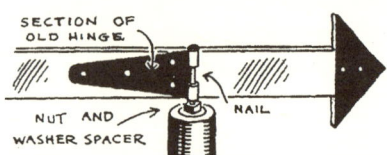

hinge. It is riveted on the arrow and a spike is driven through the loop into the top of the post. By putting on a nut and washer the bearing is held up from the post far enough to allow freedom of motion.

Repairing Drill Wheels

The wheels on my seed drill were badly worn, but with very little expense I was able to repair them. I took the tires and rims off. With a hollow auger the spokes were cut down to remove the decayed parts. Large washers

were put on as shown and the rims were placed on again. The tire should now fit very tightly, but if not, put two washers instead of one, on the spoke ends. My repair job has lasted two years now and looks good for another couple of years.—B. H. Markosky, Innisfree, Alta.

Barrel Doghouse

If you have a good barrel around the place it can be remodelled into a good doghouse. It is rain proof, snow proof, warm in winter and cool in summer. Any long haired dog can be kept in it in cold weather. The door has double acting hinges at the top to swing in or out. The visor is optional and corks can be put in the holes under it in cold weather.

Handy Depth Marker for Bit

Where several holes are to be bored the same depth a strip of adhesive tape wrapped tightly around the bit or drill at the right place will show when the hole is the proper depth and will not slip.—I.W.D.

Tools from Fork Tines

Most farms have a broken old fork or two in the scrap pile. Here are some useful tools that can be made in a few minutes from the tines. The one for cleaning out oil holes has a piece of wire beat a mile. The cotter pin tool is useful in many ways. The awls are very useful in the workshop for making

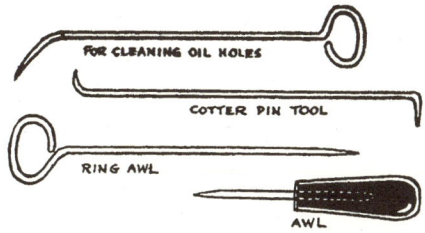

small holes and scratching lines on metal or tin. If you have no forge you can heat the tines in the stove and bend them.—George Z. Merkley, Springwater, Sask.

Leather Knife from Old Hand Saw

I made this leather knife from a piece of the blade of a discarded hand saw. I cut it out with a cold chisel and then worked it to shape with a file. I ground the cutting edge with an emery and finished the edge on an oil stone. It is very useful for cutting and trimming leather when fixing up the harness.

Good Staple Puller

An old worn-out mowing machine guard will make a good staple puller. Drive the point of the guard through the staple between wire and the post.

Awl from Valve Stem

Warren G. Parker, Birch River, Man., sent this gadget in and the artist drew the sketch from the original article. "Just remove the valve cap from a valve stem," he says, "then drop the nail

point through the valve cap and put back on the valve stem again. Sharpen the nail, put on a handle, and you have a very useful awl."

An Improvised Level

This level is made from a short block of clear white pine, two knitting needles or straight wires and a 3-ft. length of ¼ by 1½-inch lattice strip. The white pine board is 20 inches long. The sketch shows the manner of assembly. This outfit, once checked to make sure the lattice strip extends parallel with the pine board, will float level on any container of water and will be surprisingly accurate. The top strip may be moved slightly up or down one wire to maintain accuracy.—Dale Van Horn.

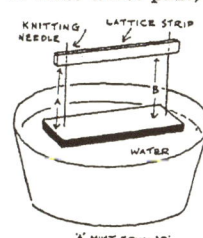

Sawbuck That Will Not Pinch

Here is a non-pinching sawbuck which will be found very convenient for sawing cordwood or posts, as these can be partly sawn through in the middle notch, and then are properly supported when pushed out so the stick can be cut completely through. Try this, as you will find it a big improvement over the old-fashioned type.

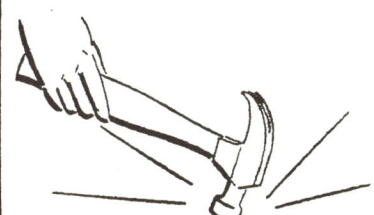

WE WANT TO HAMMER THIS HOME!

If you want top value in tools—automotive tools for use on your tractor, truck, car—or hammers, pliers, screwdrivers, wrenches—

ASK YOUR DEALER FOR

GRAY-BONNEY TOOLS

They have that "extra something" which any man who appreciates better tools can spot in a minute . . . precision, machining, perfect balance, strength, toughness, lightness.

Steel Fingers By
GREY-BONNEY include
Chrome Vanadium
Drop Forged

HAMMERS . WRENCHES
SCREWDRIVERS . PLIERS
and a
COMPLETE LINE OF
AUTOMOTIVE TOOLS

Made in Canada

GRAY-BONNEY
TOOL CO. LIMITED
St. Clarens & Royce Avenues,
Toronto, Canada

Canadian Made Quality Line • Finest Money Can Buy

Magnet as Pincushion

In the U.G.G. elevator I noticed an excellent pincushion. It is a magnet from an old Ford car hanging on the wall. No need to stick the pins in it. Just throw them at the magnet. — Henry D. Falconer, Glentworth, Sask.

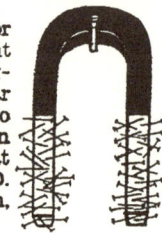

Improvised Pipe Wrench

A couple of pieces of scrap iron and a length of bicycle or motor-cycle chain are all that is required to make this pipe wrench. The two pieces of iron should be about 15 inches long and the chain about the same. The chain is passed

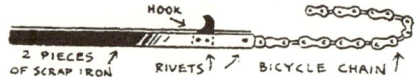

around the pipe and then anchored on the hook. — Paul Dannewald, Stettler, Alberta.

Solder Into Ribbons

To make solder into thin ribbons which are a substitute for wire solder for fine work, place the ladle of molten solder against a revolving wheel and pour. The slower you pour the finer the ribbon. — Grant Macleod.

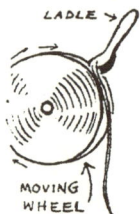

Wood-Carrying Sling

Here is a handy device which enables me to carry twice as much stovewood as with arms alone. The sling is made of gunnysack or canvas fastened to two round sticks, one of which serves as a handle, while a strap fastened to each end of the other goes over the shoulder and is adjusted to suit the user. — I.W.D.

One-Man Saw Buck

Make out of one-fourth inch or heavier rod a hook large enough to go over the largest log you are likely to saw, fasten the hook eye to one end of

a cultivator or other fairly strong spring, and the other end of spring with a rod or chain to a rod or 1x4 stick hinged as shown to a brace on the far side of the buck. Drive one or more spikes or pegs into the lower part of the buck leg on the near side, and adjust the length of chain so that pushing the 1x4 down and under the peg or spike will stretch the spring and hold down on the log. It sure works fine.

To Keep the Wood from Flying

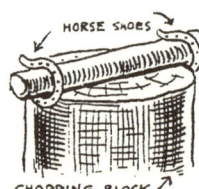

To keep wood from flying in your face when you are cutting it, fasten two horse shoes to your chopping block as shown. The wood will not fly and the danger of causing damage to your person is eliminated.

Homemade Dusting Machines

The first consideration in making a dusting machine for treating grain for smut is to have it dustproof. Three types are shown, each mounted on a pair of saw-horses, 32 inches high.

The box is about two feet square, which will hold a charge of a bushel and requires 30 revolutions per minute for 1½ minutes per treatment. The axle is 1¼-inch pipe six feet long. To make the bearings take a block, preferably of hardwood, 2x4x8-inches and bore a 1½-inch hole in the centre of the 4-inch face. Then rip the block lengthwise through the centre of the hole so that each half is 2x2x8 and has a half round bearing surface across its face for the axle. Bolt these to the saw-horses. The washers are held against one bearing with cotter pins passing through holes drilled in the axle. For the opening, cut away about nine inches from one of the free corners. The door is held in

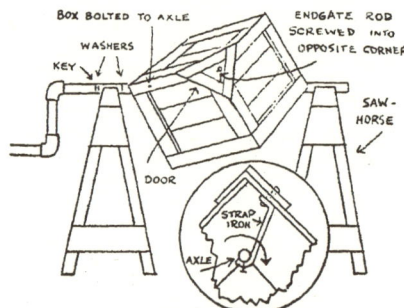

place by an end-gate rod screwed into the opposite corner and comes through the door about three inches off centre so that the door can be rotated out of the way when filling. Two pieces of ¼x1-inch strap iron are bent and attached to the box and the axle as shown in the insert illustration. These force the box to turn with the axle. The bolts which attach the box to the axle are about six inches from the corner of the box where the axle runs through it.

This duster is made from a clean wooden oil barrel. Instructions for

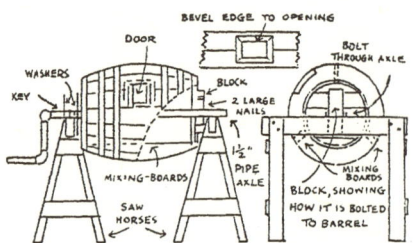

mounting the box duster apply largely in mounting this type. The axle is four feet long. Three bits of 1x4-inch board, the length of the inside of the barrel, and shaped to fit the bulge, are spaced equally around the circumference with the door in the middle of one space. They are fastened in place with iron brackets. The barrel is anchored to the axle by bolting a piece of 2x4 to each end of the barrel and then putting a bolt through the 2x4 and the axle. For the door opening select two wide staves and cut out, leaving a bevel for the door to fit against and also leaving about an inch of each stave for rigidity. Tack pieces of discarded inner tube around the opening to make it dustproof when the door is closed.

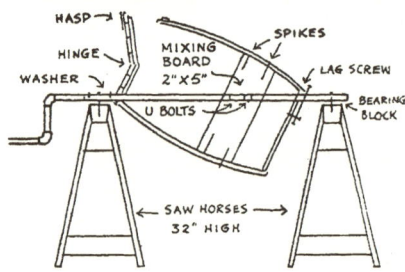

This duster is made from an old steel oil drum or a 50-gallon barrel. One half of one end is cut out. A wooden head is made hinged in the centre and attached as shown. Make dustproof with strips of old inner tubes. The door is held closed by a hinged hasp. Bolt a 2-inch reinforcing block to the opposite end for the axle to pass through. The drum is anchored to the axle by two large lag screws six inches long, screwed into the false heads through holes drilled through the barrel and axle. A baffle board, 2x8 inches, is spiked across the inside of the drum about two-thirds of the way back from the door, placing it so that it will be up and down when the drum is in filling position. The bearings are the same as in the other types and no top half is necessary.

These types of dusting machines are designed by the Dominion experimental farms system and are therefore approved by it.

Brush Hook from Shovel

An old shovel of the round nosed type can be made into a useful brush hook for clearing land along fence rows and similar work. The blade is cut away as shown by the dotted lines and the resulting sickle is sharpened. It makes a useful tool.

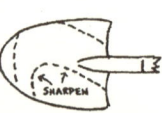

Repairing Harness

The first thing to learn in mending leather is how to make a wax end. For harness stitching No. 10 linen thread is used. A three-foot length can be taken. Long tapering ends are needed. Untwist the thread by rolling it on the thigh and then tease apart. For heavy work, like tugs, six or seven threads are waxed together. They are assembled, with the end of each succeeding thread slightly beyond the other. Place the assembled threads over a nail in the middle. Place a bit of shoemakers' wax on a piece of leather and put it near the stove so that the wax will melt on the leather. Then draw it over the teased out ends of the threads. The ends of the assembled threads are first waxed. The threads themselves are then twisted and waxed.

A harness makers' awl is used. The spacing can be done with a spacer,

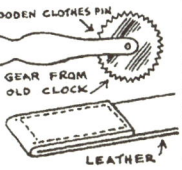

Fig. 1. Leather marker.

which can be purchased at the hardware store or, the idea in fig. 1 can be utilized. It was sent in by Paul Tremblay, St. Paul, Alta. A gear from an old clock and a clothespin are used. The teeth can be filed to any size needed.

A simple harness clamp can be made as shown in fig. 2. The dimensions of A and B are given. C is ¾-inch square, slightly bevelled. At the bottom is a piece of leather. The clamp is put in an ordinary vise when being used.

When threading the needles, which are placed one at each end of the thread, draw the tapered end through

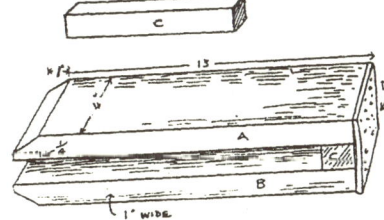

Fig. 2 Simple harness clamp.

the eye two or three inches, bend the thread back and twist between the thumb and finger.

Fig. 3 shows how to cut the ends of a strap for splicing. First cut each end square. The hair, or smooth side is much stronger than the flesh side, therefore

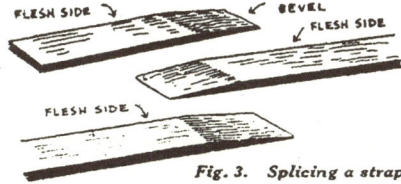

Fig. 3. Splicing a strap.

in bevelling always cut away the flesh side as shown in the upper figure in the cut. The lower two pieces are put together as shown, with the straight side of one strap placed in contact with the bevelled side of the other.

You Can SAVE MONEY with PERFECT CIRCLES

because they help your Tractor, Car, Truck, run BETTER and CHEAPER!

SMOKE from the exhaust of a Tractor that burns gasoline (or from your car or truck) is a sign of an oil pumping engine. Such an engine *always* wastes gas and oil . . . *wastes your money!*

Perfect Circle Piston Rings are specially made to stop oil pumping in tractor, car and truck engines . . . to restore power and smoothness . . . to conserve gas and oil . . . *to save you money!*

Perfect Circles are *custom made* . . . designed to fit every engine, personally — tailored to fit *yours*. Ask your Doctor of Motors to install them . . . He can—

Restore Power . . . Save Oil . . . Save Gas with

Fig. 4 shows how the stitching is done, with the two ends of strap held in place by the clamp. After marking, with a spacer or a ruler, the holes are punched with the awl, puncturing from the smooth side toward the flesh side. The stitching is done by passing both needles through the holes from opposite sides and drawing each stitch very tightly. When the last stitch through both straps is made cross over as shown, whichever method is preferred, then reverse the splice in the clamp, with the smooth side still to the right, and stitch the other edge of the splice. To finish the stitching, place the left needle and thread through as usual; then place the right needle in the hole and wind the left thread twice around the right needle and draw both ends tight. The winding will lock the threads in the leather. Make another small hole one-eighth inch below the next to the last one on the splice and put in another locking twist. Then cut off both threads. The finish is shown in fig. 5.

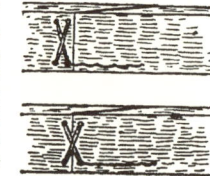

Fig. 4. The cross-over.

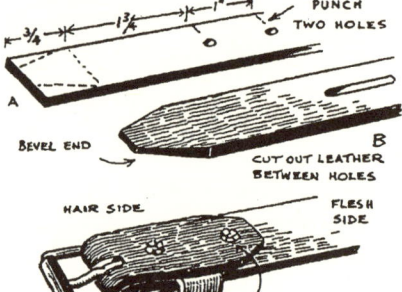

Fig. 5 Locking the threads.

To attach a buckle to a strap by riveting first cut the corners off from the end of the strap. Then mark the first hole between 2½ and 3 inches from the end of the strap. The other one is an inch further. Punch the holes and make the slot. Bevel the end for ¾-inch on the flesh side. Also bevel another piece at both ends for the insert. Place buckle and loop in position and rivet as shown. If a slide loop is used the strap should extend back three or four inches from the buckle.

Fig. 6 Riveting buckle to strap.

In mending a broken tug a plain stitched splice will do, even when the break is at a buckle hole as it generally is. The tug is shortened the length of the splice. First square the broken ends then bevel each back six or eight inches. The stitches holding the straps together are then cut for some distance so that when the bevelled trace ends are put together the straps from one end of the tug can be placed between the straps from the other end of the trace. Place the trace in a clamp and stitch.

Fig. 7 Splicing a tug.

A simple way to splice a tug is to use a metal tug splicer. The broken ends are squared. Then lay the ends together place the splicer on top and mark the rivet holes. Punch the holes, insert the splicer and then the rivets, being sure that the heads of the rivets are next to the horse.

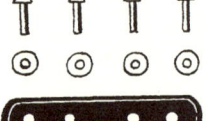

Note:— Most of the material in this article is from the Cornell bulletin, Repairing Harness.

Fig. 8. A metal tug splicer.

Protecting Barn Door Track

Sliding barn doors are usually hung on the outside, and we had much trouble with water running down the outside

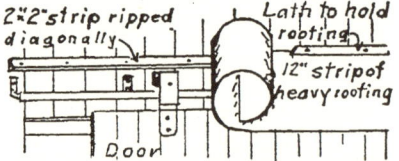

of the gable end and inside of the door, making the floor very wet in a hard driving rain. Snow also blew in badly and sleet and ice bothered in freezing weather. The diagram shows how we stopped the trouble by cutting a strip of heavy roll roofing 12 inches wide and putting it over the track and top of door. We ripped a 2x2 on the diagonal and used it under the strip at the top to throw it away from the wall and lessen the rubbing of the hangers. The top was fastened with laths so it could be removed easily.—I.W.D.

All Weather Doors for Barn

By extending the door track as far one side of the doorway as the other, the owner of this barn is thus able to use two doors. One is conventional, tight, and when used, completely closes

the barn. The other is merely a frame the size of the first door, but carrying a panel gate over the lower half while the upper half is covered with taut woven wire. This door keeps stock in or out, yet ensures ample ventilation on hot days.—Dale Van Horn.

Butting Belt Ends

Be sure that the ends are cut at a right angle. If no square is handy, employ the method shown in the drawing. Double your belt, then twist one end so that it is reversed, and cut both ends at the same time.

Quick Repair for Broken Strap

A satisfactory emergency repair for a broken strap can be made as shown. Of

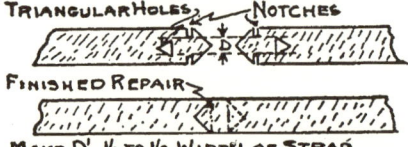

course it should be sewed or riveted at the first opportunity—I. W. Dickerson.

Another Door Catch

Seeing a door catch described in your Fall Work in the Workshop, I am enclosing a sketch of one I have used for quite a few years. It has an advantage over some door catches because it is used half way up the door and never becomes clogged with snow or ice. A bit of light spring, a short piece of strap iron and two quarter-inch bolts are all that are necessary to make it. The

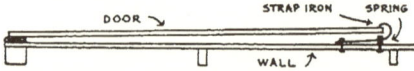

door is caught automatically when the door is pushed open wide and a slight pull back on the strap iron releases it. The catch will last as long as the building.—John R. Anderson, Rocanville, Sask.

Repairing Hay Ropes

When a rope breaks, and you have two ropes instead of one, it is well to know how to put them together again, end to end. You lengthen the life of the rope, and new rope isn't easy to get. But first

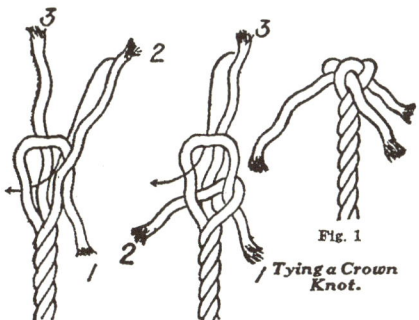
Fig. 1 — Tying a Crown Knot.

let us learn how to make a knot on the end of the rope. The following instructions are from an Iowa bulletin on knots and splices.

To make a crown knot, (fig. 1) bring strand (1) down between strands (2) and (3), pass strand (2) across the loop as shown and pass strand (3) through the loop. Then pull each strand until they are all tight.

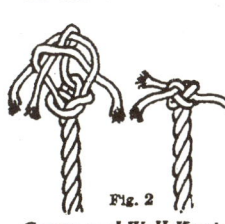
Fig. 2 — Crown and Wall Knot.

The crown knot can be made more secure by adding the wall knot as shown here (fig. 2).

The Short Splice

If there is a straight pull and the rope doesn't have to pass through a pulley, a short splice will do. The steps in making this splice are shown in (figs. 3, 4, 5 and 6). The two ends are unwound and then locked together so that those from one end pass alternately between those from the other end. Taking two strands from opposite sides, (a) and (1), (fig. 3) tie a simple knot (fig. 4). Repeat with (b) and (2) and with (c) and (3). Draw the knots tightly, then passing each strand of the rope diagonally to the left, tuck the ends under the strands of rope (y). Then turn the rope end for end and repeat by splicing down the strands of rope (y) (fig. 5). When the splice is completed each strand from both ropes should be spliced under at least two places, depending on the size of the rope and the strain it has to carry. The splice can be made, of course, without

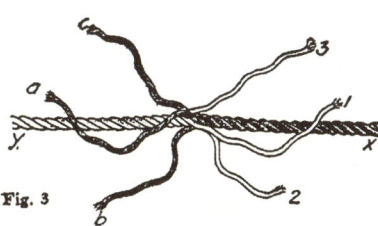
Fig. 3

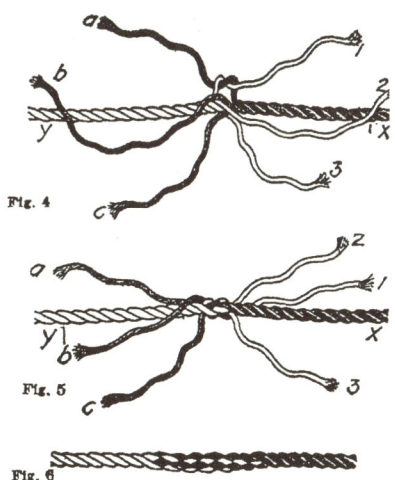
Fig. 4
Fig. 5
Fig. 6
Steps in making a Short Splice.

beginning with the overhand knots. The finished splice is shown in (fig. 6).

The Long Splice

For a long splice, a ⅜-inch rope will require a free end of about 18 inches and an inch rope, of 36 inches. The strands are locked together as when beginning a short splice, making sure that the strands are properly paired. The strands of two pairs, as (b) and (2) and (c) and (3) (fig. 7), are then tied together, leaving (a) and (1) free. Unlay strand (a) one turn and follow it by relaying strand (1) in its place, drawing it firmly and keeping it twisted tightly. Continue until 6 or 8 inches from the end of the relaid strand (1) and tie as in (fig. 11).

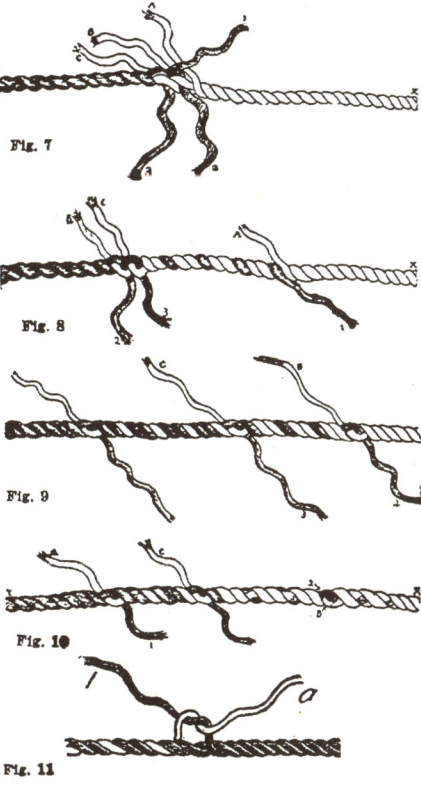
Fig. 7
Fig. 8
Fig. 9
Fig. 10
Fig. 11

Beatty Pumps

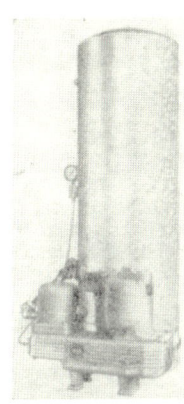

BEATTY DIRECT DRIVE SHALLOW WELL PRESSURE SYSTEM, electric or gas driven. Ideal for farm or large home. Capacity 360 wine. gals. per hour. This system has direct drive — all parts are enclosed — no belts — safe for children — takes less power to operate.

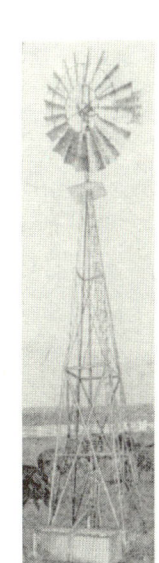
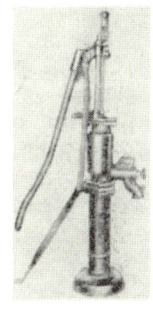

BEATTY DOMINION STOCK PUMP—Five times as many of these Pumps purchased as in our best years before the war. It's a favorite for deep well, hand or windmill use. Large capacity. Simple to re-leather. BEATTY PUMPER—Pumps in lighter breezes—Pulls in — Stormproof — Mechanism oil tight, water, snow and dust proof.

BEATTY BROS. LTD.
WINNIPEG - EDMONTON

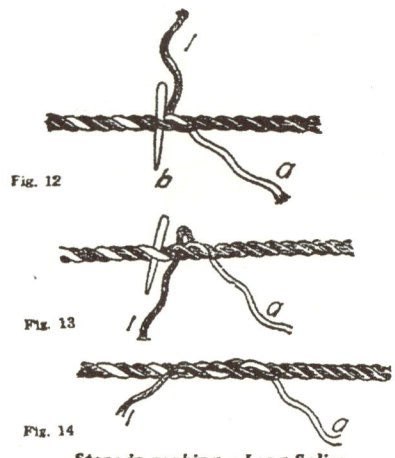

Steps in making a Long Splice.

now appear as in (fig. 9), the ties should be separated the same distance and each strand coming from (x) should be placed in front of the strand from (y) and tied. Crossing the strands otherwise, as behind (1), a mistake often made, makes it impossible to complete the splice properly.

The splice is completed as follows: With the ends properly tied, (fig. 11) with the right hand overhand knot, draw down firmly into the rope (fig. 12). The end (1) is now spliced down by being passed over the first strand (a) and under the second (b) as shown by the marline spike (fig. 12), then over the third (c) and under the fourth (a) (fig. 13). Draw down end (1) and cut it off, leaving it ¼-inch long (fig. 14). In identically the same manner splice down and cut off each of the remaining strands. The splice is finished by pounding down the uneven parts and rolling it on the floor under the foot. The finished long splice appears as in (fig. 15).

Then turn the rope end for end and unfasten (b) and (2) (fig. 8). Repeat as with the first pair and tie. The rope will

mow. The nuisance can be easily avoided by attaching this double hinged section on the bottom. Whether the door is being opened or closed one of the sets of hinges opens and prevents binding. The same idea could be adapted to a gate to make it easier to open and shut when the snow is deep.—Martin S. Walder, Raley, Alta.

New Use for Old Hoe

An ordinary garden hoe which is worn out and will not work well any more can be made into a handy hand cultivator by chiselling it into the shape indicated by the dotted lines. When the young plants are just above the ground a person can hoe them on both sides of the row and stir up the soil around and between them.—Bernard Schick, Carmel, Sask.

Fig. 15. The Long Splice as it appears when completed.

Mending a Broken Strand

When it is necessary to mend a broken strand of rope, or replace a piece of strand that is badly worn or frayed, unlay the strand back both ways, procure a new strand of sufficient length and relay it. The directions for doing this are the same as those given for making a long splice.

Improvised Garden Hoe

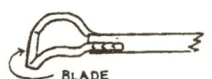

This is a homemade garden tool which I have proved myself to be very handy and useful, at the same time being cheap and easy to make. Take a piece of strap iron and bend it in the manner shown. Then fasten it to the handle. Sharpen the blade on both sides and work the hoe backward and forward. A big patch can be hoed in a short time.—Henry A. Jantz, Langham, Sask.

Inside Rolling Door

Under some conditions this kind of a barn door can be used to advantage.

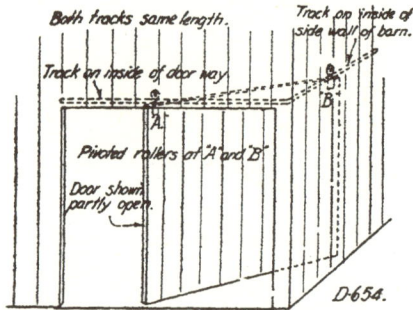

—Barn Door That Won't Blow Down—

The tracks are on the inside and both are of the same length. The pulleys have to be pivoted and must be close to the edges of the door. One advantage of inside doors is that they are not blocked by snow or manure.—I.W.D.

Barn or Garage Door Stop

Nothing is more annoying than to have the wind slam a barn or garage door just when you don't want it to shut. Here is a simple but effective stop

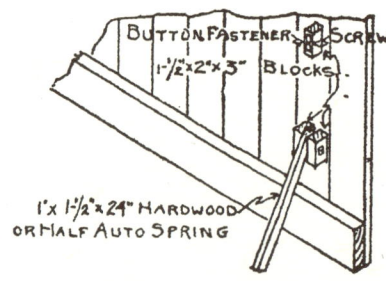

which is always at hand and will hold the door where you want it. Half an auto or buggy spring is fine for the prop.

Prevents Door from Binding

Heavy swinging doors, such as garage or barn doors, often bind at the bot-

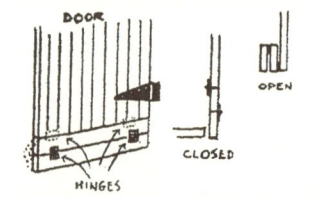

tom in winter time on the ice or packed

Briar Cutter

To make a briar cutter get a worn-out mower blade and cut off a length that has four sections or teeth. Rivet this to the eye of a scuffler hoe, or weld it to the blade of an old grubbing hoe. Sharpen the knives with a file or emery and you have a good tool to cut briars as well as an excellent fire-fighting tool.

A Harrow For a Dollar

This little harrow costs but little and can be used for cultivating between the rows of garden stuff or potatoes. It could be so constructed as to straddle the rows when the stuff is small by raising the hinges with blocks. Use

four-inch spikes as they are much heavier than the 3½-inch spikes generally used around the farm. Old plow or cultivator handles can be used, but if none are available they can easily be shaped out of some 2x2 stuff. The little affair will save a lot of hoeing.

Garage or Stable Door Stop

A door stop can be quickly and easily made from an old T-hinge. It is screwed to the bottom of the door and bent a

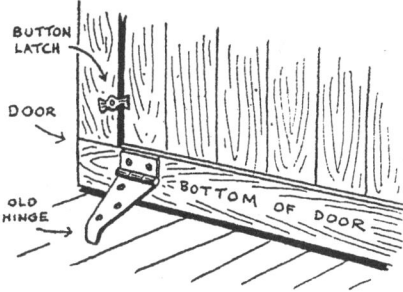

little at the other end to make it grip the floor. A small button latch is fastened above the hinge to hold it up when the door is closed.—Paul Tremblay, St. Paul, Alta.

Barn Door Closer

In barns having double doors closing in the centre trouble is sometimes caused by harness catching, or by two animals trying to go out through one half of the door while the other half is hooked. To overcome the difficulty I fixed up this arrangement. I took a wagon seat coil spring and fastened a six-inch hook to one end of it. This hooks into a staple in the door near the top. The other end of the spring is fastened to a joist by means of a wire. Another piece of wire keeps it from falling down when not in use. This automatically closes the door when it is pulled open by an animal pushing against it.—M.H.S., Calder, Sask.

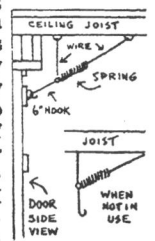

Rake from Fork

Needing a rake about the workshop I got an old fork with five tines, heated the tines and bent them over in a curve so that all the points were in line. Then I found that the rake was also useful in cleaning up the yard or cleaning stables.

Garage or Barn Door Stop

A piece of flat iron 18 inches long with a hole bored five inches from the one end and fitted into a stake will hold any door on hinges. The stake must be driven into the ground far enough so that the door will pass over it.—A. T. Glosser, Hepburn, Sask.

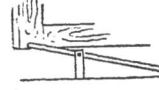

Home-made Push Hoe

Following the lead of a Manitoba orchardist, I have designed and tried out half a dozen push hoes of various angles, shapes and sizes. The one I like best, and intend to use henceforth, is light and not too wide, quite pointed, and with strongly sloping sides, though blunt on the very tip. The long slope gives one a chance to cut through even stout weeds with little force, and the blunted end enables one to hit square-

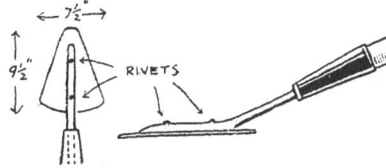

ly upon a particularly stubborn plant. The metal from the feeder knives of a grain separator makes very good material for a push hoe. No trouble should be spared to obtain a good, smooth handle, and its angle with the blade should be that best suited to the man who uses it oftenest.—Percy H. Wright, Moose Range, Sask.

A Wheel Hoe

The knife of this cultivator was made from an old car spring, shaped and tempered by a blacksmith. The handles were made from 2x2 banister stock. The braces are ordinary band iron. When laying out the garden have the rows of vegetables a little wider than the cultivator blade so that you have to go down the row only once for each cultivation.—A. T. Gossen, Hepburn, Sask.

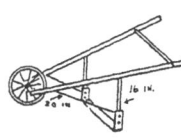

Steering Wheel Auger

A car steering wheel placed on the common post hole digger will greatly improve it, especially when it is necessary to make a post hole near another post, as the ordinary type of handle has a tendency to get caught and often results in the injury of the operator's hands. The drawing is quite self explanatory. If the hole in the steering wheel is too big a bushing will remedy this.

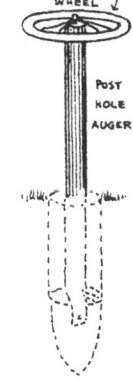

Brush Axe from Rolling Coulter

Take an old rolling coulter from a gang plow, heat it and halve it. Next drill two holes as shown in the illustration. From two pieces of 1¼-inch iron make two loops for the handle. Make the handle, bowed as shown, and saw a slot in it for the straight end edge of the blade to get greater firmness. Then rivet the loops on as shown. The cutting edge

of this axe, being wide and thin, cuts many kinds of brush or even small trees

Protect your FENCE POSTS with "Osmose"

SPECIAL FENCE POST MIXTURE

Costs only 3c to 4c per post, makes them last 3 to 5 times longer. Simply applied like paint. Preserves all kinds of posts, green or dry—Poplar, Pine, Spruce, Willow, Tamarack, Cedar.

Just paint from 4 ins. above to 8 or 10 ins. below the groundline. "Osmose" preservatives penetrate up to 2 ins. in the wood and remain there a lifetime.

Over 90 Canadian Power Companies use this treatment. Over 2,000,000 Power, Telephone and Telegraph poles have been "Osmose" treated in U.S. and Canada—ample proof of "Osmose" effectiveness.

Gallon $3.95—treats 80-150 posts.

PENTOX
PRIMER • SEALER • PRESERVER
FOR NEW WOOD
STOPS... Warping-Shrinkage
SAVES... One Coat of Paint

Use it for sash, doors, millwork, plywood, lumber, shingles, floors, farm implements, boats, fabrics, etc. Comes liquid ready to use—prevents warping, swelling, sticking and rot. Applied to wood panelling, and flooring, Pentox brings out the full beauty of the grain.

Gallon—Clear Varnish
   Type $3.20
Gallon—Dark Green or
   Dark Brown $3.45

See your local dealer, or write direct for complete illustrated literature.

Write: Department 10

OSMOSE WOOD PRESERVING COMPANY OF CANADA LTD.
83 Union Building • Calgary

Thousands of Parts To Fit All Tractors

SAVE MONEY ON TRACTOR REPAIRS

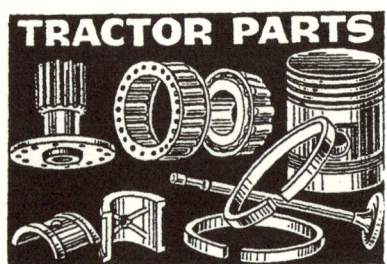

We Sell No Used or Worn Parts

Guaranteed NEW parts: thousands of them, to fit all makes! Our low prices put money in your pocket—many of our customers say that our tractor parts prices save up to 40 per cent or more. A Sask. farmer writes: "... I am well pleased with the tractor parts I ordered. Every part fitted 100 per cent and shows a saving that is truly wonderful."

Highest Quality--Perfect Fit

Perfect fit guaranteed — bearings, gears, fan belts, cylinder sleeve assemblies, valves, piston rings, connecting rod bearings, gaskets, oil filters, air cleaners, mufflers, ignition parts, carburetor parts, clutch facings, headlights and other parts to fit all tractors. Complete stock at Winnipeg, Regina, Saskatoon, Calgary, Edmonton. All new goods—we sell no used or worn parts. See our latest catalogue for prices, etc. (copy free on request), or visit our nearest retail store.

Consult Your Macleod Catalogue— You cannot afford to be without a MACLEOD catalogue. It is the most complete listing of implement replacement parts and farm supplies that is published anywhere. If you have not a copy handy, write for one TODAY

MACLEOD'S LIMITED

Five Mail Order Warehouses:
WINNIPEG REGINA SASKATOON
CALGARY EDMONTON

and 32 Retail Stores:

Winnipeg	Melville	Ponoka
P. la Prairie	Regina	Edmonton
Brandon	Weyburn	Wetaskiwin
Carman	Humboldt	Camrose
Neepawa	Melfort	Red Deer
Virden	Prince Albert	Stettler
Swan River	Saskatoon	Olds
Dauphin	Rosetown	Calgary
Kamsack	N. Battleford	Lethbridge
Canora	Lloydminster	Medicine Hat
Yorkton	Vermilion	

---

much better than an ordinary axe. Two or three may be cut off with a single stroke. I have used this axe myself, and a neighbor cut nearly 200 acres of brush with it. — Theodore Troitsky, Springdale, Alta.

Garden Time Saver

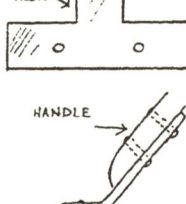

A very convenient device for hoeing between garden plants is made from an old scalloped bread knife. It eliminates hand weeding which is very tedious. Drilling holes in the blade about three or four inches apart. Take a piece of thin strap iron and make two holes to correspond with those in the blade. Note the angle of the strap iron in diagram. The bread knife blade is now attached to a handle about four inches long. This hoe is used by moving it back and forth, just as though cutting bread, as you move along the rows.—A.P., Dallas, Man.

A Push Hoe

A piece cut from a gas barrel top, two pieces of heavy strap iron and the stem of a small spruce tree or other piece of wood are all that are necessary to make this handy push hoe. A push hoe is an excellent implement in keeping the garden clean. — C. Leder, Neerlandia, Alta.

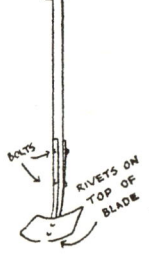

Auger from Two Shovels

A post hole auger can be made from two old shovels. The shovel blades are cut as shown and riveted to the ends of a piece of stout flat iron rod. This works all right in clay or soil that is not full of stones.— John H. Schab, Winnipeg, Man.

Shovel from Ford Fender

That perpetual rattle from a Model T fender can be converted into a sturdy fire shovel. I think the illustration is sufficient without any instructions how to make it. — Wm. Kowalchuk.

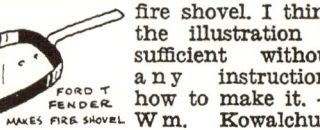

Wire Snipper

I found this a very handy tool for snipping wire. A pair of scissors that can no longer be used in the house can be converted into a good pair of wire cutters. Cut off the scissors the desired

length and then grind out a semi-circle in one blade, sharpen with a small whet stone and lubricating oil. The blades can be hardened by heating cherry red and plunging into old cylinder oil.— C. Leder, Neerlandia, Alta.

Water or Gasoline Pail

A serviceable water or gasoline pail can be made by cutting a hole in the side near the top of a pail. A piece of flexible metal hose or a 3-inch length of 1-inch galvanized pipe is soldered on

at the hole. If the pail is intended to be used only for filling radiators with water, the metal hose or pipe may be omitted and the water allowed to flow through a hole near the top of the pail. This hole should be from 1 inch to 1½ inches in diameter, depending on the size of the radiator opening.

Door Hook From Old Pail Handle

Old pails should not be thrown away. The sheet metal may come in useful for many purposes. The metal bale and lugs can be used for making a door hook.

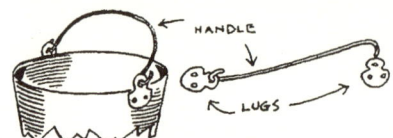

The bale is left attached to one lug and cut the right length and bent into a hook. The other lug then is placed where needed. Each lug has to be bent at a right angle.—Paul Tremblay, St. Paul, Alta.

Sharpening Shears

Sharpening shears or scissors is not a difficult job. A good, sharp, fine-toothed flat file and ordinary good care are the chief requirements. Secure the shears firmly in a vise in a horizontal position. Light, firm, even strokes in the cutting direction are required. Lift the file at the end of each cutting stroke. Do not draw the file backwards on the

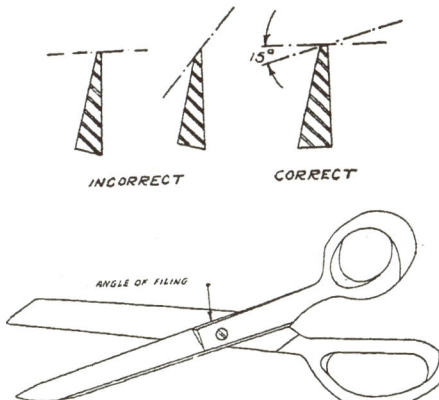

shear edge before applying the next stroke. Drawing the file backwards will dull both the file and the shear. The cutting edge of the shear blade must be filed at the correct angle, which is about 15 degrees. This is often indicated by a bevel near the screw as shown in the sketch. Do not file away any more metal than is necessary. If the shear blades are loose due to wear of screw seat, place a very thin washer below the head

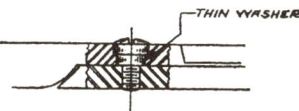

of the screw, sufficient to take up the slack. Do not attempt to remedy loose scissor blades by undue tightening of the screw, unless the screw is simply loose.

Hinge for Small Box

I found these hinges very suitable for a small box and very easy to make. They are very serviceable and work

just as well as other elbow hinges. For a very small light box tin from any tin can is plenty strong enough. For a larger box, sturdier sheet metal is needed.

Hot Water Bottle Holder

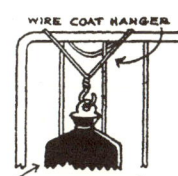

A wire clothes hanger bent to fit over the head of a bed comes in handy in the sick room for hanging such things as a hot water bottle on. It is within easy reach of the patient in the bed. It will work equally well for a well person who uses the bottle as a bed warmer.—M. Lambert.

Repair for Pail Handle

When a handle lug on a pail breaks drill two ¼-inch holes just below the rim of pail about 1¼ inches apart. Bend a piece of ¼-inch round iron into a "U" shape so that the space between the two arms of the "U" is one inch. Then turn the two ends to partly form two eyes, about 1 inch in diameter to complete the repair link. Place the repair link on the handle of the pail and then slip the ends of the open eyes through the two ¼-inch holes in the pail. Now the two eyes can be closed with a hammer to complete repair.

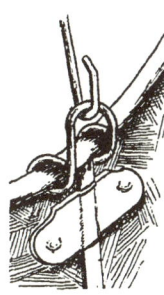

Battery Cap Lifter

We find this easily made gadget very handy in lifting the corroded caps from a car battery. Material one-eighth inch thick or thinner can be bent cold but as this is rather thin, use a nut on the inside of the gadget for the bolt to go through. Slip the points under the clips and turn the bolt and off they come.—Sidney Ransom, Sr., Mountainside, Man.

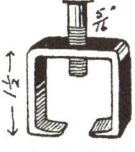

Sack Chute and Holder

This sack filler and holder can be made to be fastened on a wall or on a portable stand. When attached to a stand, the sack holder can be placed on

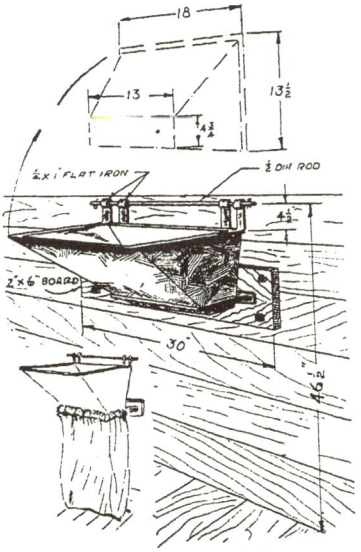

a platform scale so that seed or feed grain or other seed can be weighed. The chute or hopper is made of sheet metal and reinforced with 1x¼-inch flat iron at the top edge. The hopper is hinged on a piece of ½-inch rod as shown in the sketch. The hopper is first lifted and

FARMERS!
DO YOU NEED...

NEW EQUIPMENT

—tractors, binders, plows, pumps, choppers, etc.?

FARM ELECTRIC SERVICE

—Installations, alterations, improvements?

BETTER HERDS

—higher quality cattle, sheep, swine?

NEW BUILDINGS

—improvements, additions, alterations?

The Farm Improvement Loan Act of 1945 enables you to do all these or any other work around the farm which will increase its efficiency and productivity. Loans are repayable over periods up to ten years at 5% interest.

—*Now is the Time*—
See Your Imperial Bank Branch Manager Today

IMPERIAL BANK
OF CANADA
"The Bank For You"

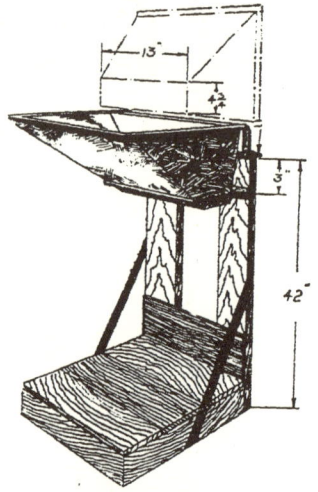

the sack is then hung on the ⅝-inch round iron sack holder. The front and side edges of the sack opening are folded over the sack holder and the rear edge is caught on a sharp pin (not shown) in the frame at the back. The hopper is then allowed to drop down and holds the sack firmly while at the same time it provides ample room for filling the sack with a scoop or a pail without spilling the grain. The height for the hopper is made to suit standard size grain sacks and measurements shown in the sketch should be followed accordingly. If sacks are short, pieces of plank on the floor or on the stand can be used to take the weight of sacks as they are being filled.

Parts Rack From Disc Harrow

An old disc harrow gang can be used as a rack for bolts, nuts, washers and screws. A discarded disc from a one-way disc machine can be used to make a good wide base. If only old discs are available, then pieces of 1½ or 2-inch pipe can be used as spacers. Both ends of each piece of pipe are plugged with wood. Holes are drilled through the wood plugs to fit a ⅝ or ¾-inch round iron rod. The rod is threaded at both ends and fitted with nuts and washers.

This rod is used to hold the discs and pipe spaces lightly together instead of the original square shaft or arbor.

Handy Single Wheel Cart

One 18-inch farm implement wheel, two light handles and a minimum of framework provides a handy cart around the farm for transporting such things as sacks of feed, plants in flat boxes for the garden, milk and cream cans, drinking water, wood for the stove, etc. Any handy farm boy would find pleasure in making this simple cart and make his farm chores lighter for additional reward. The handles, as well as the framework can be made of 1x4-inch lumber. Hardwood is preferable. The platform can be made of ordinary

boards. A piece of strap iron 2x⅛ inches is bent 2 inches at each end and used to support the front end of the platform, while a 1x3-inch wood cross piece will support the rear of the platform.

Barrel Plug Wrench

An old plow lever can be used to make a wrench to fit oil barrel plugs. Such a

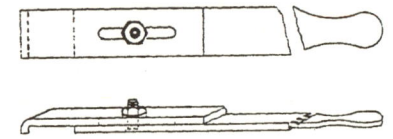

wrench can be made adjustable as shown in the sketch. When it is not necessary to make the wrench adjustable, the slot and plow bolt can be omitted and the two parts can be either welded or rivetted together to make a wrench of permanent size to fit the standard size barrel plug.

Protect The Fingers

When cutting paper or clipping magazines with an old safety razor blade only the corner of the blade is used to do the cutting so why have the whole edge exposed with danger to the fingers? Just fold a piece of adhesive tape right over the blade. The corner will cut through it the first time it is used but the rest of the edge will remain covered.

When Painting the Garden Fence

When doing it, the trouble is to find a holder for the paint can. Here is an easy way to overcome the difficulty. Just take an empty can, punch a hole in the side of it, make a wire hook and fasten to it. Then when painting just catch it on the handiest picket with the paint can hung in the hook.—D.C.R.

To Remove Road Tar Spots

Linseed oil or kerosene oil will remove road tar without damaging the finish of your automobile. Saturate the spots of tar with the oil. Let the oil stand for a few minutes to allow it to penetrate the tar. Heavier spots require more time. Then remove each spot with cheesecloth. The oil softens the tar so that it can be easily rubbed off. Then wash with water in the usual way.

Saw Horse Accessory

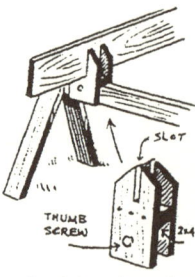

The difficulty of planing the edge of a wide board is overcome by this device. Two 1x6 pieces are notched as shown and nailed to a block so as to straddle the horse. Two of them are used. If necessary a thumb screw can be inserted to hold the gadget in place but for most work it is unnecessary.—D.C.R.

Home-made Belt Dressing

To make a hammer mill belt pull good in winter, roll up a piece of an old inner tube and tie it with a piece of wire long enough to make a handle, pour on a few drops of gasoline, and light with a match. This is held over the running belt for a few minutes, and the melting rubber gives the belt a long lasting tacky surface which prevents snow from bothering.—I.W.D.

Hold Handle Hacksaw

The handle is made of a pipe, which is bent to the desired shape. Two small holes are bored in the saw frame and in the pipe, the holes to coincide with each other. The pipe where it fits on

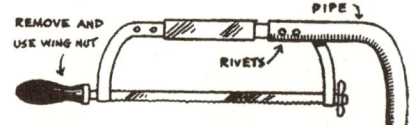

the frame has to be sawn lengthwise, so as to fit on each side of the hacksaw frame. The handle and the pipe are riveted together. The wooden handle is taken off and a wing nut replaces it to keep the blade at the correct tension.—Thos. Wishart, Starbuck, Man.

Tube Repair Rack

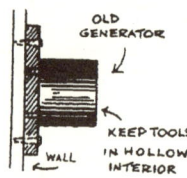

Take an old generator and discard all except the heavy casing. Mount on a block of wood measuring 12 x 12 x 1½ inches. The complete rack is then screwed fast to the side of the garage where it is ready for constant use. In making repairs to inner tubes, the tube is hung over the rack and held down with the foot, and the repair patch applied.—Dorland A. Hotz, St. Boswells, Sask.

Oil Barrel Cart

This oil barrel cart was designed by the North Dakota extension service en-

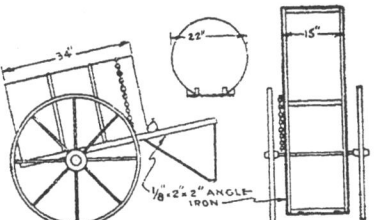

gineering department. The barrels are loaded in the same manner as a load is picked up with a warehouse truck. Tilt the barrel over and push the nose of the cart under. Then fasten the length of chain around the barrel and pull down on the handle. The frame is made 13 inches wide which will take a 22-inch barrel. The details are available by studying the sketch.

Knives From Old Files

Old flat files are useful for making various kinds of knives, such as hoof knives, poultry killing knives, scrapers and wood chisels. The old file should be first annealed by heating to cherry red

and allowed to cool in wood or coal ashes. The old teeth then can be ground off on an emery stone. Any shape of blade can then be forged on an anvil. After retempering, the blade is then ground smooth and sharp similar to the hoof knife or poultry killing knife shown in the sketch.—H.J.K.

Supports For Peonies

Peonies and other plants with large heavy blooms frequently need good support to keep the flowers off the ground and appear their best. Wire fencing makes ideal supports when formed as shown in the sketch. Supports for peonies should be 30 inches high and 18 inches in diameter. Place the supports over the peony plants early in the spring. Keep the foliage inside of the fence until the flower buds are well formed. Allow the flower buds to protrude through the openings in the fence so that they get ample sunshine. When the plant is fully developed the fence will not be seen and the entire

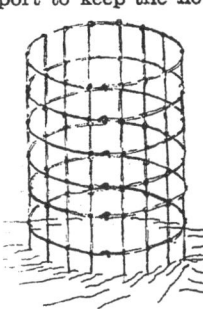

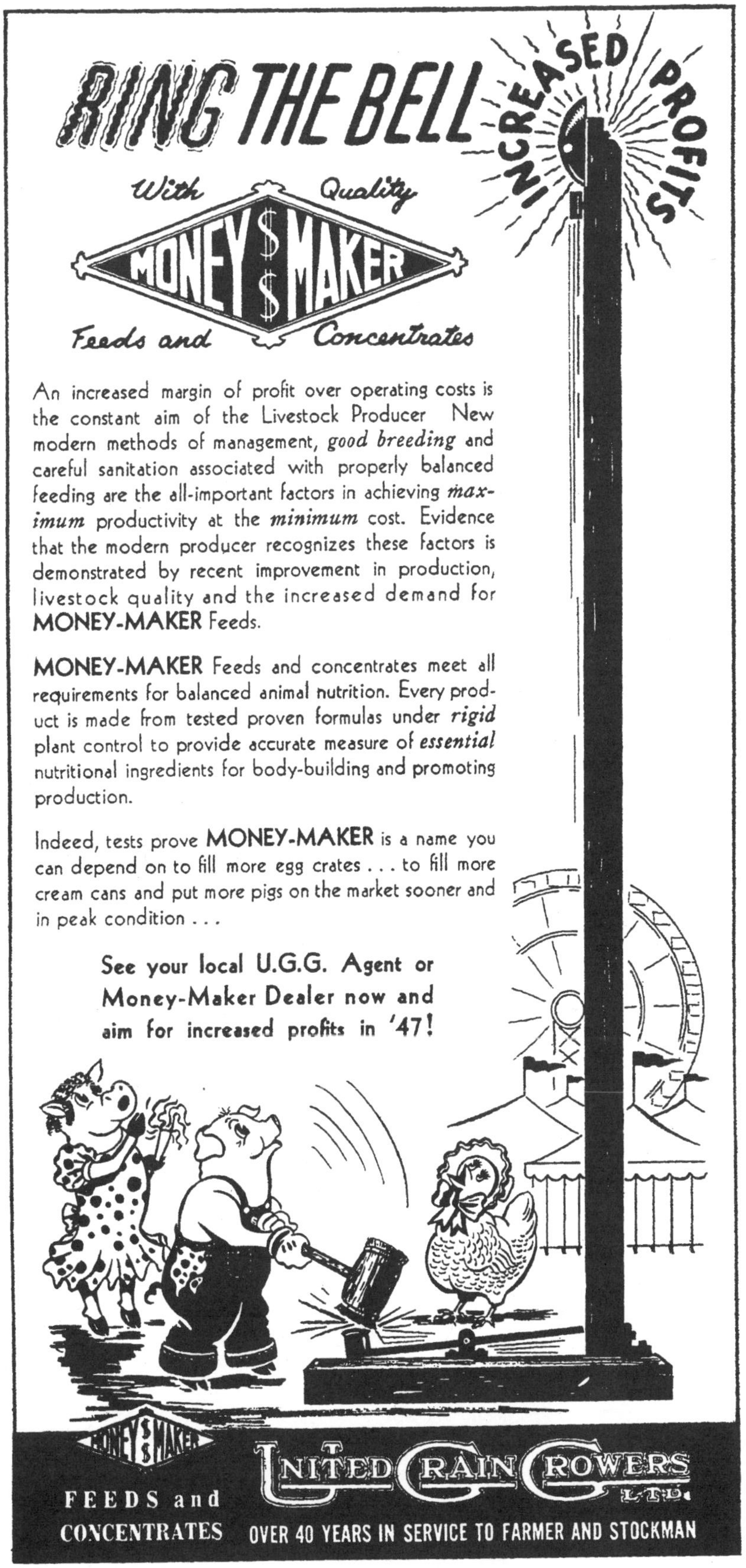

RING THE BELL

With Quality MONEY $ MAKER Feeds and Concentrates

INCREASED PROFITS

An increased margin of profit over operating costs is the constant aim of the Livestock Producer. New modern methods of management, *good breeding* and careful sanitation associated with properly balanced feeding are the all-important factors in achieving *maximum* productivity at the *minimum* cost. Evidence that the modern producer recognizes these factors is demonstrated by recent improvement in production, livestock quality and the increased demand for **MONEY-MAKER** Feeds.

MONEY-MAKER Feeds and concentrates meet all requirements for balanced animal nutrition. Every product is made from tested proven formulas under *rigid* plant control to provide accurate measure of *essential* nutritional ingredients for body-building and promoting production.

Indeed, tests prove **MONEY-MAKER** is a name you can depend on to fill more egg crates... to fill more cream cans and put more pigs on the market sooner and in peak condition...

See your local U.G.G. Agent or Money-Maker Dealer now and aim for increased profits in '47!

UNITED GRAIN GROWERS LTD.
FEEDS and CONCENTRATES OVER 40 YEARS IN SERVICE TO FARMER AND STOCKMAN

plant with all its blooms will be shown to advantage and be protected from strong winds. Try these supports in various sizes for other plants. Delphiniums eight feet high can be supported by such means.—H.J.K.

Pump Jack Pin

Frequently a pump jack pin gets mislaid or lost and causes some delay in getting the pump hooked up for operation. A pin bent to shape as shown in the sketch is easily inserted or taken out. The bent end keeps the pin from falling out when the pump is being operated. A piece of lace leather or a light chain on the pin and fastened to the pump or pump jack will insure this handy pin from being missing when it

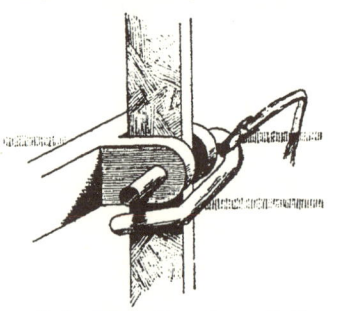

is needed. The diameter of the pin should fit the holes in the pump rod and jack as closely as possible to prevent undue vibration and wear to the entire pump and jack assembly.

Fodder Shock Binder

This diagram shows one way of tying fodder shocks so they will not tumble down, even when the corn is very tall. Use a small rope with a ring or a loop in one end. Stick a large wooden pin into the shock about where you wish the tie to come. Lay the ring or loop on this large pin, extend rope around shock and pass over large pin. Put a smaller

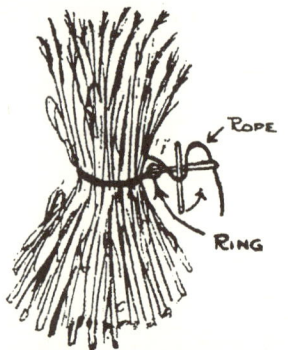

pin under rope and through the ring or loop at right of large pin. Bring end of rope under large pin and loop over end of small pin. With the small pin begin to wind the rope around the large pin, keeping the end of the rope to the outside. Wind tight enough to bring the shock into shape, then tie with binder twine, and remove rope. This makes the pressure even all around and does not pull the shock out of shape.—I.W.D.

Another Use for a Rubber Ball

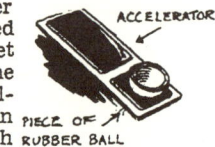

A woman driver with high heeled shoes should get the man of the house to cut a hollow rubber ball in two and attach one half of it to the accelerator upside down. The high heel fits nicely into the socket thus made. Soft rubber will not damage the heel in any way.—D.C.R.

Sawing Granary Door Boards

I noticed the postmaster's assistant in our town sawing the door boards for a granary he was building at a slight angle. On enquiry it was explained and demonstrated that, cut like this, they lift out without difficulty. It isn't necessary to slip or pound them to the top to remove them.—H. D. Falconer, Glentworth, Sask.

Stand for Salt Block

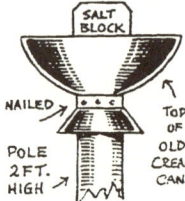

I saw this stand in a field I was passing. Take an old cream can and cut the top part off. The post is just big enough to fit into the neck of the can. The top is then nailed to the post, the whole being about 2½ feet high. The handles are removed so that the animals will not hook behind them. The top does not fit the post so tightly as to prevent the drainage of rain water.—Edwin Unger, Mayfair, Sask.

Hanging Shelves

This diagram shows how to use old auto skid chains to make good supports for hanging shelves in a basement. The chains are suspended from bolts in the floor joists, and the shelves are placed

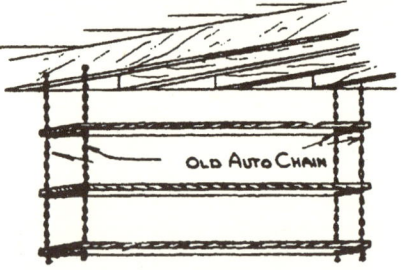

on the cross chains. Be sure that all of the links are in good condition, so that if the shelves are loaded heavily they will not break.

Generator Fails at Higher Speed

When the car generator charges normally at lower speeds but falls off to zero at 35 miles or more, the trouble is very likely due to badly worn brushes, weak brush springs, rough commutator, or perhaps a combination of these troubles. Better have these conditions checked, as neglect may require a new generator.

Pressure Grease Cup

This is a diagram of a handy home made pressure grease cup. It consists of a common grease cup, through the cap of which is drilled a hole large enough for a discarded auto tire valve stem to go through, with a valve stem nut on each side of the cap. The valve stem takes the grease gun nicely, but the valve core must be used to hold in the pressure. In case a 45 degree bend is needed, the valve core can be removed, the stem heated and bent to the shape wanted, the valve core inserted, and the stem then locked in the grease cup cap with the two nuts. This would be of especial value in overhauling old discs and other machines equipped with large grease cups.

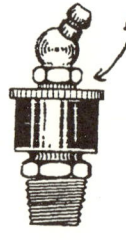

In the other diagram we showed how one reader made a pressure lubricator by soldering an auto tire valve stem into the cover of an ordinary grease cup. This works fine in summer, but is likely to give trouble when the grease gets stiff. This trouble can be avoided by screwing or soldering a regular Alemite or Zerk fitting into the cap so a pressure gun can be used for forcing the grease in under pressure. —I.W.D.

Disc Scrapes Cement Floor

Take an old blade from a disc harrow and mount on a suitable handle. Inch iron piping would make a good handle. Bend it so that when standing the disc fits closely on the floor. This does not take the place of a flat scraper but is good for lifting stuff that has become stuck to the floor.—D.C.R.

Assorted Door Handles

Here is an assortment of door handles. A is made from a bit of inch pipe and two elbows. Bolt through where shown. B is especially suited for a barn door and gives a novel effect. It is made from a harness hame and two old bells. It is fastened to the door with a couple of bolts. C is made up from an old fork handle. It can be fastened on with nails, clinched on the inside of the door.

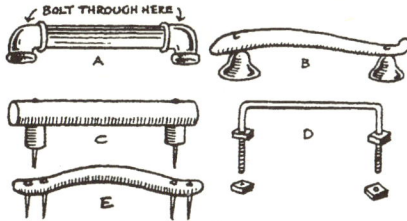

D is made from a piece of round iron stock threaded at both ends and bolted through the door. Go to the bush for the material for E. It is a natural bend or curve simply nailed to the door and the nails clinched on the other side of it.—Paul Tremblay, St. Paul, Alta.

Rope Assembly for Cream Can

Take a short piece of rope about three feet long and splice it into another rope seven feet long about 2½ feet from the end. The ends may be tied to the handle of the cream can or better still hooks provided to grab the handles. The other end of the long rope is fastened to the top of the ice well.—Fraser Robin, Inglis, Man.

Felt Hat Fuel Strainer

An old felt hat placed inside of your truck or tractor funnel will prevent no end of fuel system and carburetor troubles. Just be sure that it is clean and has no holes or perforations in the crown. Rub a film of new cylinder oil inside of it at the start. You will find it superior to several dollars worth of chamois as it holds back all the dirt and water and lets the fuel through twice as fast.—Robt. J. Roder, Reist, Alta.

Loosening Rusty Nut

This substitute for penetrating oil will come in handy when loosening rusty nuts. Put some strong vinegar in a clean can and squirt on the bolt and around the nut. In a few minutes the nut can be readily removed.—A. S. Wurz, Rockyford, Alta.

Starting Cold Tractor

One method of starting a very cold engine or tractor is to pour one cup of high test gasoline and one-fourth cup of alcohol into a bottle. A few drops of this mixture, poured into the priming cups or into the carburetor air intake usually wakes up a seemingly dead engine. Most auto supply stores have similar cold weather starting materials on sale at moderate prices.—I.W.D.

Repairing Pitchfork

To put a new handle on a three-tine pitchfork, slip a piece of pipe over the centre tine, then with a hammer drive the fork into the handle. This avoids the danger of bending or breaking the fork and insures its being driven in straight.—I.W.D.

Bridle Hooks

Do not throw broken shovel handles away. They can be turned into some-

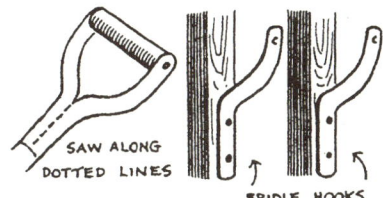

thing useful. A shovel handle of the type shown here makes two good hooks for bridles or straps in the horse stable. Saw along the dotted line and nail to the wall.—Paul Tremblay, St. Paul, Alta.

New Use for Hydraulic Jack

The hydraulic bumper jack makes one of the easiest and quickest post-pullers to be found on the farm. Here we see a stubborn corner post set three feet deep being pulled with ease. All that is needed is the jack, a halter chain with snaps, and a piece of flat iron long and wide enough to prevent the base of the jack sinking into the ground. Try it the next time instead of starting up the tractor and lugging around a heavy chain and post-puller.—J. L. Strang, Flowing Well Farm, Claresholm, Alta.

Folding Door

In a kitchen which is small and the door to the basement entrance opens against the wall where the oil stove was placed, the door was sawed in half and the two parts hinged as shown. They could then be folded together so as to occupy much less space than the single door. Three heavy butt hinges were used for this purpose, as the door will stay in shape much longer than with only two.—I.W.D.

WE HAVE WHAT YOU NEED

in

Belting (Leather and Rubber)
Briggs Oil Clarifiers
Cellar Drainers
Compressors
Contractor's Equipment
Domestic Water Systems
Electric Motors (A.C. and D.C.)
Engines, Gas and Diesel
Fire-Fighting Equipment
Generators (A.C. and D.C.)
Grain Cleaning Equipment
Jacks (Screw and Hydraulic)
Lighting Plants
Power Tools
Pumps
Saws
Saw Mill Machinery
Thermometers
Wire Rope
Woodworking Machinery

●

Write for Prices and Particulars

Mumford Medland Limited

Winnipeg, Man.
and Saskatoon, Sask.

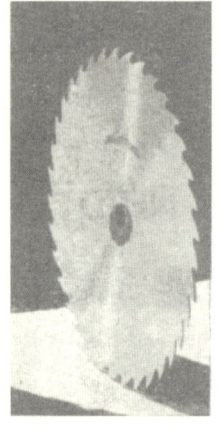

RED CENTER Circular Saws

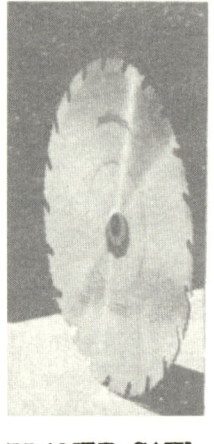

PLANER SAW For Smooth Cutting

SIMONDS
CANADA SAW CO. LTD.

SAWS *and* FILES

RED TANG FILES
For machine shop and saw filing.

BAND SAWS
For cutting wood and metal.

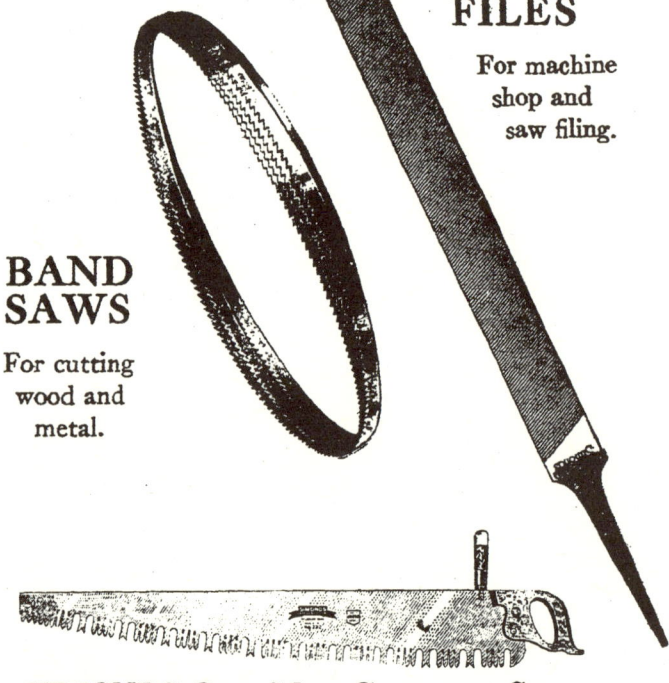

SIMONDS One-Man Cross-cut Saws
Made in various styles and lengths for general farm sawing.

- Circular and Band Saws for your shop.
- One and two-man Cross-cut Saws for the woods.
- Hack Saw Blades and Files for general use.

Get these first-quality, long-wearing Saws and Files from your store.

Simonds Canada Saw Co., Ltd.

595 St. Remi St., MONTREAL, Que.
1550 Dundas St. W., TORONTO, Ont.
42 Water Street, SAINT JOHN, N.B.
554 Beatty Street, VANCOUVER, B.C.

RED END Hack Saw Blades

CRESCENT GROUND CROSS-CUT SAWS

Improved Clothes Hooks

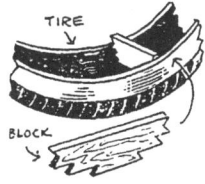

I was unable to buy the wire kind so I used old fashioned clothes pins. I took a piece of 1x4 and bored the holes the right size. Then the pins were driven in the proper distance and cut off flush with the back of the board. Wedges held the pins in position without glue or screws. The pins will not come through knitted goods as a wire hook will do.—A. S. Wurz, Rockyford, Alta.

Tire Spreader

This simple device or something like it is widely used and saves considerable trouble when making minor repairs to a tire. Simply take a bit of inch stuff and cut the notches in it to give spreads of three, five and seven inches.—J. B. Glenn, Winnipeg.

Holds Tops on Stacks

I have found this plan to save a lot of trouble and hay. It keeps the fall gales from blowing the tops off stacks. I save the old twine from my green feed

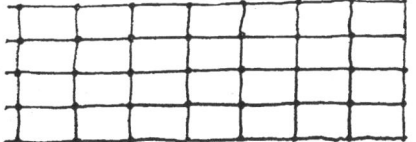

bundles for the job. Just tie them together again and make the net. It should be four feet longer than the stack, so that two feet of it hangs down over each end. A rail is hung on each side. It does not make channels down the side of a stack as a pole does if left for some time.—Horace H. Clarke, Evansburgh, Alta.

Supports Log While Hewing

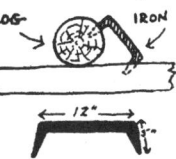

It is difficult while hewing small logs to make them stay put, but not when you use this simple device. Two short bits of old iron are sharpened at the ends and then bent as shown. Then attach them to the log and the skids, both on the side way from the hewer. The log will stay put all right.—Ralph Boesem, Fisherton, Man.

Boys' Pencil Sharpener

First take a piece of wood about ⅜ of an inch thick, 5 inches long and 1½ inches wide. Shape the handle and round off the corners. Then cut out a piece of sand paper and glue it on the holder. By

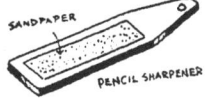

holding the holder in one hand and rubbing the pencil on the sandpaper with the other you can make a nice pointed lead. — Ernest Lavrolette, Box 208, Kamloops, Alta.

Starting a Screw Nail

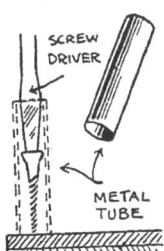

A short piece of metal tubing will help to start screw nails if it is placed around the screw and the driving end of the screwdriver. It will prevent the screwdriver from slipping off the slot in the head of the screw nail and will hold the screw plumb. The tubing is held with the fingers of one hand while the other manipulates the screwdriver. A piece of tubing of the proper size can be made from an empty rifle shell if sawed at the proper places with a hack saw.—Paul Tremblay, St. Paul, Alta.

Preserve Manure Spreader

Here's another use for used transmission oil drained from your car. Coat the chains on your manure spreader with it. They will work much more smoothly and lengthen the life of the chains.

File Pipe Wrench

This shows how a monkey wrench and a file will hold a pipe when you do not have a pipe wrench at hand. Place the end of a

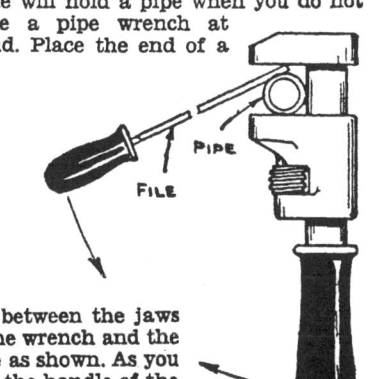

file between the jaws of the wrench and the pipe as shown. As you pull the handle of the wrench in a clockwise direction pull on the handle of the file in the opposite direction as indicated. This will tighten the grip of both these tools on the pipe.—I.W.D.

Handy File Handles

Most farm shops are sadly lacking in decent file handles. This shows one which can be made in 10 minutes and which will fit any file and last a lifetime. Cut off a piece of broom handle about 5½ inches long, hacksaw a ferrule from a half-inch pipe or light tubing, trim one end of the handle down, and drive the ferrule on tightly. Next bore a 3-16-inch hole about ¾-inch deep, a ⅛-inch hole about ¾-inch further, and then round off the other end so it will be comfortable to the

---

When Buying TOOLS for the

FARM - HOME or WORKSHOP

consult your local

Hardware Dealer

His experience will guide you right in getting the tools, hardware or paints to do a better job.

"DIAMOND A"
ABSOLUTELY PURE
PAINTS

Will protect and preserve your property efficiently and economically.

"Diamond A" House Paint

for interior and exterior surfaces

"Diamond A" Wagon, Truck and Tractor Paint

for your Carts, Trucks and Implements.

"Diamond A" Barn Paint

for all your Outbuildings.

See your "Diamond A" Paint dealer. Tell him of the job you have in mind . . . he will gladly help you . . . Color Cards are available.

We are also distributors of

Fine Varnishes and Self-Smoothing Enamels

The J. H. Ashdown Hardware Co. Limited

WINNIPEG
REGINA - SASKATOON - CALGARY
EDMONTON

MONARCH MACHINERY

DESIGNERS
MANUFACTURERS
DISTRIBUTORS

of

Monarch

GRAIN GRINDERS
6 Models
1. Speedy 8".
2. Speedy 10".
3. Heavy Duty 10".
4. Model 210 10".
5. Model 212 12".
6. Model 310 10".

Monarch

GRAIN ELEVATORS
2 types
16, 18 and 20-ft. sizes.
1. Portable Wooden
2. Portable All-Steel.

Monarch

PUMPS
All Types
1. Lift Well Pumps and Cylinders.
2. Barrel Pumps.
3. Semi-Rotary Pumps.
4. Centrifugal Pumps.
5. Self-priming Pumps.
6. Turbine Pumps.
7. Water Pressure Systems.

Distributors of
CLIMAX WASHING MACHINES
Gasoline and Electric

Producers of
GREY IRON, BRASS
AND
ALUMINUM CASTINGS

For information on the above and other farm equipment, write, phone or wire

MONARCH MACHINERY
COMPANY LIMITED
WINNIPEG, MANITOBA

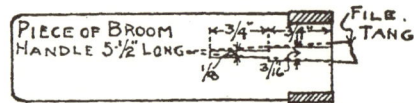

hand. Serviceable handles can even be made from a piece of corn cob if one end is trimmed down and a ferrule screwed on, made from the bottom shell of an electric socket or the outer shell of a garden hose coupling.—I. W. Dickerson.

Protecting Scoop Shovel

To protect the grain shovel from catching on nails on a rough floor when shovelling grain or coal, make three runners to keep the edge slightly

above the surface. These are ordinary fence staples. Simply put them on the edge and hammer them flat.—A. S. Wurz, jr., Rockyford, Alta.

Lace Leather Cutter

To start with, all you need to make this gadget is a jack knife. Hunt up a bit of stick and trim it as shown, one notch being to cut belt lacing the proper width

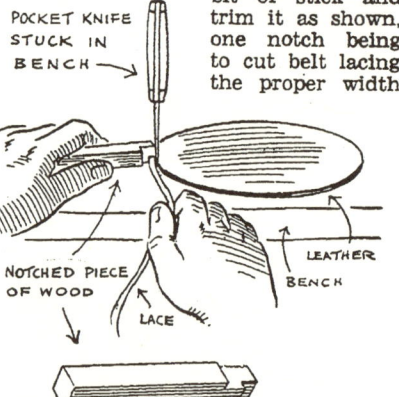

and the other to cut shoe lacing. Then stick the jack knife in the bench and go to work on the leather, which has been trimmed approximately round. You will soon get on to the knack of making laces as fast as you can count them.

Use of Monkey Wrench

When starting a nut with a monkey wrench never hold the wrench with the jaws pointing away from you. Always have them facing you. It is not a good practice to use a monkey wrench on a six-sided nut. It causes an unnecessary strain on the wrench and soon

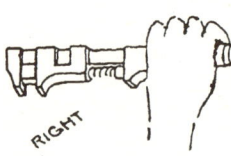

rounds off the corners of the nut. Neither does it do a monkey wrench any good to use it as a hammer. Better use a real hammer.

Loosening Tight Nuts

When carriage bolts become rusted it is difficult to remove the nuts. Here are

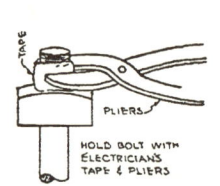 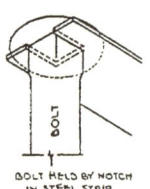

two simple suggestions to solve the difficulty in a hurry. A few drops of penetrating oil or kerosene applied and let stand a few minutes will make the job easier.

Gasoline Screen

This is a system for screening gas which I made on the end of a gas pump hose. It works very well and is easily cleaned by turning off the end fitting and blowing dirt out of the screen. If ¾-inch pipe fittings are used, then it will be easy to drain gas out of the hose into the barrel since the fitting can be put into the small bung hole of the barrel.—Cecil W. Tuininga, Neerlandia, Alta.

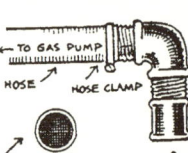

A Trap That's Always Set

Use a pail or the bottom half of a square oil can. Grease the round can well and roll it in crumbs, wheat or seeds until it is well coated. In the bottom put a few inches of water. The mice jump from the board to the can, which turns around and dumps them into the water.—Mrs. Velma Sanders, Balfour, B.C.

A Handy Sanding Block

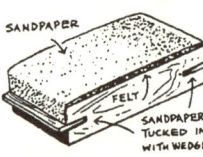

This arrangement avoids the use of the fingers and thumb to hold the sandpaper to the block. A piece of 2x4 is used and slots made in the end to take two thin wedges. The corners of the block are rounded. Thin felt glued on, to go under the sand paper will improve the working quality.

Weed Spud

Our new garden spot was badly infested with poison ivy, Canada thistle, and dock, and to get the roots out completely a spud 10 inches long was made to fit into an old spade handle. The cutting is W-shaped as shown, which keeps tough weed roots from slipping around the point. A broad spring leaf from a heavy duty truck makes a good spud if it is thickened and reinforced near the handle to give it good strength there. It is fine to remove sour dock or other perennials with long tap roots. — I.W.D.

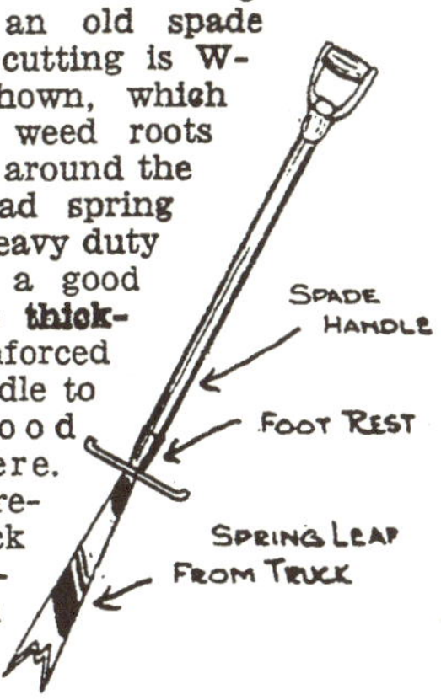

Two Ideas in One

This sketch illustrates the idea of putting a tube on a drill to gauge the depth of the hole and also a flattened drill to level the bottom of holes bored for making a key seat. To make the key seat, first centre punch the holes carefully, being sure they are in line with the shaft and right distances apart. Then drill the holes to the right

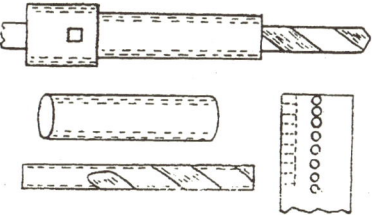

depth. It is best to drill every second hole first, then the ones in between. Then flatten the bottom, and with a sharp cold chisel, the sides of the key seat can be easily finished. I generally use a drill 1-16 inch larger than the width of the key which makes it still easier to finish. — James E. Moscrip, Major, Sask.

A Home-made Light Hoe

A light hoe can be made from an old manure fork. It can be either the usual garden hoe or a push hoe. The tines are heated, cut to length and shaped and then allowed to cool slowly. They can be split with a hack saw or drilled. The blade as made from an old scythe, or a piece of the blade of an old buck saw serves the purpose. It is either riveted on or forced into the hack saw cut.— Charles Richardson, Douglas, Man.

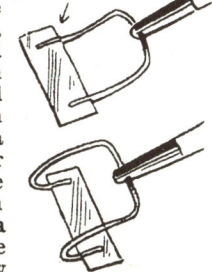

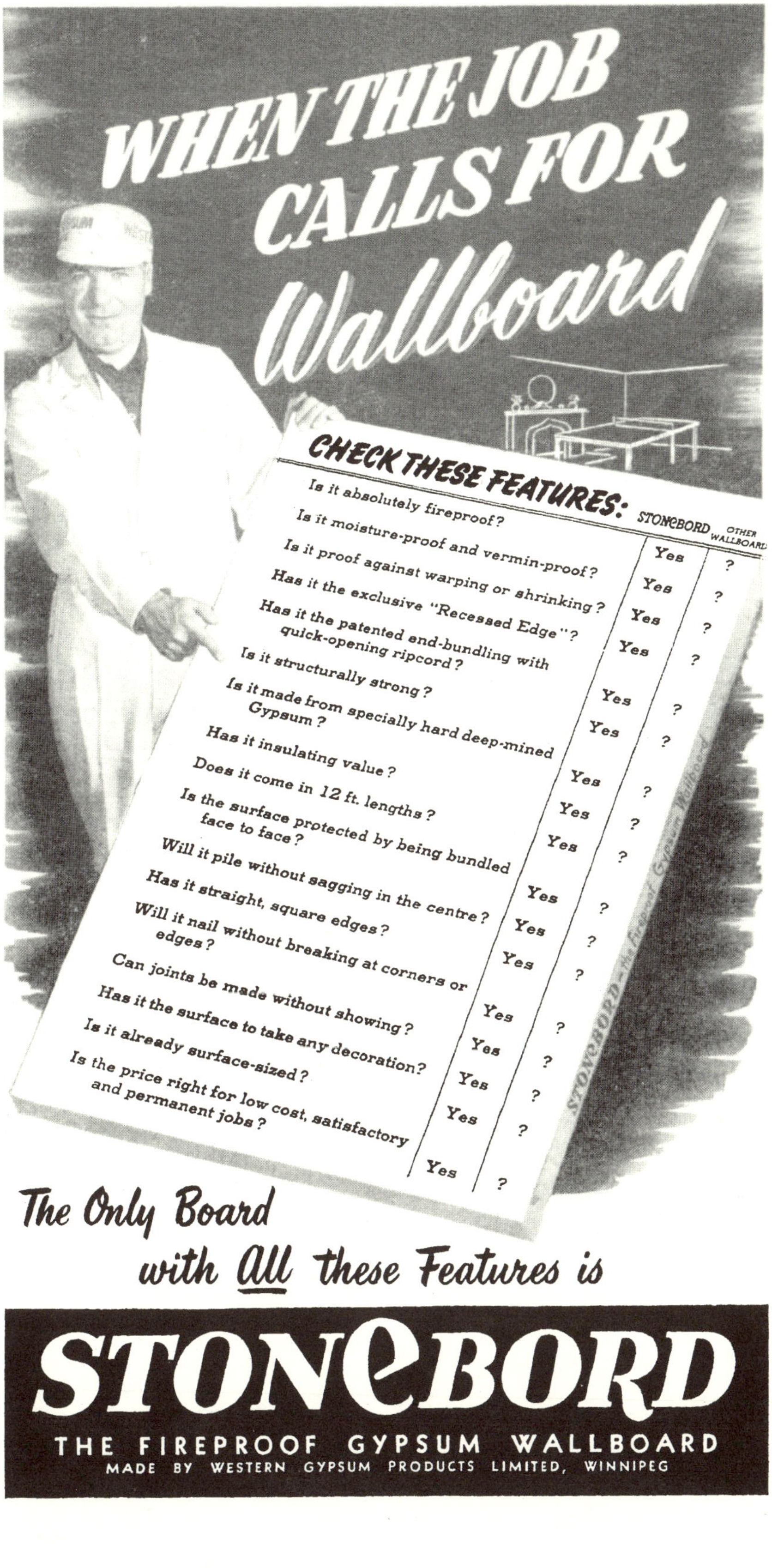

WHEN THE JOB CALLS FOR Wallboard

CHECK THESE FEATURES:

	STONEBORD	OTHER WALLBOARD
Is it absolutely fireproof?	Yes	?
Is it moisture-proof and vermin-proof?	Yes	?
Is it proof against warping or shrinking?	Yes	?
Has it the exclusive "Recessed Edge"?	Yes	?
Has it the patented end-bundling with quick-opening ripcord?	Yes	?
Is it structurally strong?	Yes	?
Is it made from specially hard deep-mined Gypsum?	Yes	?
Has it insulating value?	Yes	?
Does it come in 12 ft. lengths?	Yes	?
Is the surface protected by being bundled face to face?	Yes	?
Will it pile without sagging in the centre?	Yes	?
Has it straight, square edges?	Yes	?
Will it nail without breaking at corners or edges?	Yes	?
Can joints be made without showing?	Yes	?
Has it the surface to take any decoration?	Yes	?
Is it already surface-sized?	Yes	?
Is the price right for low cost, satisfactory and permanent jobs?	Yes	?

The Only Board with All these Features is

STONEBORD
THE FIREPROOF GYPSUM WALLBOARD
MADE BY WESTERN GYPSUM PRODUCTS LIMITED, WINNIPEG

SECTION 9.
Concerning Livestock

Tank Heater
I made a tank heater from an old 8-gallon cream can and it has proved to be very satisfactory. A 7-inch stovepipe fits snugly into the mouth of the can and can easily be removed when more fuel is needed. A hole is cut as shown and a piece of old well pipe is inserted to provide the draft. It comes to within about four inches of the bottom. I burn wood, with a few pieces of coal placed on the fire in the evening to keep it going till morning. I am very well pleased with the service I am getting from this heater.—Harlan Petersen, Dickson, Alta.

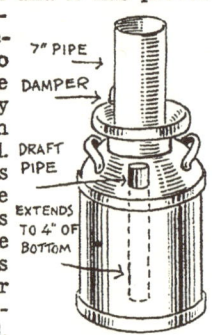

Breeding Rack
This breeding rack for cattle has been found to give satisfactory service. It is strong and adjustable to different lengths of animals. It is not difficult to build. In case the figures are a little hard to decipher they will be repeated.

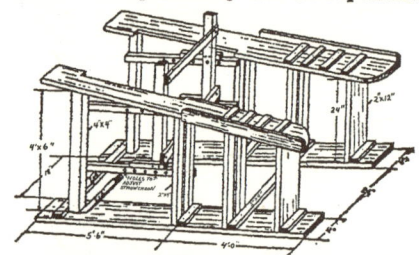

The sills are divided as shown, 4 feet and 5 feet 6 inches and 9 feet 6 inches long and are of 2x12. The front legs are 24 inches high and are also of 2x12. The rear legs are 4x4 and 4 feet 6 inches high. The clearance between sills is 24 inches in front and 18 inches at the rear. The holes to adjust the stauncheon are shown. The upright supports and the cross pieces are of 2x4 stuff.—I.W.D.

Protects Mash Feeder
Any size or length of trough can be used. Drive four stakes into the ground, depending on the width and length of the trough. Then nail cross pieces as shown for the trough to rest on about 12 inches from the ground and extending about 12 inches out on each side of the trough. Then put a 1x4 along each side for the fowls to stand on while eating. Then cut fairly heavy wire into about 24-inch lengths, bend them into a U-shape and tap them into holes drilled about two to three inches apart in the edges of the trough. If desired, it can be made portable by nailing and bracing the legs and cross pieces to the trough itself.—I.W.D.

Dehorning and Branding Chute
The chute shown is recommended and the construction is clearly indicated. Round posts are most often used

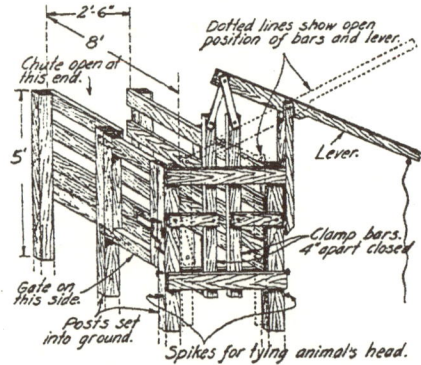

and poles and slabs may be used for the sides, provided they are free of knots and splinters. The gate shown at the side is for castrating, branding, etc.

Oil Can Grit Feeder
The diagram shows how to solve the problem of feeding oyster shells and minerals to poultry with no cash outlay. Use a discarded oil can, cut and bent as shown. When completed, this can is nailed against the wall just high enough

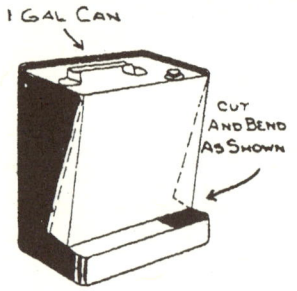

so a chicken can reach in. Several of these may be used to feed the different minerals and grits. It can be made much more convenient by cutting all around the can about half an inch below the top to make a removable lid for easy filling. The top could be made to fit over the can either by slitting the corners of the top and spreading it slightly or by pinching in the corners of the can. Cut and force in only about four or five inches of the side.

A Cow Poke
To make a satisfactory cow poke take a piece of iron 36 inches long, and for the cross piece, a piece of iron of the dimensions shown. The length from the nose ring to the cross bar is 15 inches, which leaves the top end the longest. The top is bent forward somewhat. The lower end is doubled over and riveted to hold the nose ring. The rings which go over the horns are held by holes in the cross bar. The ring is 3½ inches in diameter

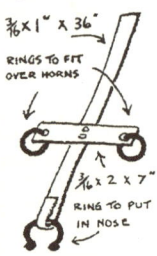

and fitted into the nose but not through it. Fit the poke loosely on the head so as not to injure the nostrils.—H. G. Classen, Box 2, Aberdeen, Sask.

Stop Car Chasing
The diagram shows how to break a young dog from chasing cars and chickens. Put a strap around his neck and to it fasten a light chain just long enough to reach his hind legs with a large ring three or four inches in diameter fastened at the end. When he

runs after cars or chickens, he steps in the ring and falls head over heels. This soon cures him of the habit and he gives no more trouble.—I.W.D.

Animal Catcher
Here is a device for catching sheep, calves, pigs, etc. It consists of a wood stick or small pipe about four feet long, with a rope or sash cord five or six feet

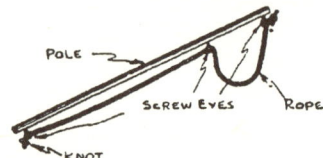

long fastened to it at each end and through a heavy screw eye about a foot from one end. Throw the loop over his head and pull up the slack and you have him securely. This is especially handy for use in the barn or stalls at marking, shearing, vaccinating, etc.

Shelter for Salting Trough
So that his sheep could have access to salt at all times, and to prevent waste, one farmer con-

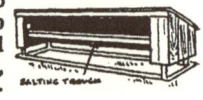

structed this salting trough shelter. It is just high enough to let the sheep at the trough which is sheltered from rain or snow. The size of the shelter can be varied according to the size of the flock. A small one will do on most western farms.

Hoof-trimming Rack

Here is a diagram of a handy rack for trimming bull and cattle hoofs. The only materials needed are four posts 8½x9 feet long, five eight-foot 2x6's, a short post, a barbed wire stretcher, and a sling. The posts are set solidly in the ground about four feet apart and the stall constructed as shown.

To lift the hind foot, hang the wire stretcher from the beam above the animal's hind quarters. Take a heavy strap with a ring in each end (the breast strap from work harness will do) and

put a sling around the animal's leg just above the hock, hooking the rings on the lower end of the wire stretcher. Now when you pull up on the stretcher, the foot is brought into position where you can do a good job of trimming it with a pair of hoof nippers. The animal may kick a little, but will soon quiet down when it finds it does no good.

To lift the front foot, attach the sling to the post across the top of the stall and around the fore leg just below the knee, and lift as before with the fence stretcher.—I.W.D.

Pig Catcher

Ever make a pig catcher? This one is made from an old fork handle and bit of wire rope—that is what a clothes line is. The rope is anchored to the nut end of a ¼-inch eye bolt and runs through three eye bolts on the other side of the handle. At the upper end of the handle a ¼-inch bolt, long enough to take 1½ inches of ½-inch pipe on each side of the handle, is used to wrap the other end of the clothesline on when you have snared your pig.

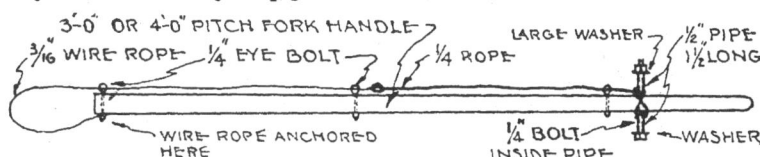

Holds Dehorning Pencil

Instead of using an old glove when applying the caustic potash dehorning pencil you will find the following method

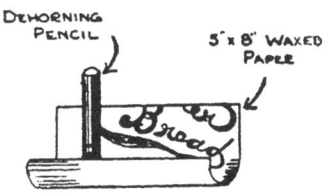

much better. Cut a strip 5 x 8 inches from waxed paper, such as that used in wrapping bread. Lay pencil on strip so bottom edge can be turned up a little, then roll or wrap around the pencil. In this way one does not get anything on his hands and no gloves are needed. When done, slip pencil into bottle and burn paper.—I.W.D.

Safety Device

Stock often acquire a habit of nosing a gate hook open. The habit can be effectively cured by attaching a metal

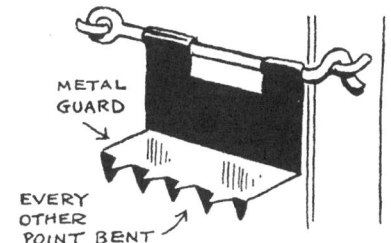

guard on the hook as shown. The bottom edge of the metal is serrated and every other point bent down.—A. S. Wurz, jr., Rockyford, Alta.

Guards Poultry Fountain

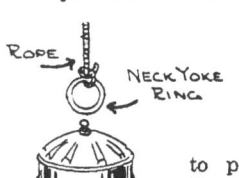

If you have trouble with chickens roosting on your water fountain and getting droppings in the water, rig up this device to prevent it. Take the centre ring from an old neckyoke and hang it about four inches directly above the cover. When chickens try to fly up on the cover, the weight will knock them off. This will save a lot of time and trouble in cleaning the fountain.

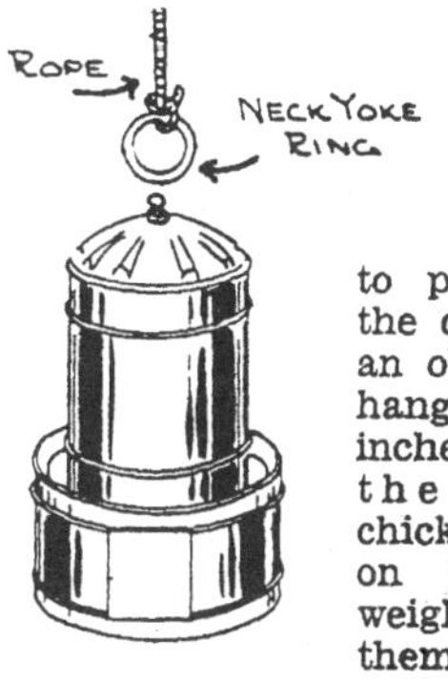

BELTING for every purpose ... ELECTRIC MOTORS

- Pulleys Shafting Saw Mandrels Saws
- Vee Pulleys and Vee Belts Hangers
- Contractors' and Mining Equipment
- Pumps Rubber Hose

ENDLESS THRESHER RUBBER BELTS

Length Feet	Width	Ply	Net Price
30	6	4	$16.65
40	6	4	22.50
50	6	4	27.70
75	6	4	38.00
100	6	4	50.45
120	6	4	60.50
100	6	5	63.70
120	6	5	75.65
50	7	4	32.80
75	7	4	46.50
100	7	4	59.45
120	7	4	70.50
140	7	4	82.15
150	7	4	88.00
100	7	5	74.40
120	7	5	88.60
100	8	4	64.10
120	8	4	76.20
120	8	5	95.25
150	8	5	118.70

SAFETY BELT HOOKS		STEEL HINGE LACING	
	Per Box		Per Box
No. 2	$1.79	No. 1	$1.65
No. 3½	1.79	No. 2	1.65
No. 4½	1.90	No. 3	3.30
No. 5	2.11	No. 4	3.52
No. 6	2.59	No. 5 (long)	2.09
		No. 6	2.97
		No. 7	3.30

FRICTION RUBBER BELTING TRANSMISSION BELTING

Width Inch	3-Ply	4-Ply	5-Ply
1	.08	.10	----
1¼	.11	.12	----
1½	.13	.15	.18
1¾	.15	.17	.22
2	.16	.19	.24
2½	.20	.23	.29
3	.23	.27	.33
3½	.27	.31	.39
4	.29	.34	.42
4½	.32	.38	.47
5	.36	.42	.52
6	.43	.50	.63
7	.50	.59	.73
8	.54	.63	.79

ENDLESS HAMMERMILL BELTS

Feet	Inch	Ply	Price
50	6	4	$33.35
75	6	4	46.45
50	7	4	39.05
75	7	4	54.40

N. SMITH Belting Works Ltd.

Established 1886

138-140-142 York St., Toronto 2, Can.

Phones: Adelaide 1437, 1438 - Kingsdale 1379

Hog Scalder

Joe Maze, of Khedive, Sask., has an improvement over a barrel for scalding hogs. The trouble with a big hog, he says, is that when you try to pull it out of the barrel the barrel comes with it. He took a 45 or 50-gallon drum, and cut it in two lengthwise by starting at the four seams with a hack saw and then taking an old hatchet to complete the cutting. One end of the drum was removed and the two halves riveted together with a four-inch lap, using sickle rivets and packing the joint with asbestos. Then he took the angle iron from an old bedstead, cut it into the proper lengths and formed a top. Two pieces of 2x4s, a foot longer than the trough, formed grip handles for carrying it.

When scalding a hog he doubles a log chain, lays it across the trough and rolls the hog on top of it. One man stands on each side of the trough, pulling alternately on the chain. The hog rolls in the hot water. Two men can easily handle the heaviest hog.

Novel Fountain Support

We built a new poultry house recently and have been looking for good ideas for a fountain stand. We built one like that in your handy hints column, and the idea was good but it took up too much floor room. We finally worked out the plan shown by burying a nail keg about half way in the ground with the

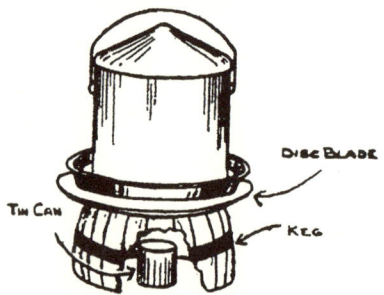

upper end open. A gallon can is set inside of the keg, a discard disc blade set concave side up on top, and then the fountain on top. Any water spilled out runs down through the centre hole into the can and does not get on the litter. When the can is full, it can be emptied. The keg can be sawed into two bases for use on concrete floors.

Magpie Trap

THE framework is made of 2x4's and 2x2's, with 1x8 boards around the bottom. All sides and the top sections are covered with wire cloth or small mesh poultry netting. A section on one side is hinged so it can be opened easily for entering and killing the pests and liberating any desirable species. The entrance is about three feet from the ground between the sloping top sections, being a slot about four inches wide extending across the six-foot width. On it are nailed several crosspieces on which

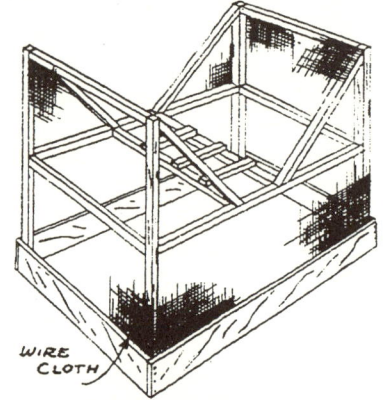

the magpies can rest before dropping to the bait below. Entrails or exposed meat make the best bait, and the trapped birds fluttering around trying to escape attract other birds. Even crows might be caught if one side is hinged up until they get used to feeding inside and then closed with a long wire.—I. W. DICKERSON.

A Loading Chute

Here is a sketch of a cattle and hog loading chute which explains itself. Now when all scrap iron and wheels are shipped away for scrap this chute is mounted on skids.—J. H. Holman, Lougheed, Alta.

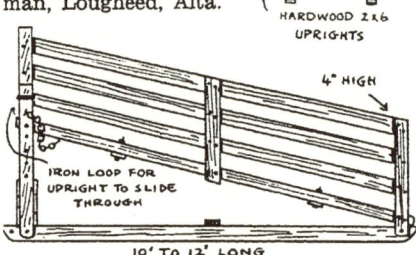

Anti-Jumper

Here is a simple but very effective anti-jumping device which keeps the

cows in the pasture when everything else fails. Take a light pole four or five feet long, and bore a hole through it seven or eight inches from one end. Put one end of a tie chain through the hole and the other end around the cow's neck. This is effective even when the fence is almost down.—I.W.D.

Board-Paper Tank Insulation

The diagram shows one method of insulating an outside concrete water tank from freezing. Instead of banking it up a couple of feet thick all around, put one or more layers of heavy waterproof roofing paper on the outside of the concrete, then build a wooden box on the outside of the layers of paper. The boards can be nailed to each other as shown at the corners, or better, a 2x4 placed at each corner and the boards

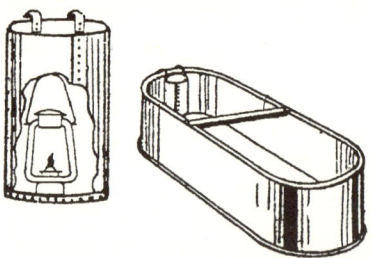

nailed to it, as the air space helps give insulating value.

The tank heater is placed near the middle of the tank with the pipe coming up through a sheet iron cover to avoid danger of fire. The top lids at each end are covered similar to the sides, one end being hinged to lift up so the stock can drink. This also keeps the water cooler in summer and will last for years.

Lantern to Heat Stock Water

Here is a very convenient and satisfactory way for keeping ice out of a tank in a barn where cattle or sheep are run during the winter.

The small tank helps to keep the temperature of the water higher. Also bank the tank with six or eight inches of manure and have a cover to drop over the tank during extremely cold weather. Take an old can about eight inches in diameter and 18 inches long, remove

one end, and have hooks attached to the open end so it will hang inside the water tank. It is necessary to weight this open tank down. A lighted lantern set in this open tank will help keep ice from forming on the water, with the help of the cover in very cold weather, but provision would have to be made for the fumes to escape. A lantern will burn from 24 to 36 hours, so the cost isn't much.—I.W.D.

Combination Inside Self-feeder

This is a design of a combination inside self-feeder for hogs. The top illustration shows one-half of a cross-section of the building, which is 16 feet wide. Both sides, of course, are constructed the same. The hopper capacity on each side is 750 cubic feet, or 1,500 cubic feet in all, holding about 1,200 bushels. The outside studs are cut seven feet long. Note the manhole at the top for filling the hopper. Also the detailed drawing at the right, showing how the regulating board is adjusted. There is a row of studs down the centre, which is boarded up three feet to form a partition. The measurements for the diffeernt parts are given.

The lower illustration shows the ground floor. Note that the troughs are along the sides. At one end is a water tank, with a tank heater for winter use. An automatic cut-off keeps the water trough from overflowing. At the other end is the main door in the middle, swinging outward. On each side of it is a pivoted flap which automatically closes, but which the pigs soon learn to open. It swings freely both ways. A swinging gate gives access by the attendant to each side of the pen when cleaning it out, etc.

Light is admitted by two windows, one over each pivoted flap, and another sash in the upper part of the main door. Two small sash can be fitted between the studs at the other end for additional light. The feeding troughs are 14 feet long in an 18-foot building.

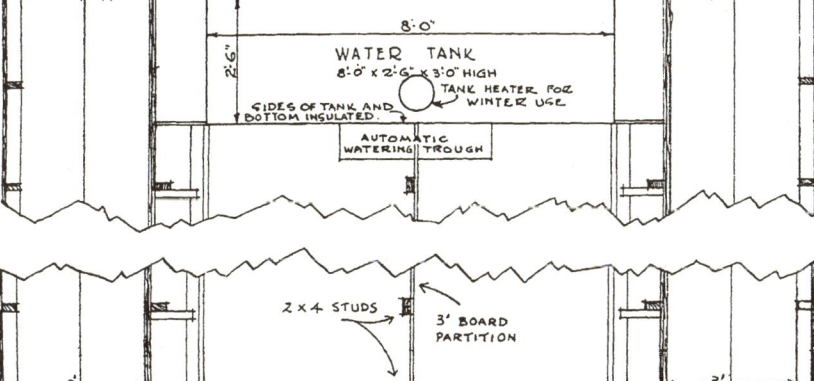

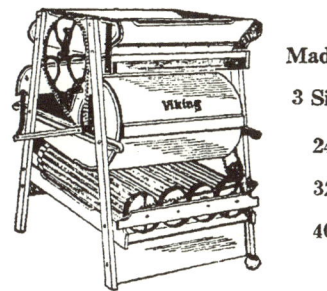

Canada's Largest Manufacturers of Grain Cleaning Equipment

We can help you with your grain cleaning problems by advising the proper equipment to use, or the mill most suitable for your individual cleaning requirements. This is a FREE SERVICE. Circulars on request.

THE VIKING FANNING MILL
3 MILLS IN 1

Built to fit the need of the Farmer who is particular.

Made in 3 Sizes
24"
32"
40"

Also

BULL DOG Fanning Mills—
BULL DOG Smut Machines—
BULL DOG Wild Oat Separators
—CARTER Disc Cleaners —
VIKING Combination Mills —
HART Corn Graders.

Now...
Complete Stocks of Repairs

HART Perfection Low Elevators for Threshers — HART and GARDEN CITY Feeder and Weigher Repairs—Wire Cloth and Perforated Zinc, all Meshes, for Fanning Mills and Thresher Screens.

Write for full Information.

GENUINE ORIGINAL PARTS

THE HART-EMERSON CO. LTD.
WINNIPEG, MAN.   SASKATOON, SASK.

Field Self-feeder for Hogs

When 10 or more pigs are kept the labor-saving feature of a self-feeder merits attention, says the Alberta department of agriculture in explaining this easily constructed self-feeder for hogs. It saves 60 per cent of the labor, reduces waste and gives the weaker pigs a chance, though it is best to keep hogs of the same size together. Trough space is saved and since grain is before the hogs all the time, digestive trouble from overloading the stomach is reduced.

The illustration gives most of the construction details. The skids are 10 feet long and the sills are placed three feet apart. The main floor is covered with matched lumber and the V-shaped part is best covered with galvanized iron to make the grain slide down more readily. The 2x2 braces are a foot apart and are let into the front side of the trough. For outside use lids can be hinged over the trough to keep out the rain and prevent the dry feed from blowing. Hogs soon learn to lift the lids. A separate lid is made for each one-foot section.

Sides, ends and roof are covered with shiplap and the roof covered with ready roofing. A canvas strip nailed over the ridge prevents rain from entering. The overhang of the roof is to prevent rain from dripping into the trough.

A feature of the trough is the adjustable slide. It consists of a 6-inch board, held by flexible wooden slats ⅜"x1½"x28". They are nailed firmly to the slide board as shown. The slots at the top are in the wall of the feeder and provide for the adjustment of the slide board. A feeder six feet long will accommodate 24 pigs.

The lower cuts show how the feed capacity of the self-feeder can be increased. With the extension shown on the left a six-foot feeder will hold approximately three-quarters of a triple wagon-box load of ground grain. For inside use 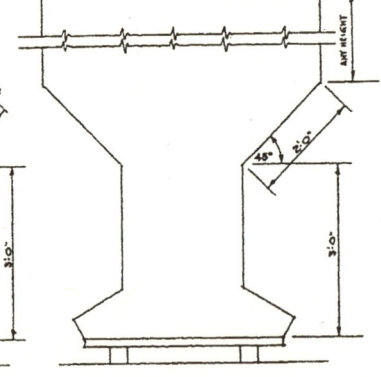 the extension shown on the right can be carried to any convenient height.

Water Tank Cover

A small shack of old lumber can be built over and around a tank to hold in 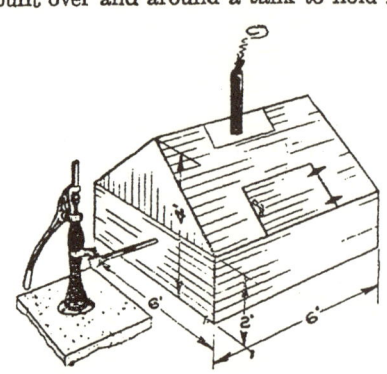 the heat from the tank heater. There are two openings for the stock to drink from, one on each side of the roof, as the tank is half in one lot and half in another, with covers which can be laid back on the roof during the mild part of the day. There will be no trouble with ice in the tank, and it is light enough so two men can lift it off in the spring and put it back in the fall.—I.W.D.

Grain Self-feeder for Cattle

An A-1 self-feeder for cattle can be converted from an old granary, according to this plan, also by the Alberta department of agriculture. The building is blocked up 20 to 24 inches above the ground. Leave the first 8-inch board on the bottom of two sides and an end and remove the next three boards above it. This leaves an opening two feet deep.

Next construct baffle boards on two sides and one end of the interior, as shown in this cross-section view. Frame in the struts of 2x4 stuff, one at each stud. It would be safer to have the bottom of the strut come directly above the joist as there is considerable weight on the baffle board. Cover the struts with shiplap as shown and the job is done. The illustration shows the framing and gives an interior view of the back end. Do not be confused about the bottom of the baffleboard at the back. It might be mistaken in the cut for a false floor. There is an opening 4 or 5 inches deep at the bottom of the baffleboard to allow the grain ration to flow within reach of the cattle.

Calf Weaner

From an old car, we got a sheet of strong sheet metal from which we cut a strip nine inches by ten inches. The strip was bent to fit the calf's muzzle. Holes were punched in the tin, with the jagged ends outward, to help the calf breathe and to make it uncomfortable for the mother when it tries to suck. The weaner extends about two inches past the calf's mouth. It is best suited to big, strong calves. This is simply a new use for the old Model T.—Wilfred Brewer, Ashville, Man.

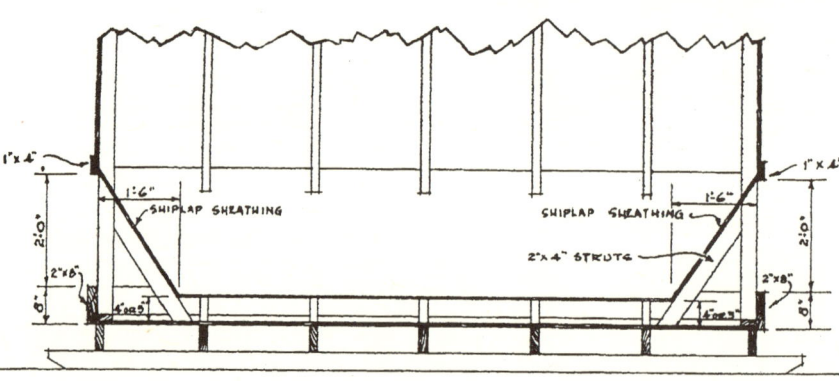

Cow Stanchion

A few pieces of oak or other strong hardwood, two pieces of chain, a couple of bolts and some nails are all you need to make this cow stanchion. It is swung on the chains and is best fastened permanently at both top and bottom. It will hold any cow and last a long time. The measurements are given in the diagram.

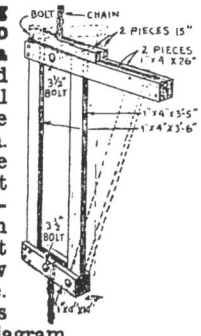

Simple Control Valve

This is a simple way to make a valve to control the flow of water from a tank to a trough. It brings in another use for the well-known inner tube. A bit of

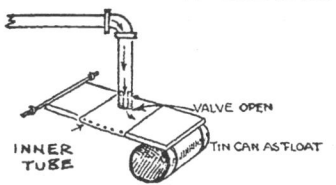

board is hinged as shown. To it is attached an empty tin can of say a gallon capacity. This presses the rubber against the end of the pipe and shuts off the water when the trough is full.

Breaking Sucking Cows

Some cows can be successfully broken of the habit of sucking themselves by putting a horse collar around the neck. Try the treatment for a period; then remove the collar. If the cow attempts to return to her old habit, replace the collar. Repeat until cure is effected or cow indicates a hopeless case.

Better Tank Heater Pipe

With this device an air jacket around the pipe does away with water condensing on the inside and dripping down into the firebox. The inner smoke pipe is a five-foot section of 4-inch soil pipe, which replaces the original smoke stack and fits closely over the lower section. An old cream can cover has a hole cut in it to slip over the lower 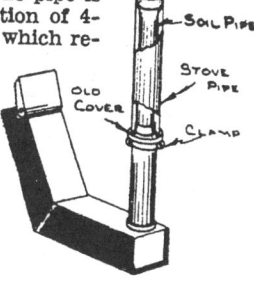 smoke pipe, while a clamp made of heavy strap iron fits around the lower pipe to hold the upper pipe at any desired height by means of bolts through the ends of the collar. Six-inch stove pipe is slipped down outside the soil pipe so as to make practically air-tight and water-tight joints both at the top and bottom. This will need careful fitting, but should last for years.—I.W.D.

Red Head PRODUCTS

DEPENDABLE QUALITY AT LESS COST!

Red Head MOTOR OIL

Refined for Western Canada to meet the needs of farmers, ranchers, commercial vehicle operators. Gives maximum performance and protection at minimum expense. Available in quart and gallon cans ... also bulk in drums.

Red Head BATTERIES

Highest quality materials and construction. Unexcelled for quick power ... dependability. Fibre glass "dual insulation" assures up to 50% longer life! Red Head wet and dry cell batteries are made for every purpose ... carry an Iron-clad guarantee.

Red Head PAINTS

One quality only ... the BEST! Best for barns ... buildings ... furniture ... machinery. Available in a wide range of colors and finishes ... in quart cans or bulk. Red Head paints beautify and protect ... increase the value of your property.

GREAT WEST DISTRIBUTORS LTD.
CALGARY · EDMONTON · LETHBRIDGE · SASKATOON · WINNIPEG

SECTION 10.
Hitches -- Tractor and Horse

Tractor Hitch

A V-shaped device as pictured in the drawing for hitching a binder or other

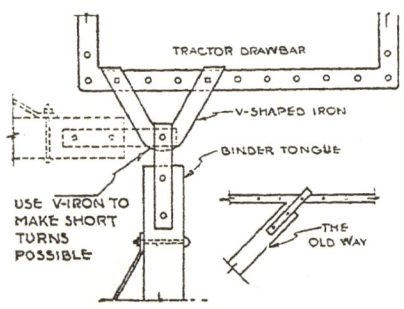

short tongue implements to a tractor will go a long way to prevent breaking the tongue when making short turns in the field. The new hitch allows a short right angle turn. With the old hitch this is not possible without breaking the tongue.

Three-horse Evener

Here are four different ways to make a three-horse evener. Fig. 1-A shows how it can be done by using a short evener under the main evener with the end chained back to the king bolt. Fig. 1-B gives the detail of the short evener as attached to the pole. Fig. 2-A shows another method with straps bolted to the tongue, and Fig. 2-B the general design of the neck yoke equalizer. Fig. 3 utilizes a chain and pulley to give the single horse an even draft with the other two. In Fig. 4 notice that the main evener is in two parts pivoted on the draw bolt, while the draft all comes on the short rear evener. The measurements are all given, including the dimensions of the different eveners.

Three-Implement Hitch

A three-implement hitch developed by the Montana agricultural engineering department. With the present short-

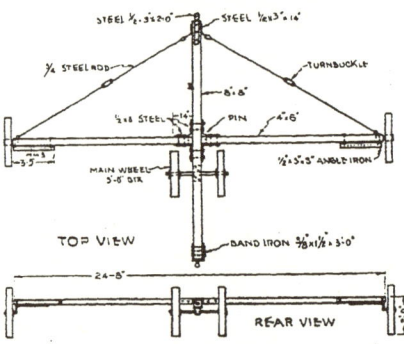

age of manpower, and because of high operating costs, it is important for every farm operator to make sure his tractor is pulling close to its rated load. However, do not overload the tractor and cause unnecessary wear.

Hook-up for Tandem Discs

Hook-up (Fig. 5) can be made by anyone the least bit handy with tools. A shows the forecarriages of the discs, B and C show the left disc and D and E the right disc, F is a logging chain, G an eight-feet oak evener and H another logging chain. The small drawing shows the clevis arrangement at each end of the spreader bar G. Adjustments to get proper alignment are easily made. In coupling up such an assembly three discs should overlap. This avoids ridging to some extent and you are sure of cutting all the land. Care will have to be exercised in turning corners so as not to get the chains caught in the lugs of the

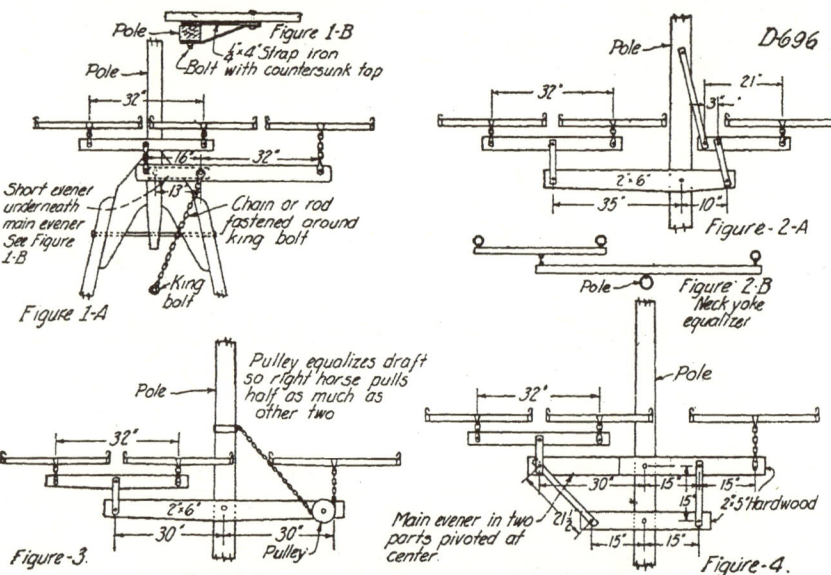

Four different ways of making a three-horse evener.

tractor. Take it all the way through it is a fairly satisfactory outfit for tractors. Some misses will be made at the ends but these can be gone over when the

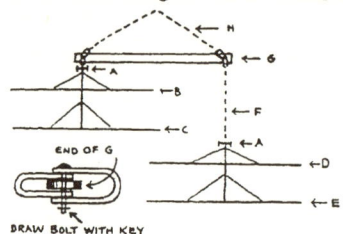

Fig. 5

field has been disced. When working on stony ground this is a satisfactory hook-up because when one disc strikes a stone the equalizer bar will swing and the obstructed disc is drawn over the stone rather gently then both discs swing into parallel position.—George W. Caldwell, Ridgedale, Sask.

This is a tractor hitch for weeders or drag harrows that is the acme of simplicity but which can be used only on quite level land. It was developed in Colorado and is recommended by the Montana

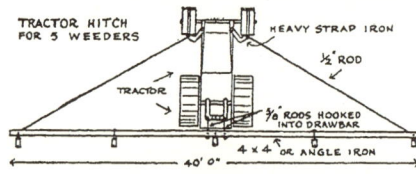

Fig. 6

authorities. A reader at Rouleau, Sask., writes that he has used this hitch successfully with drag harrows. He has a stay rod across from the middle of the cable to the draw bar.

Fig. 7 (above right), is a three-implement hitch, also a Montana design. It is a complete unit in itself and no additional rear extension truck is needed. It is very flexible for use on uneven or rough ground. The turnbuckles are not essential.

Fig. 8 (on next page), is a 10-horse hitch for a one-way from a drawing by Prof. Evan Hardy, of Saskatoon. It is so designed that the centre of the draft is in the centre of the eight-foot implement. The main doubletree is seven feet eight inches long with the clevis in the middle. Attached to it are two doubletrees, 50 inches between clevises, divided 40 inches and 10 inches for one and four-horse draft respectively.

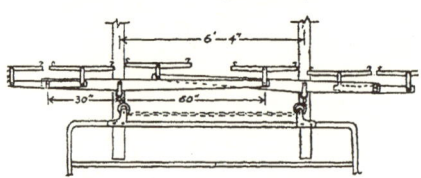

Six Horse Hitch for Drill or Packer

Above is a simple six-horse hitch for a drill or packer. The main doubletrees are seven feet six inches long, divided 30 inches and 60 inches. The chain and pulley attachment is used.

Below is a six-horse tandem hitch for a 14-inch gang. It and the one following are from drawings by Prof. Hardy. In this one the horses are three abreast and the side draft is three inches. Note that the centre of resistance of the plow is 19 inches from the edge of the land while the centre of draft from the horses is 19 plus 3 or 22 inches. The main doubletree is 45 inches net, divided 30 inches and 15 inches, while the singletrees are 30 inches between centres.

The last hitch shown is a seven-horse hitch for a 14-inch triple gang. The centre of resistance is 26 inches from the edge of the land and the side draft is 3½ inches. On the main doubletree four horses pull against three and the division is 30 and 22½ inches. On the outside doubletreee two horses pull against one and the division is 20 and 10 inches. The inside doubletrees are 28 inches between centres of clevis holes.

If you want to haul three drills, have the long timber one-third as long again and hitch the third drill on the left

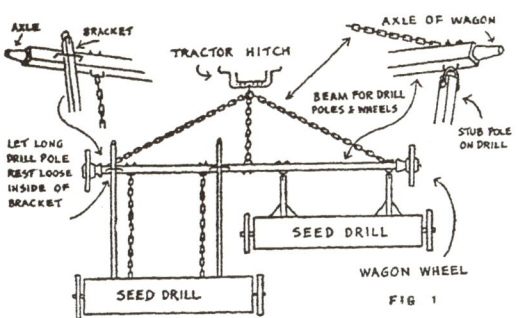

side using long tongues, while the drill with the stub tongues is hitched in the centre. In that case the long timber would have to be strengthened.—J. M. Holman, Lougheed, Alta.

For 28 and 20-Run Drills

This is a hitch (fig. 2) that we have used for several years with complete satisfaction. Our drills are 28 and 20-run and once when seeding breaking we adjusted an ordinary chain from the tractor draw bar to the frame of the front drill to steady it. They can be hitched very close as the wheels may rub the frame or running board on the corners without doing the slightest harm. The hitch is made of 4x6 beams joined by ½x3-inch irons. — Harvey Bros., Flaxcombe, Sask.

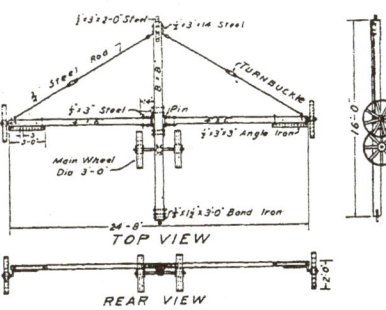

Fig. 7

The response to the request of The Guide for designs of homemade hitches for linking up two or more horse-drawn machines behind a tractor was generous. Many sketches with explanations were received and four of them, of particular value at this season, are published herewith. They have all been put to the test of practical use in the field and have given satisfaction. Other designs will be published in due course, including several binder hitches.

Two-Drill Hitch

Take two wagon wheels and a piece of timber wide enough for two drills (fig. 1). Then bolt the axles of a wagon to each end of the long pole or timber and put on the two wagon wheels. If you have not as many chains as are shown in the sketch you can use a long iron rod with a few links in each end so that the rod will not break.

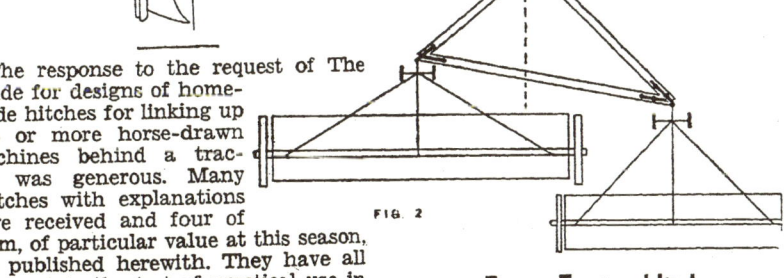

Frame Tractor Hitch

Here is a sketch of a frame tractor hitch I have found to be quite satisfac-

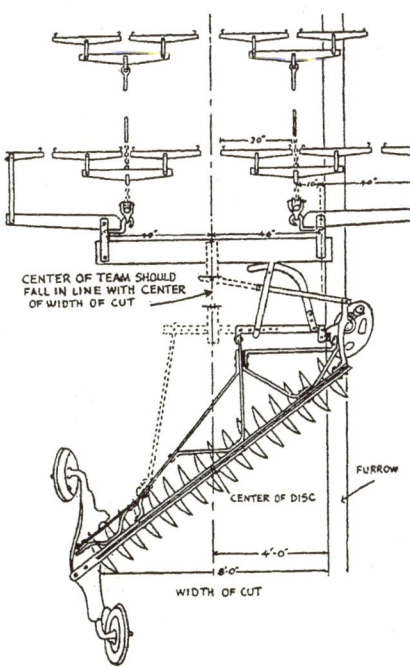

Fig. 8

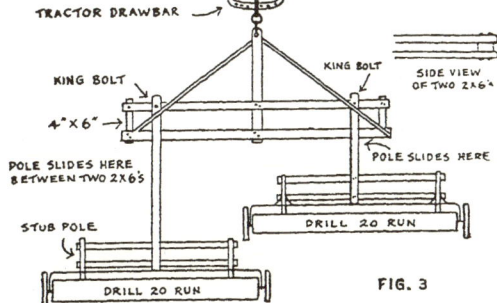

tory and simple to make (fig. 3). No wheels are necessary, but with a little alteration it could be changed into a truck with the addition of two wheels if desired. As shown, the frame work is composed of four 2x6's 14 feet long for crosspieces and pieces of 4x6 each two feet long for the ends. The short pole is a five-foot 4x6, and the braces are two pieces of angle-iron each eight feet nine inches long. The machines are hitched to the frame by means of king bolts on which the tongues pivot. The poles slide between the two rear 2x6's when turning. The frame is attached to the tractor draw-bar by means of a ring and two clevises. An iron step or leg is fastened underneath the front end of the frame pole so that it doesn't have to be raised to couple or uncouple from the tractor draw-bar. This leg should be about eight inches long.—J. R. Duncan, McKenzie Island, Ont.

Drill After Tiller

The accompanying diagram (fig. 4) shows a very successful hitch for drawing a drill after a tiller. The drill is hitched behind the tiller as closely as it will work. Use a stub tongue in the drill about two or three feet long. On top of this put a 2x4 hardwood tongue (birch will do) that reaches to the frame of the tiller. Put a bolt through the 2x4 and the stub tongue so as to form a pivot that will turn easily. Fasten a U-shaped iron over the frame of the tiller so that it will allow the 2x4 to slide freely. Hook a cable from the left corner of the drill frame to the right corner of the tiller frame and another from the right corner of the drill frame to the left corner of the tiller. The drill will then make short corners. The drill we used is the same width as the tiller. We used this hitch last year and it worked perfectly.

The diagram shows how the hitch is applied and from it any farmer will know just how to go about making it. Very little new material is required for

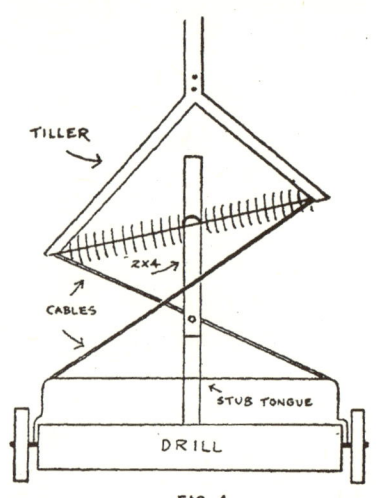

FIG. 4

making the hook-up.—Carl A. Tatroe, Sedgewick, Alta.

Four, Five, Six and Eight-Horse Hitches

The four-horse hitch shown in fig. 1 is for a 14 or 16-inch sulky plow, with the teams strung out tandem and two of the horses walking in the furrow. This makes it easy to handle the horses. The hitch is in the true centre of the draft, eliminating side draft on the plow and reducing the power required.

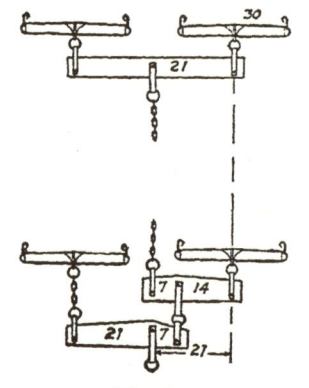

Fig. 1

A 42-inch wagon evener is used in the lead and the chain is long enough so that the wheel team will not bump their knees on the lead evener. The hitch works equally well on a wagon and can be used on a binder. On the wagon, buckstraps and tie chains should not be used. Use lines on both lead and wheel teams.

Fig. 2 shows how to make a simple five-horse abreast evener. The main evener is 78 inches between clevis holes. To it are hitched two eveners 45 inches long, divided nine inches and 36 inches since the middle horse has to pull against the other four. A space of six inches is left between these two eveners. The iron rods are made long to bring the middle horse as far forward as the others.

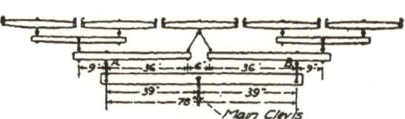

Fig. 2

A five-horse tandem evener is shown in fig. 3. The rear evener is divided six and nine inches, as shown at A. since the lead team pulls against the other three horses. The middle horse of the wheel three pulls against the two outside wheel horses. To equalize the draft between these three horses two pairs of steel bars are bolted on the wooden evener at the outer end as shown at B, but left to swing freely. The division of the bars is nine and 18 inches. The rear whiffletrees are 28 inches long and the front ones 30 inches. The evener for the lead team C, is 36 inches long. The cable or chain, D, is just long enough to prevent interference.

In the six-horse tandem evener, fig. 4, the long evener, A-C, is 56 inches between clevis holes and is divided evenly, with B in the exact middle. The rear evener, E, is 30 inches long, divided 10

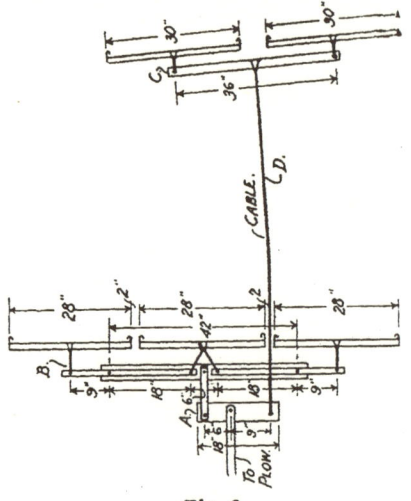

Fig. 3

and 20. The singletrees are all 28 inches long, and the doubletrees the same length, 28 inches.

Here, in fig. 5, is an arrangement for stringing three teams out tandem. The front evener is an ordinary one. Working back the next one is 22½ inches

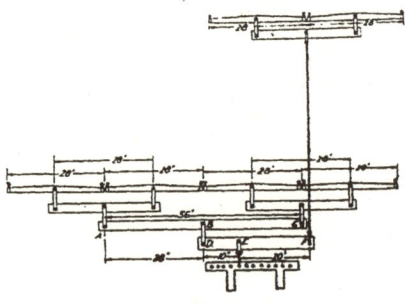

Fig. 4

long, divided 7½ and 15 inches, with one horse pulling against two. On the next one, three horses pull against one and the division is 7½ and 22½ inches. Then these four pull against one with the next evener, 4½ and 18 inches. Finally the five pull against the sixth,

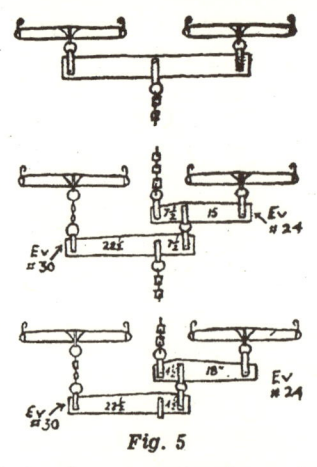

Fig. 5

which is given an advantage of 22½ inches to 4½ inches. The advantages of tandem hitches are that they are cooler on the horses and side draft is not a problem.

Finally we have an eight-horse hitch in fig. 6, where the horses are arranged four and four. There is little side draft and only one chain or cable is required. Note, where the main hitch is connected with the plow, that short chains making a crotch extend back to small clevises to facilitate turning the plow at

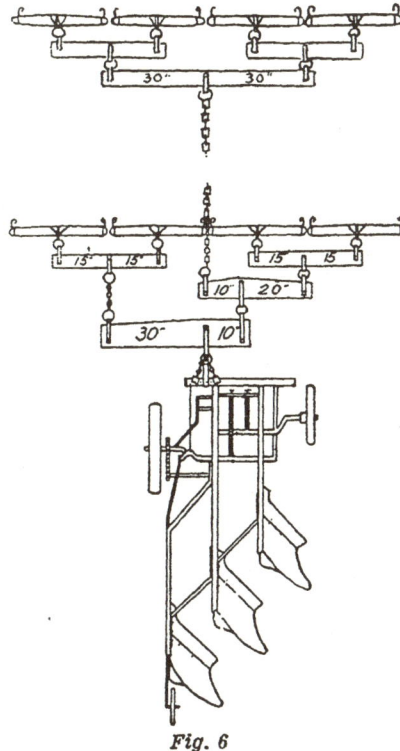

Fig. 6

the corners. The lines go to the lead horses but a short strap is attached from the bits of the two outside rear horses to the lines. Lead lines extend from the bits of the rear horses to the singletrees of the lead horses. The lengths used in making the eveners are all shown clearly in the illustration.

Tandem Plow Hitch

Someone asked for a hitch for hauling a two-furrow plow behind a three-furrow tractor plow. I have used these implements by attaching a cable or long chain from the left-hand side of the tractor draw bar through a ring wired to the left beam of the three-furrow plow, through a clevis bolted to the tongue of the two-furrow plow and through to the draw clevis of the second plow. If it is desired to hitch the two-furrow plow up closer to the three-furrow plow discard the wood tongue and make a V-shaped devise from angle iron to haul and guide the two-furrow plow. If both plows are the power-lift type the usual turns can be made on the headlands. Otherwise it will not be possible to make short turns at the ends of the fields.—T.N.S., Cadogan, Alta.

Steers from Chain

This tractor hitch for two binders takes only a few pieces of flat iron and a 14-foot logging chain. The diagram shows how it is attached to a McCormick-Deering tractor, but of course it can be hitched after other makes.

A crossbar is bolted to the drawbar of the tractor behind the differential. This crossbar is drilled in the centre and a stout piece of flat iron about three feet long is drilled and bolted to the crossbar and the side hole in the tractor drawbar. This being on an angle will enable the tractor to follow the grainside while the first binder takes the full swath.

In place of the tongue in the first binder a stub pole about two feet long, with flat clevises on the end for hitching to the tractor, replaces it.

The second binder also has a two-foot stub tongue. The hitch for the horses is left on this binder; only the eveners removed. A logging chain at least 14-feet long is attached in place

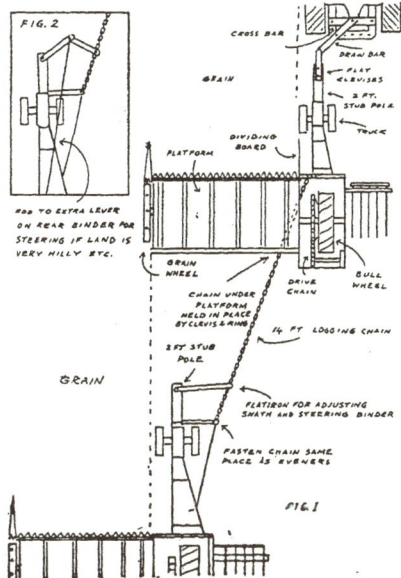

of the eveners. The chain is then passed under the first binder and held in place by a clevis and ring fastened to the frame of the binder where the truck of the binder is attached to the binder proper.

On the stub tongue of the second binder is a flat iron 2½ feet long bolted to the front end of it, with the other end bolted to the chain. This iron steers the truck of the second binder. It adjusts the swath by turning the binder in or out. A study of the diagram will clarify how it works. In turning corners turn the tractor when the first binder is two or three feet out of the grain.

If the field is very hilly or irregular and the second binder has a tendency to go off its course a contrivance for steering it can be made, Fig. 2. We have used this hitch on hilly land and on quite small fields.

To move the outfit remove the chain

from the front frame of the first binder and attach it to the rear part of the frame. The chain on the second binder is attached to the end of the stub tongue. This makes the two binders follow each other.

Using Steering and Spreading Bars

When two binders are hitched behind a tractor there is considerable side draft. Do not expect too much from a homemade hitch in keeping the corners square. They should be made somewhat round.

The hitch on the first binder must be well braced as it has to pull the second binder. It is advisable to take the truck off the first binder as it simplifies turning the corners. The rear binder is drawn by a chain or cable which is fastened to the front binder at A. The chain passes under the frame and almost directly under the sprocket which drives the canvas. It is held down by a block so it will not interfere with the sprocket and should be held in place by a ring or wire loop at B, to prevent it swinging at the corners. It is fastened to the rear binder frame approximately at C. The entire length of the chain is nearly 18 feet when the binders are spaced properly.

The success of the hitch depends on getting the binders spaced properly and on having the steering bars the right length. The length is given at GF. The length of the spacer bar must be adjusted to have the rear binder cut nearly

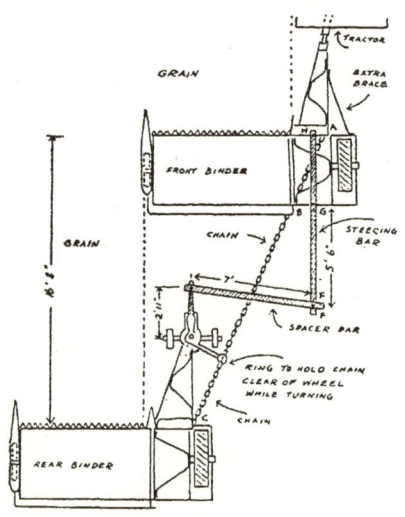

the full swath. It is about seven feet long. The length of the spacer bar will vary according to the angle the trucks make when turning. This varies with different makes of binders.

The spacer arm hinges at both ends, where it is fastened to the steering bar and to the stub tongue of the rear binder truck. The steering bar can be clamped to the frame of the front binder with U-bolts at H and G. A straight grained 4x4, or a strong piece of angle iron will serve for steering and for spacer bars.

Occasionally it is necessary to weight the rear binder truck with sand bags.

Combination Hitch

I have primarily designed this hitch for hilly land but it will work equally well on level land. After trying all kinds of hitches, none of which would work on sidehills, this one finally filled the bill. I think I can safely say this is the only hitch I have ever seen that will successfully work any place that a Wheatland itself will work.

All the material used in this hitch was salvaged from an old individual beam plow used in the early days, with the exception of one wheel. BC and CD are beams from said plow, CD being two welded together. AB and AC are levers off two beams. The wheels G are the small wheels that are in front of each bottom of the plow. The castings which hold these wheels on the draw-bar are sawed off until they match the height of the tractor draw-bar and then welded to the draw-bar as shown, with provisions being made to allow enough room for the wheels to turn freely.

Before any measurements are taken great care must be exercised in proper alignment of the machines to be used, especially that there be no appreciable amount of side draft. If these precautions are not taken trouble must certainly be expected.

Once the machines are aligned then they should be dropped to their working positions and drawn together until about one and one-half feet separate the back and front wheels. The full cut of both machines must be measured accurately and divided in two. This distance will be the draw-bar BC. Under this method any two machines may be used regardless of size, but they should preferably be the same make. However two machines with the same draft in proportion to size may be used. Next cut a board to equal the computed length of BC and lay in position as illustrated in diagram.

CD may also be a board laid at right angles to BC and corrected with a square. Enough distance should be allowed between point B and first machine for linkage. All points should be flexible except AC and AB which are welded solid to BC. Wheel F is an old salvaged wheel which is not too high. The brace ED and shaft FE will vary with different machines and may be measured as before with boards. Once the complete layout is made in boards then it may be transferred to the steel parts. AC and AB should be the same length as BC to allow the maximum turn. These parts to be solidly welded or bolted together. CD must be joined to CB with a swivel link which will allow a certain amount of movement in all directions. The same principle to be followed in connecting first machine to draw-bar. Point E should be equidistant from both wheels and be made in the form of a U so as to allow about six inches free movement of drag-bar CD ahead or back. The shaft FE to be bent at point E to form an eye which will slide on U bar. U bar can be attached to front machine, with every possible care being taken as to height and position in reference to wheels. In some cases this part may have to be redesigned to fit your particular machine. The purpose of the six inches travel is to act as a guide in balancing the load of two machines. Contrary to popular belief I find that the back machine pulls the lightest in nine cases out of ten.

I have successfully seeded two crops with my two Wheatland combines on sidehills equal to the hills of the Big Bend country of Washington. So far I have found no misses in the crops. A solid hitch has many advantages over a cable hitch on sidehills since there can be no sag in a solid bar.

Two Mowers Behind Tractor

This illustration shows a simple hitch for drawing two mowers. A hardwood scantling, 3x4, is clamped to the front mower. A fir 4x4 would do as well. On the end of it a flat iron bar carries the stub tongue of the rear mower. An iron bar, passing over the axle, is bolted to the tongue and to the scantling to hold it up. The stub tongue of the front mower extends three feet in front of the cutter bar. To prevent side swing a chain or cable runs from the drawbar of the tractor to the bar which hitches to the rear mower. By placing the two mowers in their proper relative positions the length of the different parts of the hitch can be readily calculated.

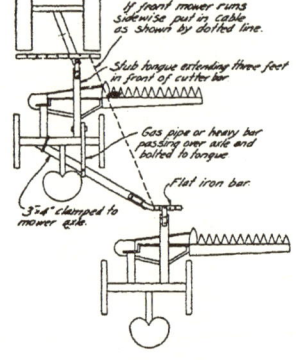

One-way Behind Binder

This illustration is from a leaflet issued by the extension department of the Manitoba department of agriculture.

It is quickly and easily constructed by extending a chain from the tractor drawbar to a point on the rear of the one-way disc that is directly in front of the stub tongue of the binder when the latter is in the correct operating position. At this point a pulley should be solidly bolted so that the chain may run on it, as it will

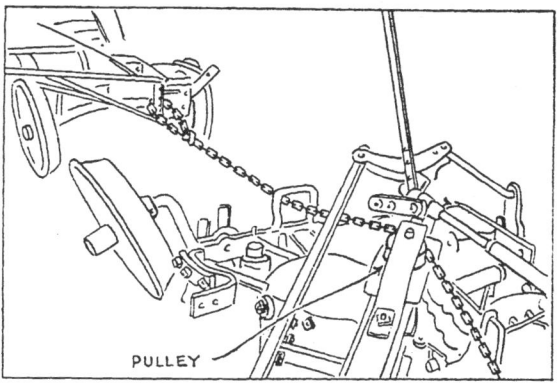

do on the corners. Carry the chain back to the stub tongue, but first passing it through the U guide bolted on the rear of the disc frame, as shown in the picture. This guide gives more positive control on corners. The chain can be secured to the truck at the base of the stub tongue.

It might be necessary on some makes of one-way discs to raise the pulley a short distance above the frame to allow the chain to clear the framework. This can be done by bolting an L-shaped standard to the frame and supporting it with braces. The pulley can be secured to the top by means of a clevis arrangement. Using this hitch it is not necessary to run the chain back through the guide, the binder following satisfactorily from the pulley.

Safer Binder Hitch

Here is a diagram of a simple attachment to the tractor drawbar for hooking up the binder, which has stopped split binder tongues for us. The V-shaped attachment shown makes possible a sharp, right-angled turn, and

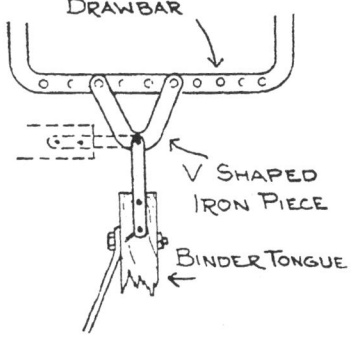

a neater job of cutting the field. The attachment is made from strap iron a little lighter than the drawbar, such as a heavy wagon tire.—I.W.D.

SECTION 11.

Water, Drainage and Heating Systems

Cheap Sewage System

This system is composed of a pipe leading from the kitchen sink, under ground to a buried oil barrel or box which, in turn, stands on several feet of rocks. Rocks are also piled around the barrel for part of its height. The barrel is covered with earth and some sacking or paper placed on the rocks before the

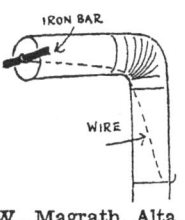

hole is re-filled. The pipe should be buried about five feet deep, on a slope, with the lower end resting on the top of the barrel or through a hole near the top. The barrel should have no bottom and the stones beneath should be covered with coarse gravel. The short piece of pipe from the sink connects with the horizontal pipe, which should be 12 feet long and four inches in diameter. Use a T-connection so that the plug can be removed and the horizontal pipe cleaned out. This system should be used to dispose of liquid waste only. The scraps are better fed to the hogs.

Stove Pipe Holder

If you are troubled with your stove pipe slipping out of the chimney opening just take a small iron bar and fasten it to a piece of wire.

Slip it into the chimney as shown and bring the wire down one length of pipe below. Hook it over the bottom of this length and it will hold the pipe firmly in place.—J.G.W., Magrath, Alta.

Easily Made Heater

This heater will keep you warm while driving in winter. It is made from an old shotgun creamery can. Open an old

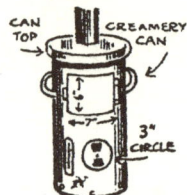

stovepipe and use it to line the inside of the can so as to make it more heat-worthy. Put two inches of earth in the bottom and then put in a disc of sheet iron as a false bottom. This prevents the solder from melting off. The draft is made of half a baking powder tin. The door is made half an inch larger than the feed hole. The pipes can be made of old stovepipes, cut down to half size, or an eavestrough conductor pipe may be used.

Cheap Pipeless Furnace

When building a home-made furnace select a section of the basement where the stove pipe can run directly into the chimney. If possible the stove should be set parallel to the joist of the floor so the registers can be easily put in place. The air ducts and 2x6 furnace frame posts can be readily placed by this type of placement.

In operation, the cold air passes through the narrow cold air grates in the floor and goes down between the inner and outer shell of the furnace casing. It then travels around the stove or drum and up through the hot air grate directly above the heating unit. The circulation is rapid and the efficiency of the unit is very good. If it is

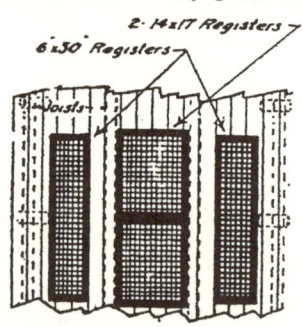

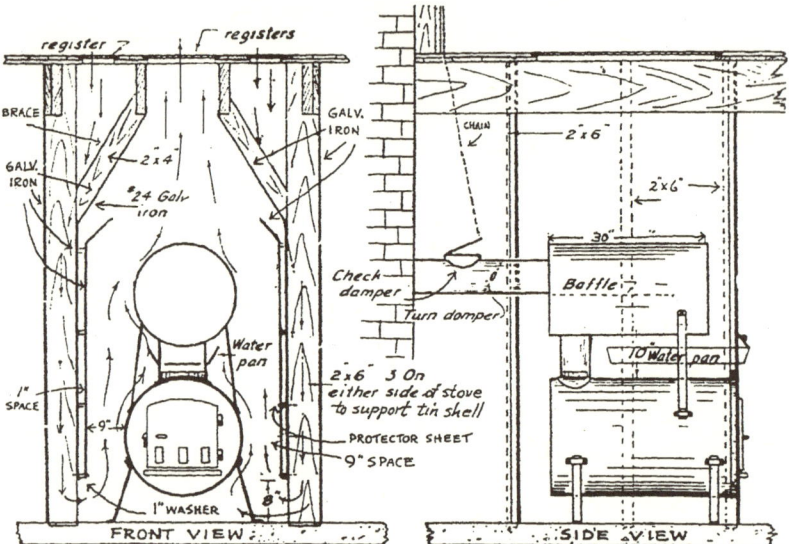

built relatively tight along the basement floor (earth or cement floor) you will find that the basement receives practically no heat from the furnace.

Construction of the furnace is very simple. Six pieces of two by six form the frame of the furnace and they are fitted to the floor or joist, according to the width of stove used. The two side walls and the back are clean sheets of galvanizd iron or sheet tin. At the front, the sheet of tin is fitted around the end of the stove so it can be fired. On the inside, the inner shell extends from the surface of the upper floor down to within eight inches of the basement floor. To protect the frame a protection sheet must be installed with a one-inch air space along the sides next to the heating drums. The protection sheet can readily be installed with nails and a shell has been completed. Cheap cold air grates can be used for both the cold air and hot air floor registers.

Heating units may consist of a large box stove or one or two old gasoline or oil drums. A blacksmith will put a door and damper in an oil drum in short order. A 24x30 cottage can be heated with a single drum unit, but a second drum will greatly increase the efficiency of the heating equipment.

By placing control chains on both the few washers after the main frame and pipe damper and pipe check damper, the heat can be conveniently controlled from the main floor.

In designing and building the home-made heating unit every effort has been made to make it fire proof and yet inexpensive. With the floor register directly above the stove it can be easily observed even though it is not on the main floor. There is but one person who can protect your property from fire and that is you.—W. Kalbfleisch.

Get the Chimney Ready

High time to be looking over the chimney and get it ready for use. Inspect it both inside and outside the attic to be sure it is firesafe. If cracked, if bricks are loose, or mortar is soft and crumbly, the chimney should be rebuilt. Clean out all soot by tying tire chains in a loose bundle to a rope and jerking it up and down so as to reach all corners. All stops and openings must be tight so as not to spoil the draft. If smoke pipes are three or four years old and a nail can be pushed through them, replace them with new ones.—I.W.D.

Mending Broken Grates

The grates in our kitchen range became so worn in the middle that the fuel was wasting. We took two pieces of

strap iron, bored holes at the ends and bolted them to good links of the grate at either end of the broken places. They work very well and appear to be durable.—Carl A. Tatroe, Sedgewick, Alta.

Stove From Oil Drum

FIRST cut a hole in the side of the carbide can or oil drum near the bottom for the stovepipe. Cut the hole about two inches less in diameter than the pipe to be used. After the hole has been cut, centre the pipe over it and mark around the pipe. Now cut slits from the hole out to the mark, and bend the strips between the slits up or out, and try the pipe to see that it slips snugly inside the row of bent strips. Cut slits about two inches long all around the end of the pipe, slip the pipe into place and from the inside bend the pipe strips out so the pipe cannot be pulled out. Drill a few holes through the drum and pipe strips and bolt the two together.

Cut a fuel hole about eight inches square. Cut from heavy galvanized iron a fuel door a little larger than the fuel hole, cut a V-shaped draft hole near

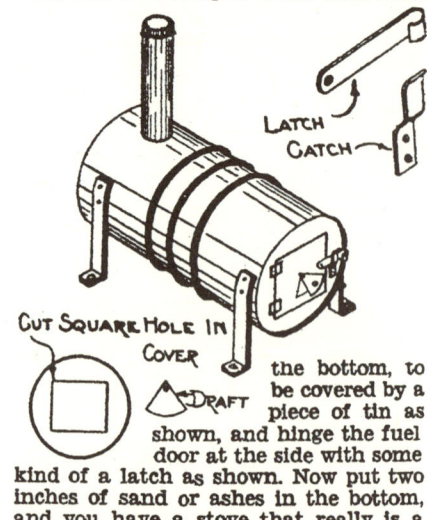

the bottom, to be covered by a piece of tin as shown, and hinge the fuel door at the side with some kind of a latch as shown. Now put two inches of sand or ashes in the bottom, and you have a stove that really is a heater.—I.W.D.

Filter for Rain Water

An effective filter should be used for the cistern, especially if the water is to

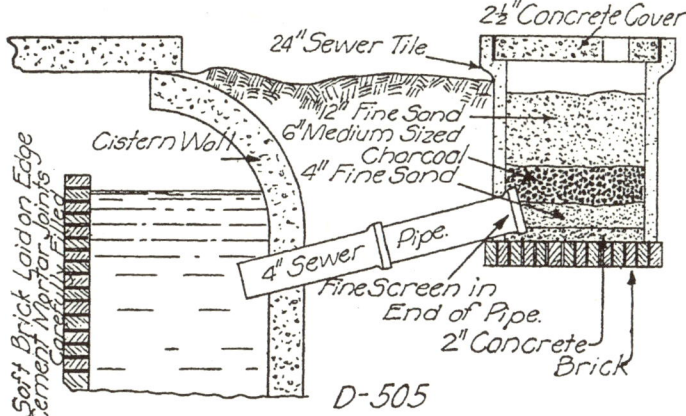

be used for drinking or cooking. A sand and charcoal filter is effective, as it removes both coarse and fine impurities and improves the color and taste of the water. The reinforced concrete box is at least three feet square and three feet deep, and with 12 inches of charcoal. Skim off and wash the top two inches of sand occasionally, and every year or two remove sand and charcoal and wash thoroughly, or better replace with new.

Cure for Noisy Water Pipes

Does the cold water pipe in your home thump and bang when a faucet is closed suddenly? The figure shows how you can use a 2 foot 6 inch length of pipe to provide an "air cushion" that will eliminate this trouble.

Cheap Water Installation

The accompanying drawing shows how this installation is made. When a barrel like this is utilized no pressure can be used. The float tells how high the water is. A tap in the connection between the pump and the barrel is closed when not pumping into the barrel and cold water is obtained by straight pumping. It is necessary also to have a tap on the cold water spout; when closed the water is forced into the barrel. Hot water is taken off by means of a T on the ascending hot water pipe.—I.W.W.

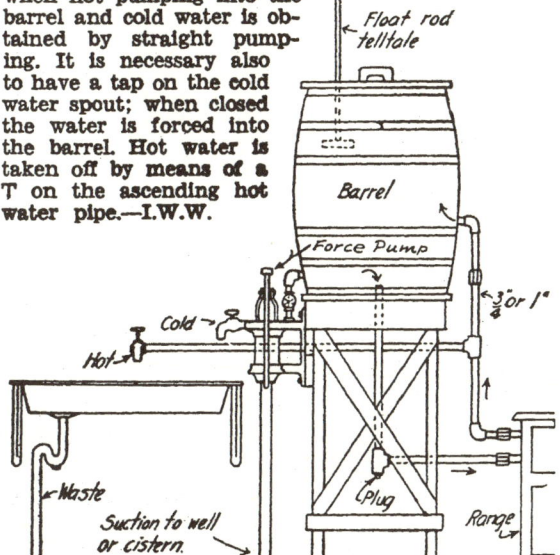

Cleaning Chimney

Burning a double handful of dry salt on a hot fire every few days will loosen and remove ordinary soot and creosote, and will gradually crumble it when caked. The zinc and manganese in old dry cells also have a similar effect when burned on a hot fire. We do not know of any other materials which would be safe or practical to use.

Keeps Pipe From Dropping Into Well

If you have trouble with your pump coming apart so that the pipe below the cylinder drops down into the well, where much time is required to fish it out, you can remedy the trouble by twisting a heavy wire several times around the upper end of the lower pipe, and then looping it around the drop pipe above the cylinder. It would be even better to fasten a hose clamp tightly around the lower pipe, so that there would be no possible chance for the pipe to slip through the lower wire loop. —I. W. Dickerson.

Insulating Furnace Pipes

A thorough study of heat wastes in the cellar brought out some rather surprising things. For example, it was found that a great deal of "insulation" that is put on warm-air furnace pipes is actually worse than no covering at all for these pipes. Careful measurements of heat losses shows that bare, bright, clean tin pipes lost a smaller amount of heat than the same pipes did if covered with one or even two layers of thin asbestos paper. It took at least a ¼-inch thickness of asbestos insulation to keep the heat in as well as a bright metal surface.

WOOD'S ELECTRIC FARM EQUIPMENT

"I Save a Half-Day Every Week -- thanks to my WOOD'S Feed Grinder"

WOOD'S Electric Feed Grinder

Entirely eliminates wasted days hauling grain to the mill. GET FRESH FEED DAILY—Grind grain in your own barn ... AUTOMATICALLY. The Wood's Feed Grinder turns out the quantity and grade of chop you require, while you get ahead with other work. CUTS GRINDING COSTS 90%. Wood's 16 years experience building sturdy electric farm equipment makes the Wood's Feed Grinder the best for your job. Heavy duty motor will also operate other equipment in the barn.

WOOD'S ELECTRIC MILKER

CUTS MILKING TIME IN HALF — One-piece seamless pail. CAN'T RUST. Easy-fitting one-piece inflations. Attractive LOW PRICE.

The WOOD'S Line

MILKERS — MILK COOLERS — FEED GRINDERS — OAT ROLLERS — ELECTRIC FENCERS — FARM FREEZERS — WATER HEATERS — WATER SYSTEMS — HAY DRIERS — FARM VENTILATORS

Write for Catalogs

The W. C. WOOD CO.
LIMITED

Head Office & Factory: GUELPH, CANADA

Thawing Overhead Pipe

Pick up a number of tin cans and wire them about a foot apart up close to the pipe. Put about an inch of gasoline in each can and touch them off with a torch on a long pole. When most of the cans quit burning, start up the

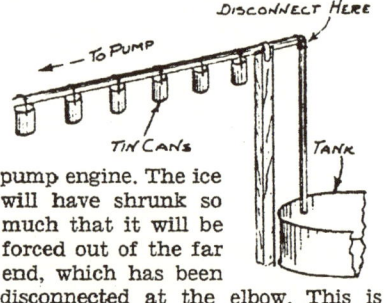

pump engine. The ice will have shrunk so much that it will be forced out of the far end, which has been disconnected at the elbow. This is a simple idea but it will save a lot of disagreeable work.—I. W. D.

Hot Water Heater

This heater can be made cheaply and will keep lots of hot water available. It is attached to the kitchen stove. Most

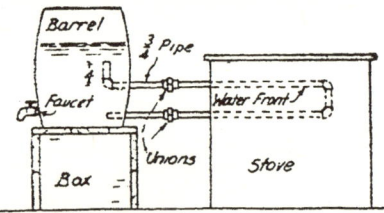

stoves are made with holes for putting in a water front. Use 3/4-inch piping and a barrel with a faucet. The unions can be secured at any hardware store.

Stove Pipe Lock

The diagram shows how one man solved a problem. "I put the pipes together as they should be with the seams turned right and so on. Then I took a one-eighth inch or smaller drill and drilled a hole on the top and bottom at each joint and turned a small self-tapping metal screw into each hole. These are much used now for sheet metal work,

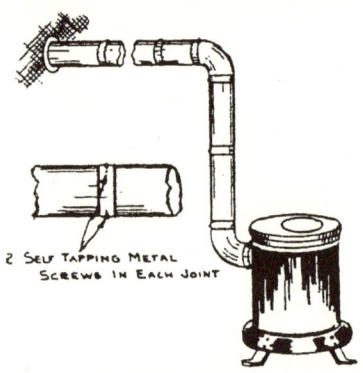

but small lag screws can be used if the others are not available. These screws should be used at each joint and at the stove connection. They are quickly put in and I find the pipes never sag, always stand straight, and can be quickly taken down for cleaning."

I believe it would be safer with a ceiling support at the upper elbow to prevent an accidental push sidewise and to hold the vertical part if it becomes necessary to remove the horizontal part for cleaning.—I.W.D.

Easier Pump Operation

In a deep well, the plunger and long pump rod may weigh a good many pounds. To avoid lifting this dead

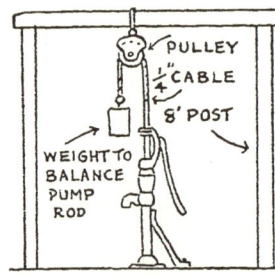

weight every stroke and letting it drop back with a jerk, many owners balance it with a rope run up over a pulley, as shown in the above diagram. This permits the use of a smaller motor and saves much wear and tear on the pump. A similar result can be secured by using heavy coil springs which are compressed on the down stroke and help to lift the plunger on the up stroke. The pulley should have ball bearings, and the weight should just about balance that of the plunger and rod.—I.W.D.

An Improved Windlass Well

This simple, pumpless well performs with surprising ease and efficiency. An 8-in. tile leads down to water which is only 14 ft. below ground. The top of the tile bell end is provided with a milk can cover with which it is closed when the well is not in use.

A rope winds about a hand turned windlass, runs up over a pulley in the top of the well house, then extends downward and is tied to the 6-in. sheet iron cylinder which is the bucket. This bucket has a flap valve in the bottom and a strap iron hoop on the top. Dropped into the water it quickly fills but the valve seals the bottom when the bucket is drawn up.

To empty it is but a single movement. The illustration shows the position of the bucket. At the far end of this trough is turned a lag bolt. Dropped on this the valve is forced upward and the water runs out, down the chute and into a pail waiting for it.—Dale Van Horn.

Repairing a Noisy Tap

A water line, tap or pipe that chatters when the water is running slowly has a loose washer (W). If the tap drips the rubber is worn out. To repair the tap, shut off the water line, unscrew the cap (A), unscrew the shaft (B-B) by screwing out the handle and shaft. Remove the screw (C) and take out the rubber washer (W).

Rubber washers can be obtained for a few cents each. They come in three sizes, so obtain the correct size. Hard rubber and soft rubber washers are readily obtainable for hot water or cold water taps.—W. K.

Drains Pump Pipe

Wire a heavy spring to the spout end of the pipe and fasten the other end of the spring to a crossbar of the windmill so that the pipe will be held up close to the spout, and even a little above when the spring is lifted out of the way for pumping. This gives drainage of the pipe at all times and there is no more trouble from pipe freezing up.—I.W.D.

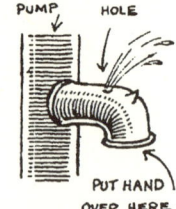

A Drinking Fountain

Holding your hand over the spout as shown, forces water up through a hole drilled in top of the pump spout, to form a sanitary drinking fountain. A cork or plug should be provided to close the hole.

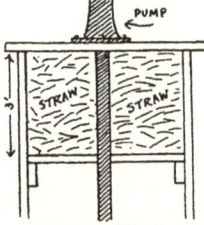

Protection from Frost

Now that zero weather is here how many farmers try methods to protect their shallow dug wells from ice. I did it this way. I put a second cover or ceiling three feet from the top by nailing 2x4's all around on the cribbing and covering with boards and paper. Then I packed in with straw. If a pump is used a hole only large enough for the cylinder needs to be left. If you bail the water out with a bucket a second lid will have to be put on, and this could be covered with sacking. In spring the straw can be removed and a fine cool place is left for storing eggs, butter, etc.—H. C. Pinnegar, Box 103, Langdon.

Test for Faulty Chimney

A smoke test will show whether a suspected chimney is safe or not. Build a small fire with kindling and when it is blazing, throw on some pieces of tar paper or asphalt shingles. When smoke pours out of the chimney, cover the top with a board or a wet sack. If a flashlight in the attic shows smoke coming through or if smoke shows in the house around the chimney except from flue stops, you may be sure the chimney is unsafe. If very bad, it should be rebuilt. Leak into the attic can be remedied by raking out loose mortar joints and then giving it two coats of rich cement mortar.—I.W.D

Warming the Pump

I am enclosing a sketch of a pump which is heated with a double burner Buckeye incubator lamp, with two one-inch wicks. The chimney on the lamp is six inches at the large diameter which just fits a six-inch stove pipe elbow.

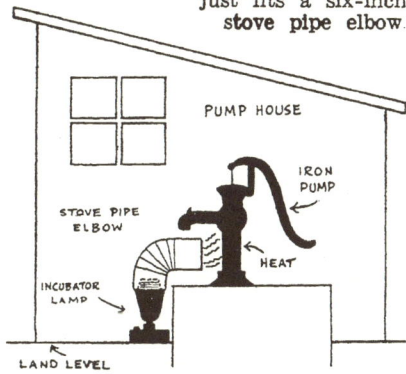

This, inside a small pump house which is insulated, will keep the pump from freezing at all times. I use just the regular iron pump with the cylinder at the top. This should help any who have trouble with their pump freezing up.

Chimney Extension

This is about as cheap an extension as you can make to the chimney. Sometimes, when the chimney is on one side of the house, and the wind blowing over the peak toward it, the current works against the draft and the chimney does not draw well or the furnace or stove may even smoke. Often another couple of feet on the chimney will completely dispose of the difficulty I had this trouble  and ordered a flue four feet long and eight inches square. Then I took one-quarter-inch iron and made two hooks two feet long, with two inches of the top bent over at right angles. The flue simply sits on the hooks and is banked around the outside with mortar to close the joint, keep the flue straight and shed the rain. A coat of paint made from iron oxide and oil brought it out to about the right color.

Insulating a Water Pipe

Where a water pipe goes up into a tank in the attic, hay mow, on the silo, and so on, it must be carefully insulated against freezing. The method shown is cheap and quite effective, provided the sawdust, rock wool, or other insulating material can be kept absolutely dry. Hence the need of careful waterproofing to keep out condensed moisture from the pipe and moisture from the outside air. The best way to insulate underground pipes is by putting them below frost. If this cannot be done, a good sized sewer tile should be laid and made waterproof. The water pipe should then be covered with waterproof commercial insulation, pushed into the tile, and then connected up at each end and at intermediate manholes.

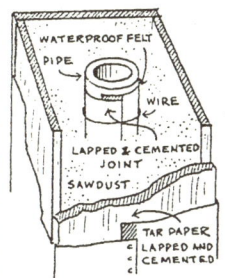

Pipe Connections to Tanks

In fixing up water tanks, cistern connections, and other work about the farm, it is often necessary to connect up a pipe so as to make a watertight

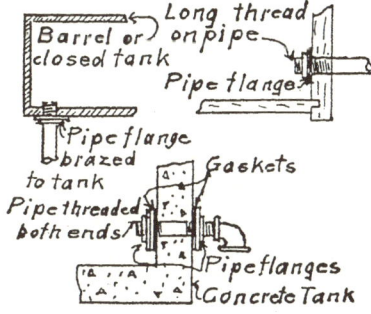

joint, as shown in the diagram. About the only solid way for a steel barrel or other closed metal tank is to braze a pipe flange on the outside and then screw in the pipe. The method shown for the concrete tank can be used equally well for any type of open tank.

Wood in Coal Burner

Many farmers find it convenient to burn wood in a coal burning heater in not so cold weather. Owing to the way the residue of hot ashes falls through the bottom grating it is not satisfactory. Suddenly the room is cold and the fire

MOTOR OILS
for
BETTER LUBRICATION
at
LOWER COST

* * *

THOUSANDS of owners of tractors, trucks and cars in Western Canada have proved in their own experience that they get better lubrication with **NORTH AMERICAN LUBRICANTS**.

These users of oils and greases have saved thousands of dollars in the operation of their engines.

If you have used North American Oils, you know they are better. If you have never used them, you should order at once, and you will be glad that you did so.

For the past twenty years, we have been saving money for farmers on their purchases of oil, grease, tires, batteries and many other guaranteed products.

Let us help you too.

Write for catalogue and free booklet on "H-D" Oil.

NORTH AMERICAN LUBRICATION
COMPANY LIMITED
NATIONAL CARTAGE BLDG., WINNIPEG

Temporary Pipe Repair

A handy way to make temporary repairs for cracked or leaky pipes is to wrap a strip of tightly stretched inner tube around the pipe over the break and fasten with friction tape, heavy cord, or soft wire. This will hold until

the pipe can be replaced or a permanent repair made, and may save throwing the water system out of use at an inconvenient time.

Thawing Frozen Pipes

For thawing out frozen water pipes in cisterns, tanks, etc., where an open flame will do no damage, a brick and wire torch serves the purpose. Take a brick, wrap the wire around it 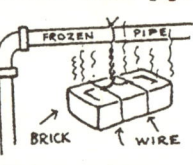 as shown and twist the wires to form a handle. Fasten to the pipe after soaking the brick with kerosene. Light the kerosene and move the brick along the pipe as it thaws.—Ernest Peterson, Chinook, Alberta.

Flue Hole Cover

A very neat flue hole cover can be quickly made by riveting a light, flat, iron, or 22-gauge tin to a common can cover. The iron shaped as shown will securely hold the cover in place. A coat of paint matching with the surrounding wall will greatly improve the appearance.—A. S. Wurz, Rockyford, Alta.

Saving Chimney Heat

A lot of heat escapes up the flue when furnace pipe connections are short and

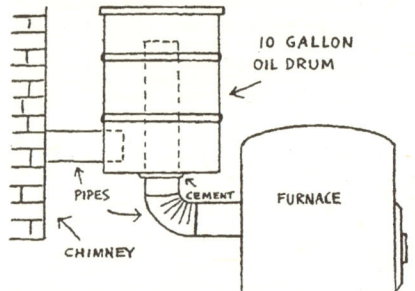

direct. A lot of it can be saved if a heat trap is made from a ten-gallon oil drum. Connect the pipe to the drum as shown. Fire cement is used to seal all connections.—A. S. Wurz, Rockyford, Alberta.

Opener for Drain Pipe

For this operation nothing serves the purpose better than an old seat spring attached to a broom stick with small wire staples. The spring is inserted 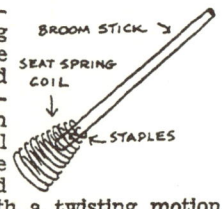 into the pipe with a twisting motion. When withdrawn the sediment is ejected without mess or bother.—Dorland A. Hotz, St. Boswells, Sask.

Old Engine Pump Jack

Here is a diagram of a home-made pump jack which has been used for three years and found very satisfactory. The pump is an ordinary standard make, while the pump jack is made from an old discarded 1½ h.p. engine with everything stripped except the base, flywheels and crankshaft, connecting rod and

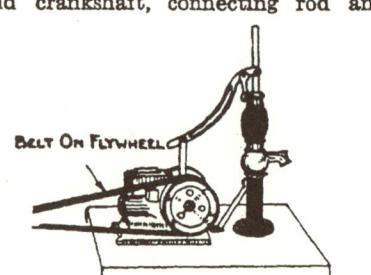

grease cups. Two strap irons are bolted through holes drilled through the connecting rod, and to a hole drilled through the pump handle. A washing machine engine can be used for power with a V-belt pulley removed and a flat belt run on the shaft alone and around the flywheel on the pump jack.—I.W.D.

Well Bucket Tipper

This simple device saves a lot of bother for those who use a bucket in their well. Take a piece of garden hose and shape it as shown. Then slip it on the rope above the bucket. It will 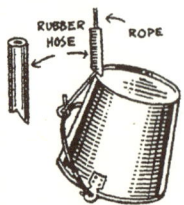 keep the bucket upside down until it hits the water when a sharp jerk loosens it.—D.C.R.

Pipe Holder

Anyone who has had pump trouble knows how hard it is to hold the pipe while making connections. All that is saved by this assembly of an old disc and an old hinge. A slot is cut in the disc to take 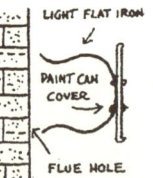 the pipe. The hinge is welded or bolted to the disc after the other end of the hinge has been notched to grip the pipe. This can be placed on the platform and as the pipe is raised, it is held while getting a new hold on it each time.

SECTION 12.

Building, Carpentry, Concrete

Seed and Feed Building, Cleaning, Grinding, Storage

This building was designed by the Dominion Experimental Station, Swift Current, Saskatchewan. It is intended

mainly for large farms, seed growers, co-operative farming projects, or as a municipal enterprise. It is not intended that a building be constructed as shown in the sketch but rather to be used as a suggestion for formulating plans that will more closely meet local requirements. Some of the features can be incorporated in a smaller seed and feed building for smaller farms.

The principal features of the plans shown here are as follows:

1. A sheltered driveway for unloading or loading.

2. A basement and stairway to permit ready access to hoppers and elevator boot for thorough cleaning purposes.

3. Deep, narrow storage bins to make best use of movement by gravity.

4. Seed bins lined with light gauge sheet metal to prevent mixing of varieties.

5. Facilities for grinding feed grain with maximum utilization of gravity to avoid excessive equipment and handling.

6. Grain elevator to convey grain into any bin in the building.

7. Unloading bins directly over the driveway for quick and easy loading into trucks and wagons.

8. A space in the driveway for installing weight scales (if such are required). An alternative would be to build an unload hopper on a scale on the unload bin floor just above the driveway.

9. Seed cleaning and grading machines located in the floor above the bins.

10. A gas engine room on the ground floor. If one large electric motor is used, this may be installed on the seed clean-

This floor is constructed with 1-inch fir flooring on shiplap with heavy paper tween to ensure no mixing of varieties in bins below the floor.

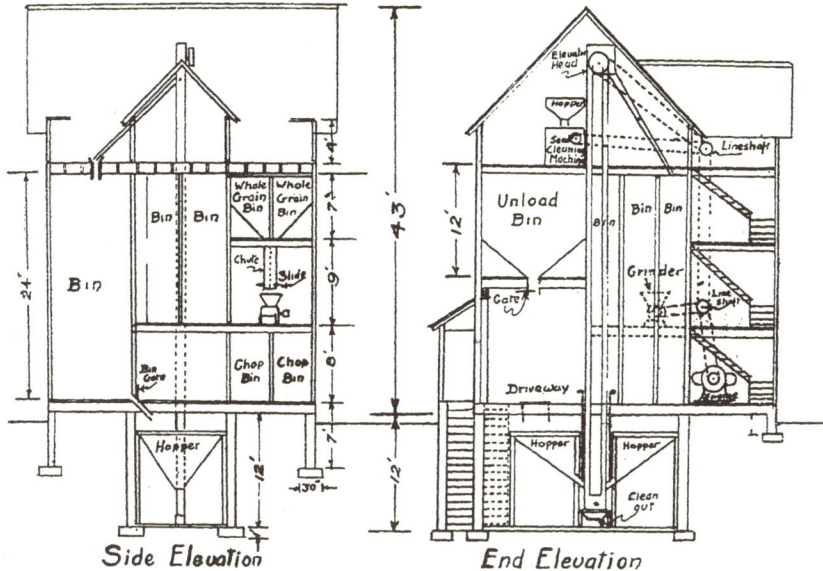

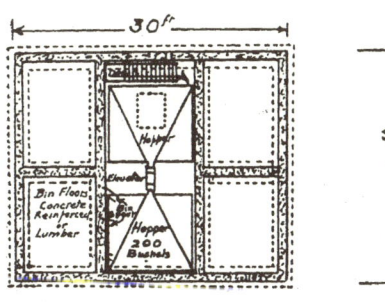

Foundation & Basement
Elevator boot and two hoppers in basement

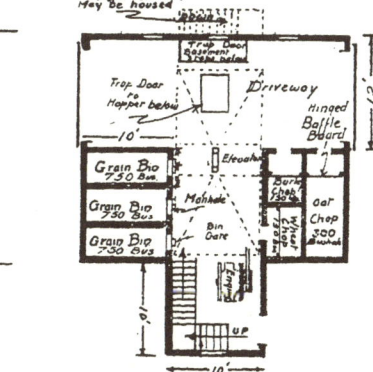

Ground Floor
Driveway-loading & unloading-Engine room

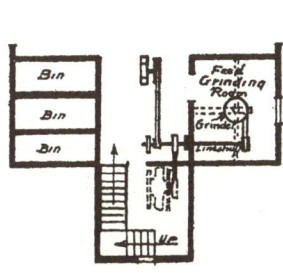

First Floor
Lineshaft-Grinder with chop bins below, whole grain bins above

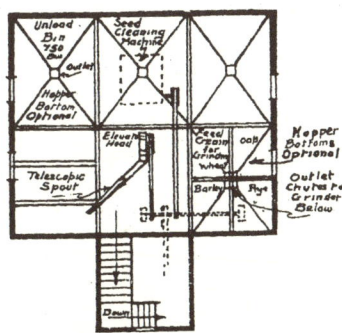

Second Floor
Elevator head, Grain distribution to bins Seed Cleaning & Treating Equipment

All the Electric Power you want

with a FAIRBANKS -MORSE ELECTRIC GENERATING PLANT

If your farm is not reached by commercial power lines, you needn't do without the advantages of electricity. A Fairbanks-Morse Electric Generating Plant will give them all. The original cost is low; the average operating cost is about one quart of fuel per kilowatt (1,000 watt) hours at rated capacity. You can have as much or as little power as you need. F-M Electric Plants range in size from those supplying simple lighting for homes, to those furnishing complete light and power for the largest farm. They are simple to install, being shipped complete in one unit, ready for service anywhere. All F-M Plants are guaranteed for one year.

THE CANADIAN FAIRBANKS-MORSE COMPANY LIMITED
WINNIPEG   FORT WILLIAM   REGINA
CALGARY   EDMONTON   VANCOUVER

If the local F-M dealer cannot supply you, mark the items in which you are interested and mail to the nearest F-M branch. We will advise you if and where the equipment may be obtained.

Water Systems.. ☐   Windmills....... ☐
Lighting Plants. ☐   Wind Charger... ☐
"Z" Engines.... ☐   Electric Fence.. ☐
Hammer Mills.. ☐   Hand Pumps.... ☐
Grain Grinders. ☐   Wash Machines. ☐
   Scales......... ☐   F.W.

Name................................

Address.............................

ing machine floor with suitable line shafting. However, a direct electric motor drive for each machine and elevators is the modern and most satisfactory method of providing power for stationary equipment.

11. A stairway provided for quick and safe access to any floor and machines.

When planning a large seed building, it is wise to have the plans drawn up by a good architect familiar with the type of such buildings. Also consult a local carpenter or contractor who may erect the structure.

A Kitchen Wood Box

This wood box was designed by Mable C. Mack, of Oregon State College and requires no further instruction to make. The height may be from 32 to 36 inches. Note that the lid is covered with linoleum. The following suggestions are made for building a wood box: It should

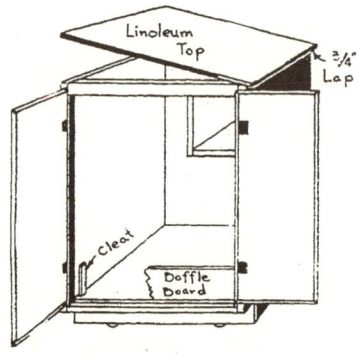

be placed at the left of the stove for convenience; large enough to hold a day's supply of wood and kindling; easy to clean; easy to put the wood and kindling into and easy to take it out; designed to save kitchen space in small kitchens; simple to design and easy to build. This wood box has been designed to meet these requirements. The top can be used for working space.

How To Frame a Roof

Before undertaking to frame a roof, make a fence for your steel square. Take a piece from a common board two inches wide and three feet long. Run a gauge line down the middle of the edges and with your rip saw run a kerf down from each end, leaving about 10 inches of solid wood in the middle of the fence. The blade of the square is inserted in one kerf and the tongue in the other. Small screws inserted in the fence will clamp it securely to the square.

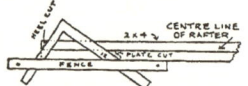

Fig. 1. Framing a rafter with fence on square

Fig. 1 shows how the fence is attached to the steel square and illustrates also how the square is applied to the rafter in getting perpendicular and horizontal cuts. The tongue of the square is on the left and the heel cut, when the rafter is in position, is exactly up and down. On the right is the blade of the square, shown as it is applied in making the plate cut, that is the part of the rafter which rests on the level plate of the building.

When you start to frame your roof, first pick out a straight scantling to make a pattern rafter. Be sure you have it framed right and then mark all the other common rafters from it. Draw a line down the middle of the dressed side. In framing the rafter always work from this line.

Here is a handy thing to know in getting the middle of a board or scantling that has an odd width. Put your square and rule on it at an angle until you get a measurement that can be easily divided in two. The scantling will be about 3¾ inches in width. Angle your ruler on the face with the end of the rule at one edge and the four-inch mark at the other. Then tick off a point at the two-inch mark and there you have the exact middle, haven't you?

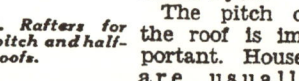

Fig. 2. Rafters for third-pitch and half-pitch roofs.

The pitch of the roof is important. Houses are usually framed with half-pitch roofs. In this case the rise or the height of the peak above the level of the plates is one-half the width of the building. In carpenter's language, the rise is the same as the run.

In getting the cuts the 12-inch mark is used on both the tongue and blade of the square. A common pitch used on small buildings is the one-third pitch. Both pitches are shown in fig. 2. In this case the rise is eight inches for each foot of run; the eight-inch mark is used on the tongue, still using the 12-inch mark on the blade. The tongue gives the upright cuts and the blade the horizontal ones.

Since we are dealing with house roofs we will assume that the half-pitch is used. Adjust the fence on the square so that the 12-inch marks on both blade and tongue come exactly on the centre line you have marked. First mark the heel cut, as shown also in fig. 1. If the projection of the roof is one foot, then

Fig. 3. Measuring a rafter by the step method.

the plate cut can also be marked. It is from the centre line to the bottom edge of the rafter.

Now make the heel cut and then rip the scantling on the centre line until you come to the mark for the plate cut. Then make the plate cut. It is always best to do the ripping first, since there is less danger, when making the other cut, of sawing in past the centre line and weakening the rafter. The lower end of the rafter is now framed.

To get the length of the rafter apply the fence to the upper edge of the scantling and step it off as shown in fig. 3. First place the 12-inch mark on the blade of the square at the intersection of the plate cut and the rip cut. Tick off the point where the 12-inch mark on the tongue lies on the centre line. You will note that this gives you one foot of run and one foot of rise. Move the square up until the 12-inch mark on the blade is on the point you have ticked off and again mark as before. The square is applied as many times as half the width of the building in feet and the ridge cut is along the tongue of the square at the last application. If a ridge board is to be inserted between the rafters cut back five-eighths of an inch to allow for it.

You now have your pattern. The other rafters are marked from it. When a pair has been framed it is best for the amateur to put them up in place and see that all the cuts are right. Be sure you are right, then go ahead, is a good motto. It is best to put the heel cuts on the four end rafters only. Then after all the rafters are in place stretch a chalk line, mark to it and then cut off the ends. This ensures a straight edge to the roof.

Fig. 4. Getting the cuts for hip or valley rafters.

If the cornice is not on the slope, but on the level, the detail may be as shown in fig. 1, How to Frame a Frame House, on page 143 In this case the roof is framed as if the body of the house extended out as far as the end of the ceiling joist.

Cutting a Hip Rafter

To get the cuts on a hip rafter, take the same rise on the tongue of the square as for the common rafter, but instead of 12 take 17 on the blade as shown in fig. 4. Frame the bottom after the rafter is up when the proper place to make the cuts can be found by carrying out the lines from the heels of the common rafters by means of a straight edge.

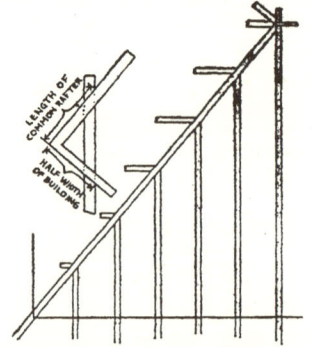

Fig. 5. Hip and jacks in position. Insert shows how to get backing for hips.

To get the length of the hip rafter step it off as you did the common rafter using the same number of steps. To get the bevel cut to fit against the common rafter take the length of the common rafter on the tongue and the length of the hip on the blade. Blade gives cut.

To get the backing of the jack rafter to fit against the hip rafter take half the width of the building on the tongue and the length of the common rafter on the blade. Blade gives cut. The novice can get these cuts by putting the rafter in place and marking by the hip. For the vertical cut at the top and the plate cut at the bottom the same bevels are used for jacks as for common rafters. To get the length of the longest jack rafter measure across from the last common rafter to the hip so that the two will be the proper distance apart and mark. Place the steel square on the common rafter and mark it directly opposite the mark you have just put on the hip. There you have the length of the longest jack.

Always nail the jack rafters on in pairs to avoid putting the hip out of line. The cuts and lengths for valley rafters and valley jacks are got in the same way as hips and hip jacks.

Precautions must be taken to have valley and hip rafters in line with the rest of the roof. Hips are placed so that the upper corners are exactly level with the other rafters. With valleys the middle of the upper edge is in line with the common rafters.

Estimating Materials

It takes about 7½ shingles, laid five inches to the weather, to cover a square foot. This will allow for waste. Where the roof is cut, broken or hipped, more will be taken. For lath and plaster, each 100 square yards, take the following: 1,450 laths; for the brown coat, 900 to 1,000 pounds of No. 1 hardwall and 1¼ yards of sand; or, 400 to 1,700 pounds of fibre plaster and ¾ yard of sand. The finishing requires 100 pounds of finish and 275 pounds of hydrated lime.

A four-inch wall takes 6½ bricks for each square foot and a double brick wall, 13 bricks. To lay 1,000 bricks will take 300 pounds of hydrated lime and ⅝ yard of sand. A chimney with a flue 8½ inches square will take 30 bricks per foot in height. With a flue 8½ by 13 it will take 35 bricks.

For 100 yards of woodwork two coats of paint will take four gallons of ready mixed paint, one gallon of linseed oil and a pint of turpentine. For each additional coat of paint allow 2½ gallons.

For framing lumber allow 20 pounds of four-inch spikes and five pounds of 2½-inch common nails. Sheeting and shiplap take 20 pounds of common nails per 1,000 feet and siding 20 pounds of siding nails. For shingling allow 3½ pounds of nails per 1,000 and for laths seven pounds per 1,000.

And here is a table for estimating materials for concrete:

Mixture	Cement Bags	Sand Cu. Ft.	Gravel Cu. Ft.
1. 1½. 3	7	10½	21
1. 2. 3	6¼	12½	20
1. 2. 4	5⅓	11	22
1. 2½. 5	4⅘	11¼	22½
1. 3. 6	3¾	11¼	22½

Building With Logs

So many requests have come to The Guide for information on the construction of log houses and other buildings that we reproduce herewith instructions written for The Nor'-West Farmer some years ago by V. W. Horwood, well known authority on farm home construction.

The first thing to do after the site is picked is to get a corner. To lay the building square use the simple rule known as the 3, 4, 5 rule and by using a common multiple it can be increased to any length, say by 2—gives 6, 8, 10. At one corner of your building set a

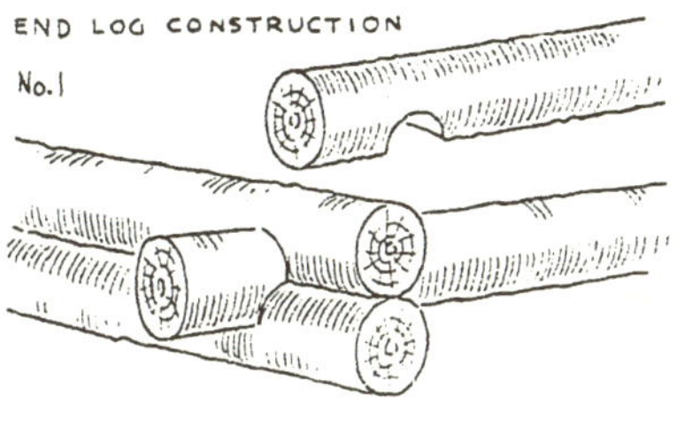

mason's square and stretch a line each way from the corner, following the square sides; then on one line measure off three feet and on the other measure four feet; adjust the two points so measured that the distance between them will measure exactly five feet.

In fig. 3, the notching of floor beams into sills is shown, floor beams being at two-feet centres. These beams are hewn on top. The construction of a notched log wall is shown in fig. 1. The construction is as follows: The first log is in place. The log above should be marked for the cut on its under side to fit over the log below. On long logs six-inch nails or pins should be driven in to prevent the log from springing—and if the log is not straight, a cut from a crosscut saw will allow it to be sprung into position. In laying the logs the butt end of one should be laid alternate to the small end of the other, so that the logs will be kept level. Often the doors and window frames are put in place and the logs spiked to the frames. An-

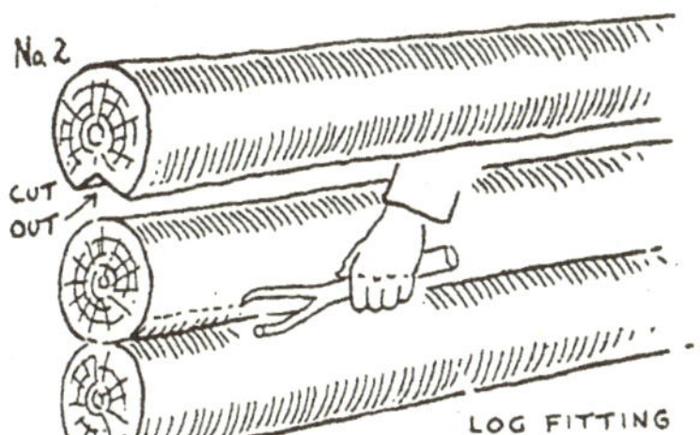

other way is when heights of doors and windows are reached to make two saw cuts in the logs at the proper width to give a starter for sawing.

The trouble in log wall construction, unless the logs are squared, is to get the upper logs to fit snugly over the under log. There is a gap to be filled with plaster or moss—a poor construction. In fig. 2, is a method which gives a snug joint. When the log is notched approximately to its position lay it on the wall. A wide crack will be between the two. Take a piece of wood about the size of the crack with a pencil on top. With the piece of wood follow the top side edge of the under log on both sides. The pencil will mark the irregularities of the under log on the top one. A scribing tool like the one shown can be purchased or made. Cut to this line, adjust the log; cut out until it rests snugly over the upper log. Put moss on, making an air-tight joint.

Fig. 4 shows dove-tailed construction with jamb logs rabbeted and the logs tenoned into them. This construction can be used on a round log construction

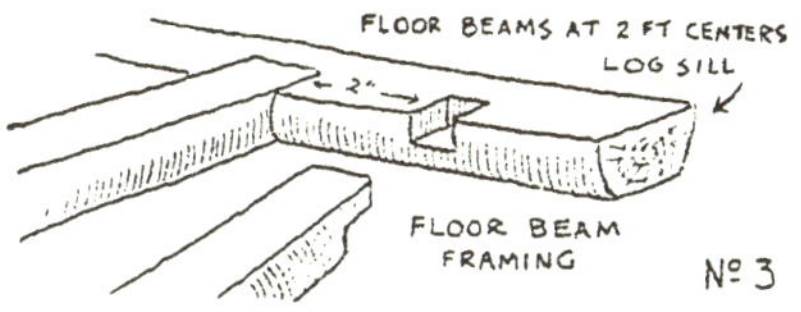

by using the method in fig 2. The jamb logs are grooved about 3x3 inches by boring holes with an auger and chiseling between to take out rabbet. Into this the wall log is tenoned. Erect these logs. Measure for the length of wall log and lay on the wall; on this measure for depth of jamb rabbet. Saw this piece out; put tenon into rabbet, then go to corners and mark the length and where the corners fit into under piece, saw and bevel to make dove tail. Fit this log into place. If your log is square it will rest on the lower log. If round you will use the method shown in fig. 2. Bore holes through log when in place and drive hardwood pins to hold logs. The construction of the roof with poles for rafters would be carried out by notching as in fig. 1.

The Hudson Bay log house is another type which is more easily and quickly put up. Spike together two two-inch planks, one as wide as the diameter of the logs to be used, the other two inches wider, the planks as long as the building is to be in height from foundation to plate. Set up a pair at each corner with

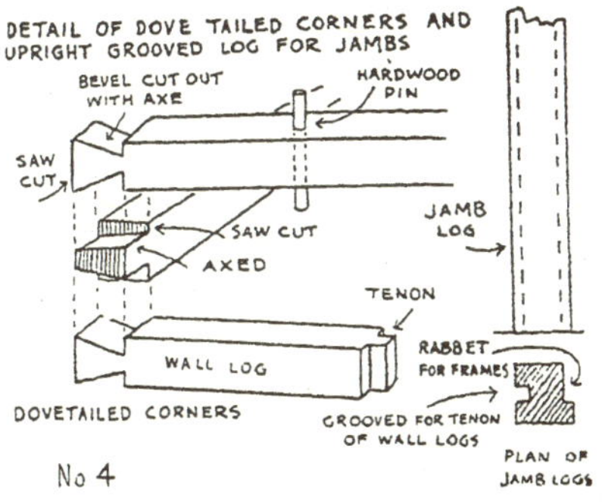

the point of the "V" placed so as to be exactly where the inside of the corner of the building is to be. Cut logs in exact lengths, place between the upright corners and spike through the planks into the log ends. Set up plank frames for door and window openings and spike firmly to the ends of the logs. For plate use a log or pole about four feet longer than the length of the building to provide for a two-foot projection at each end. Use poles for rafters. Spike a plank securely to each of the four rafters that are to hold up the gable ends. Cut end logs the proper slant and hold them in place by spiking through the plank and into the ends of the

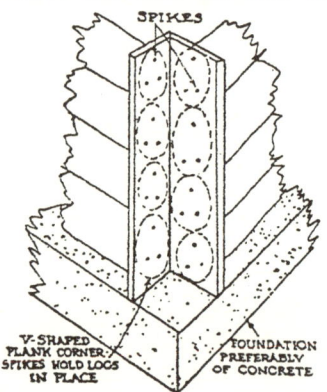

gable end logs. Roof with sheathing, shingles, or ready roofing. Floor to suit requirements. If concrete is not used for foundation, a log will serve, but concrete is best. Five or six-inch spikes are used to secure the corners. To finish off the corner, an upright log fitted into the "V" gives a better appearance but doesn't add anything to the strength of the corner. Buildings of this construction will stand for years.

The simplest type of log building is the one made of small logs or good sized poles and with notched corners. There is no trouble putting up a building of this construction except to see that the notches fit fairly snugly over the log they contact with and that the logs are laid with the notch down instead of up, this to shed rain and keep the corner from rotting. No foundation other than logs or blocks is used as a rule for buildings of this type.

For plastering a log wall use one part each of lime and sand, or one of cement to two of sand. To keep the mortar in position drive small nails into the logs at four-inch centres and staggered. Another method of keeping mortar from sliding off a log is to nail small strips, often willow limbs, on the logs just under where they meet.

Various Wood Joints

Here are some of the simpler corner joints that can be made in working

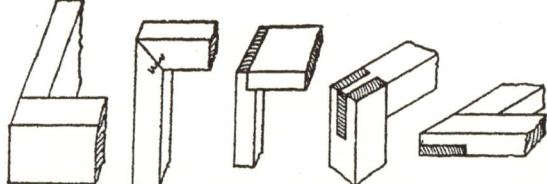

wood. The first is the simple butt joint. It is held together by nails or screws. The chief consideration is to have the connecting surfaces cut true and square. In the mitre joint the stuff is cut at 45 degrees and a mitre box should be used. Corrugated fasteners can be used. The rabbet joint has the advantage that it can be nailed from both directions. The dado joint is similar to the rabbet joint but is some distance from the end, as in making book shelves and step ladders. Crosslap joints are used in making sash or window screen frames. They are held together by small metal pegs in the case of window sash and by clinched clout nails in the case of window screen frames.

Lengthening Garage

Here is a simple way to lengthen the garage so that it will take a longer car.

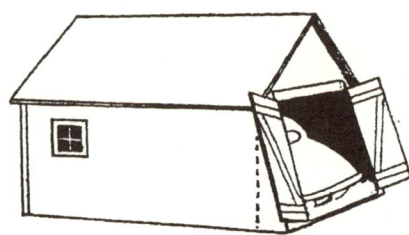

Just extend it out at the bottom, without lengthening the roof. The doors will stay either open or shut and they also lift clear of the snow. The drawing was made from a snapshot taken on the farm of John Stevens, Morrinville, Alta.

Shingling Scaffold

The sketch shows the safe and convenient scaffold for shingling the roof on a building. For the main scaffold, use a 4x4 tied to ropes which run over the peak and are tied on the other side

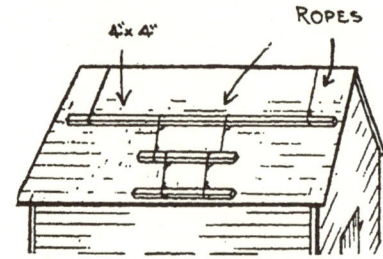

of the building. This is large and long enough to hold a large supply of shingles. Scaffolds lower down can be tied to the main one for starting the roof. When the main scaffold is reached, it can be pulled up as the shingling proceeds. After seeing an old friend fall and almost impale himself on a post, due to the splitting of a cleat nailed to the shingles, I recommend this as a simple and thoroughly safe scaffold to use.

Taming Refractory Flooring

This enclosure outlines an idea that may possibly be useful to someone, who, like myself, occasionally comes up against the problem of trying to persuade serpentine pieces of flooring to come within bounds by the aid of nothing more potent than a wood-chisel and some compound cuss-words.

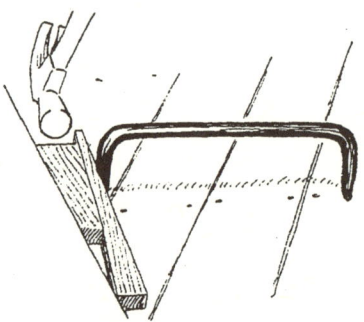

Being the sort of bird that likes to monkey around in a workshop himself, I never cease to marvel at some of the "inventions" produced from odds and ends picked up in the farm "boneyard," and which find their way to the farm workshop.—The boys are good!—R. S. MacNeill, Shelby, B.C.

How To Finish a House

In finishing a house the first thing to do is to fit the windows and outside doors to get the building enclosed. To fit a window, first trim the top sash so that it will fit closely against the blind stop. For details see figs. 6 and 7, in How to Frame a House. Remove sash and put in the top parting bead, cutting it long enough so that it will fit into the grooves at the side. Cut the side beads to fit the slope of the sill and fit them under the top bead. Replace the sash, dropping it to the sill, and mark the distance the bead comes out on the parting rail. Cut the sash to suit, replace it resting on the sill, slip in the side beads, raise the sash to position and fasten there with small cleats between the beads and the blind stop.

Trim lower sash the proper width, place against the beads, around which it is fitted. With a pair of dividers find the distance that the sash must be dropped to bring the tops of the parting rails flush on top. Scribe along the sill on the outside, rip off the excess wood at the bevel of the sill and fit to the sill with a smoothing plane. The sash is then held

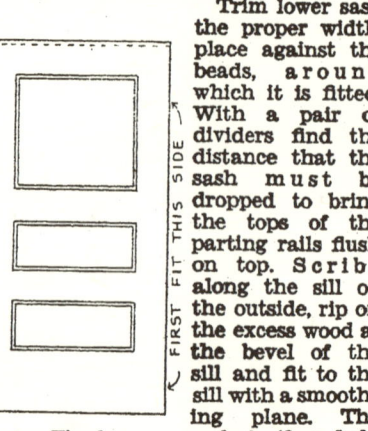

Fig. 1

in place with the window stop.

The window stool is then cut to fit against the bottom sash and is long enough to take the side trim and allow an additional 1½ inches to work a return at each end. The trim is then put on around the opening, last of all the apron, under the sill.

To fit a door, straighten the hinge side to fit the frame (fig. 1). Then scribe the top and other side. Inside doors swing half an inch free from the floor. Outside doors must be fitted to close just clear of the sill.

To hang a door, wedge into position tightly against the hinge jamb. With a half-inch chisel mark off the door and jamb 11 inches from the floor and six inches from the top. Use a hinge gauge to mark the distance the hinge is kept back from the side of the door and the corner of the rabbeting. You will thereby avoid hinge binding. Let the hinges into the wood the depth of the metal, with the bottom hinge above the lower chisel mark and the top hinge below the top one. Some adjustment is almost sure to be required to get the door to swing properly.

Locks are placed about 3 feet 2 inches from the floor. First make the mortise so that the body of the lock will fit into it snugly. The rest of the work is a matter of fitting.

In putting on the base, start with the longest span of wall and work in each direction toward the door. Fig. 2 illustrates how base with a moulding is coped at the corners of the room. Cut the piece to be fitted, as shown in the unshaded part, at an angle of 45 degrees, using a mitre box. Cut away with a coping saw to the intersection line and if the piece shown as shaded is plumb, the other will fit against it neatly at every point. This principle applies with any pattern of moulding.

To build a stair properly calls for special skill and is best done by an experienced carpenter. The rise is the distance from the top of one step to the top of the next. The run is the distance from the face of one riser to the face of the next one to it. The step is wider than this as there is a projection. A general rule regarding the proportions of the rise and run is that the two together should total about 16 inches. Fig. 3 shows a stair string, fig 4 the enlarged scale and fig. 5 the detail at the top, also enlarged. The shaded parts show the treads, risers, etc.

The first thing to ascertain is the exact height between the top of the finished downstairs floor and the top the upstairs floor. It is better done with a piece of say 1x2. Supposing it is 9 feet or 108 inches; if 15 risers are used this will work out to about 7¼ inches to the riser. Set a compass to this and keep adjusting it until you come out exactly to 15. Take a piece of board about half an inch thick or less and cut it into a triangular shape, with one side equal to the rise and the other equal to the run. After straightening the top edge of the string do some experimenting and find out exactly the point where the face of the riser, extended upward, intersects the top of the tread. Since this point will be the same distance from the top of the string

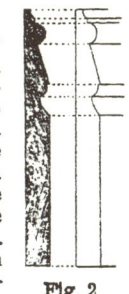

Fig. 2

SEED-QUALITY

Sometimes unseen but priceless

The quality you desire will be found in Steele Briggs' high purity graded stocks available for your selection early in the New Year.

For that new field buy Lion Brand High Grade No. 1 Seed.

ALSIKE CLOVER
RED CLOVER
ALTASWEDE CLOVER
ALFALFA
SWEET CLOVER
TIMOTHY

FIELD SEEDS

Carefully milled, grading No. 1 or Certified No. 1 Seed.

BROME
CRESTED WHEAT GRASS
SLENDER WHEAT GRASS
MEADOW FESCUE
CREEPING RED FESCUE
SUNFLOWER
SEED CORN
RAPE
MILLET
FLAX

●

Write for our 1947 Garden Seed Catalogue and Field Seed List

●

STEELE, BRIGGS SEED CO.
LIMITED
WINNIPEG, Man.   REGINA, Sask.
EDMONTON, Alta.

for each riser and tread, a line can be drawn the whole length of the string through this point. Then invert the pitchboard with the long side on the line and lay out the steps.

Now make two templates the exact size of the housings, one of the tread housing and the other of the riser housing, including the wedges in each case. The housing is at least half an inch deep and for the nosing it is made by boring a hole that depth with a bit which is the same thickness as the tread.

Always bore the nosing holes first and then do the sawing for the housing. In building the stair, wedge the risers and treads as shown after giving the wedges a generous coat of glue. Note, fig. 4, that the bottom riser differs in width from the others because it does not have to lap over the back of a tread at the bottom.

How To Frame a Barn

The weight of a barn should be taken by concrete or masonry. When concrete walls are used for a low foundation it should be two or three feet above the ground level. Bolts are inserted every six or eight feet to hold the sills (fig. 1). The wall is allowed to set well before putting the weight of the superstructure on it.

For sills select straight 2x6s. Beginning at one corner place one length of the sill on top of the bolt shanks, taking care that it is exactly above the position it will finally occupy. Hit it a smart blow with the hammer over the first bolt. This will give you the mark by which to bore the first hole. Bore it and replace. The bolt will catch in the hole if one man holds the scantling down a little. Use a short straight edge to keep the edge of the sill plumb over the side of the wall while the marking is being done. Mark, bore and replace. Continue until the sills are all fitted. Then make a batch of thin cement and sand to bed the sills. If properly done the sills will be level both lengthwise and crosswise.

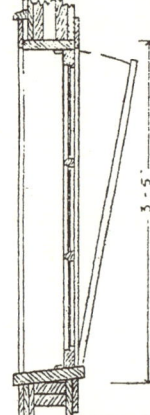
Fig. 1

The posts that support the girders stand on concrete abutments going down below the stable floor and having a good large bearing surface to take the weight. In cutting them make allowance for the depth of the girder and a corbel, if one is used. Place them in proper position and brace them perfectly plumb. On top of them place the corbels, if any, and on that build the girder, after lining them true with a chalk line.

Previous to this the studs will have been framed. In the plan, fig. 6, they are 14 feet high. They are squared at both ends and the only framing that is required is a housing in each to take the ribbing piece which supports the joist. The ribbing should be 1x6 of sound material. The top is one inch lower than the top of the girder to allow for crown.

In laying out the sills for the studs, make sure to work from the same end of the building so that the studs will be exactly opposite. Also see that the middle of the second stud from the corner is exactly two feet from the outside corner of the frame. When the side sills have been laid out, mark the ribbing pieces from them so that the studding will be plumb.

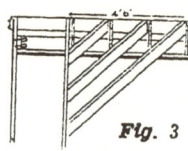
Fig. 3

Now you are ready to begin raising. The corner is built up of two 2x6s kept two inches apart by bits of 2x4. These are erected and braced to the inside of the sills. The next stud takes the other end of the first ribbing piece. A couple of joists can then be put up to tie these studs to the girder, with the outside ends resting on the ribbing and flush with the outside of the walls. The toe-nails into the girder will have to be drawn later and the heads are left out far enough to allow this to be done, not forgetting how hard it can blow in this country. Brace everything firmly, and proceed until all the side studding are in position.

The end sills and joists are then laid out to take the end studding, which are next nailed in place. *Fig. 4*

Lay out the openings for doors and windows in the stable. Fig. 2 shows how to frame a wide opening so as to prevent sagging. For nine-light stock sash with panes 9x12 the opening will be 2 feet 11 inches wide and 3 feet 10 inches high in the clear. Carry up the shiplap on the outside to the top of the joist and then straighten the walls by sighting along the top of the shiplap. With the joists all in position spike them together over the girders and toe-nail to the girders, being sure that the building is firmly tied together to prevent spreading. Fig. 4 shows how the joists are bridged to give further strength. The top plates can then be put on and the shiplap carried up to them.

Before framing the roof lay the floor of the mow. Then clear a big stretch of this floor to lay out your rafters on. Select two studs, which are exactly opposite, and strike a chalk line between them. Get the exact middle of the building and strike another chalk line up it using the first chalk line as a base. Then on each side of this centre line, and 7 feet 6 inches from the outside of the frame strike a line parallel to the centre line. Select two straight 14-foot 2x6s and two 12-foot 2x6s as rafter patterns. For the amateur it is a cut and fit proposition. Study the framing of the roof in fig. 6 carefully. The measurements are all there. Those rafters, braces and struts, as assembled in position, can be laid out on the floor. It is just a case of using your ingenuity and every single cut and length can be figured out. The heels of the rafters are scribed to fit the sides of the studs. Where the hip comes the proper cut can be found by placing the end of one rafter over the end of the other one, ticking off where the two edges of the scantlings intersect, and drawing lines for the cuts by connecting the points ticked off. The cut at the top of the rafter is along the centre line that has been chalked off. Make a good set of patterns for one side of the roof. Be sure, and doubly sure, that everything is exactly right and if they fit together on the lines you have drawn on the floor they have got to be right when they are standing in position.

Fig. 3 shows the projection of the frame to carry the hay sling track. The top piece had better be a piece of strong 2x6, framed back into the roof of the barn. The lower piece is also 2x6 fastened to the collar tie by iron brackets and hung from the ridge in iron stirrups and stayed to the outlook rafter by iron rods to keep it in line. It is a wise precaution to tie the top of the walls to the joist by pieces of 1x6 to prevent spreading.

The window shown in fig. 5 is the one commonly used to provide additional ventilation in warm weather, and at the same time avoiding drafts. It swings out at the top in a frame, which can be either of lumber or of galvanized sheet iron so that the inflow of air is over the top.

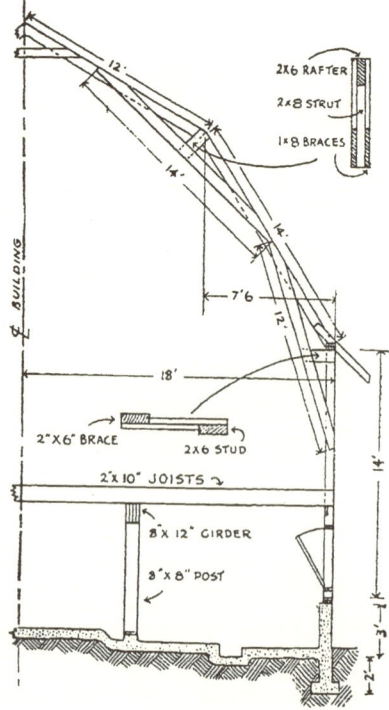

Fig. 5 (two straight 14-foot 2x6s)

How to Frame a Frame House

The illustrations, figs. 1, 2 and 3, show cross sections of the wall of a two-storey house and fig. 4 shows a cross section of the roof and wall where they join a one and a half storey house.

Beginning at the bottom, fig. 3, note that the wall plate is imbedded in the concrete and is flush with the outside. The bottom plate, to which the studding is nailed, is directly above it. This brings the sheeting and siding outside the perpendicular line formed by the outside of the basement wall and the outer edge of the studding. The mud sill, which should be at least 6x8 inches or better still 8x8 inches, is not shown in the drawings. It carries the inside ends of the ground floor joists and its ends are imbedded in the basement wall while it is supported inside the basement by one or more posts. A vital point regarding these posts is that the concrete basement floor should not be brought up around them. If it is they will rot and disastrous settling will result. They should be on a concrete base which is built up four inches above the level of the basement floor.

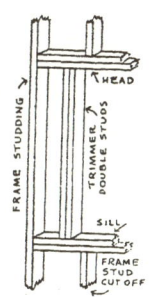

Fig. 1

After the joists are in position put on the rough flooring. All the bottom plates, both of walls and partitions, are placed on top of the rough floor. The outside finish at this point is a thickness of sheeting, preferably shiplap, a base board and drip cap and above that the siding. Two thicknesses of building paper, the inner one white and the outer one tar paper, are placed under the siding. Where the framework sits on the basement wall both outside sheeting and base board with the paper between, are brought down well below the wall plate. With the beam filling added you have a warm, windproof job.

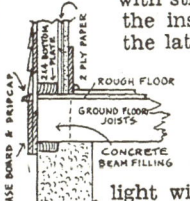

Fig. 2

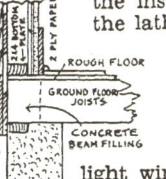

Fig. 3

The inner finish shows a thickness of shiplap siding on the studding. It is covered with two thicknesses of white building paper, then strapped and lathed and plastered. Instead of building paper on the sheeting, felt or sheet insulation can be used. The wall between the studding may be filled with shavings, moss or other bulk insulation material.

At fig. 2, is shown the wall at the first floor ceiling. A 1x4-inch ribbon is let into the studding to support the joists. The gains are made in the studding before they are put in place.

Fig. 1 shows the construction where the roof rests on the wall. The ceiling joists are carried out the width of the eave to support the rafters. The frieze board is nailed in place and the siding finishes up to it. The soffit may be plain board or V-joint. The facia is nailed on the end of the ceiling joint. It takes the eave-trough and projects down about ¾-inch below the soffit. A layer of white building paper, covered with a layer of tar paper goes on beneath the shingles.

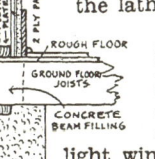

Fig. 4

Fig. 4 shows the construction of a storey-and-a-half house where the roof meets the walls. The collar ties are 2x4 inches and are nailed to the rafters. The soffit is on the angle of the roof though outlooks may be nailed in and the soffit put on the level. If it is, returns have to be worked in at the corners of the building.

For the upstairs ceiling two types of insulation are available. If bulk insulation is used all that is necessary is to lath as in fig. 1, directly on the joist or rafters, between which the insulation is placed. If blanket or sheet insulation is preferred, the construction is as in fig. 4 with strapping used to hold the insulation and to take the lath.

Fig. 5

Fig. 5 shows how an opening for a window is framed. If the opening is for a door the trimming is the same except that the side studs come down to the floor. Where two-light windows are used the opening is trimmed seven inches wider in the clear than the width of the glass to allow for the sash, the window frame and some play for plumbing the frame. The depth of the openings are the depth of the two panes of glass plus nine inches to allow for the sash and the top and sill of the window frame. The trimmed opening is of double 2x4's all round to give wood for nailing the finish to. In cutting the studs additional allowance of 3½ inches has to be made for the top and bottom trimmers. From where the bottom cut is made to the top of the finished stool is about seven inches.

When trimming an opening for an outside door leave it four inches more in height and about three inches greater in width that the dimensions of the door. In addition 3½ inches must be allowed when cutting for the height, to allow for the double trimmer.

Fig. 6 shows a cross-section of a two-light window. It includes the outside

Fig. 6

Standard Quality Lines

Your Best Buy Anytime for Economy — Lasting Performance — and Satisfaction

"DOMO" "STANDARD" Cream Separators

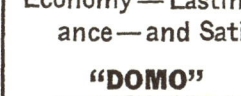

A size and style for every need. Prices as low as **$22.50** and up. A dozen sizes to choose from.

FORDS Electric and Gas Engine Milkers, and Fords Hammer Mills

Easy to install—no pipelines—fits all barns. Finest performance—lasting satisfaction. Prices within reach of all farmers.
Inquire about the **FORDS MILKER** before you buy. Write us or see our dealers.

"STANDARD" GRAIN CRUSHERS

with patented feed control — roller bearings. Highest capacity. A marvel in performance. 9¼, 10 or 10½ size.
$59.50 f.o.b. Winnipeg.

"GILSON"

Furnaces, Fans, Oil Burners, Stokers, Chick Brooders, etc.

Canada's Leading Heating Line

for any size home, school or church. We supply all fittings, piping, registers, etc.

Write for full information, prices.

GILSON "Snowbird" Power Washing Machines

Canada's outstanding washer value. Four electric; four engine models.

★

GILSON

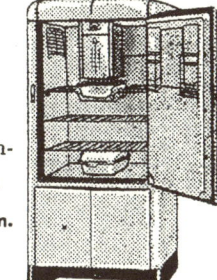

Domestic — Commercial Refrigerators.

Six sizes to choose from.

★

Write us for complete price lists and literature.

STANDARD IMP. & SALES CO.

78 Princess Street   9851 Jasper Ave.
WINNIPEG, MAN.   EDMONTON, ALTA.

and inside finish. The details of the construction of a window frame are shown in fig. 7. The inside of the frame is four inches wider than the glass. The head, and the sill at the shoulder, are cut ¾-inch longer than this as each end is let into the side ⅜ inch. The sill is given sufficient pitch to drain the water off. The depth, inside measurement, is six inches more than the depth of the two glass panes.

In fig. 7 the detail is shown. Imagine one side of the assembled frame to be sawn through after being nailed in position and that you are looking down on the cross-section, which is shown in the middle of the cut. The 2x4 is the studding in the wall, which is usually double. The frame is held in position by being nailed to the studding through the

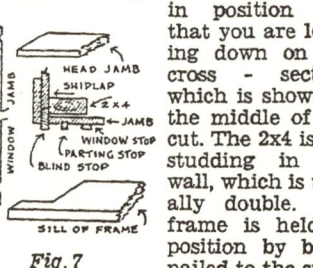

Fig. 7

blind stop. The top sash fits between the blind stop and the parting stop and the lower sash slides between the parting stop and the window stop. Outside the blind stop the outside casing is shown. The side jamb projects in past the studding far enough to allow for the sheeting, strapping and lath and plaster. The plaster, therefore finishes flush with the frame so that the inside casing lies flat with the plaster.

The cellar frame is the easiest of the outside frames to make. The details are shown in fig. 8. The amateur carpenter had better take his measurements direct from the sash. Be sure that the sides fit in between the head and sill so that the frame will stand up under the weight that

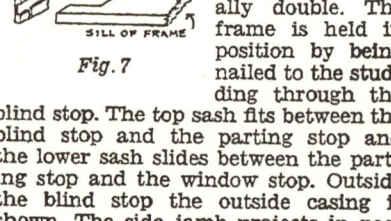

Fig. 8

may be put on it by the building above. It is rabbetted as shown to take the sash on the inside and the storm sash or screen on the outside.

Building with Stonebord

Here are some helps in putting on Stonebord or other kinds of wallboard. In using the sidewall helper, butt lower board tightly against the upper one, using the lever helper shown, so baseboard covers any open space. Span bottom of window openings. Maintain at least one-inch space between board and basement floors.

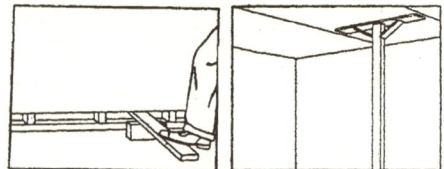

Over cracked plaster, rout out loose plaster down to wood lath. Apply furring strips 16 inches o.c., flush with old plaster surface. Nail through old surface to wood studs or joists.

Forming arches, face boards are cut with a saw and sanded. Soffits are curved by scoring back paper at intervals of approximately ½-inch depend-

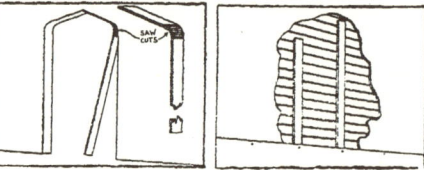

ing on radius. Then nail to curved contour and sand edges.

A Ceiling "Helper"—Apply ceiling boards first using the handy "T" helper shown. Over cracked plaster ceilings, furring strips may be used. Start nailing at centre of the board and work towards ends.

The edge of this board provides a channel for applying the tape cement. It assures a level surface without "feather edging" over a wide area and provides for perfect concealment of the tape.

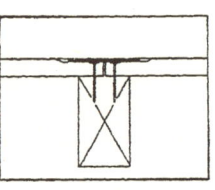

Improvement to Ice Cellar

Our ice cellar is 7-ft. x 7-ft. square and 7-ft. deep. The cribbing extends about one foot above ground level. It is banked up on the outside to keep out surface water. The ceiling of the ice well is a wooden platform at the ground level and above it the cribbing

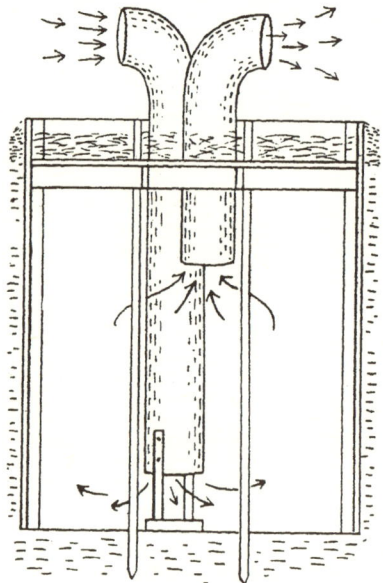

is filled with straw. Our problem has been to freeze the ground from the inside of the cellar for 6 or 8 feet in all directions so the ice will not melt so rapidly on the sides and bottom. By placing a couple of pieces of old blower pipe through the trap door as shown in the sketch, this can be accomplished. This wind will generally be blowing into one pipe and out the other and if the cellar is kept clear of snow it should be well frozen by March. The ice should be cut earlier in the winter and piled near this ice cellar and covered with a load of straw. Toward spring the ice is put in the ice cellar and straw is spread on the ground all around to keep the frost in, near the ice cellar, as long as possible.—Jas. E. Moscrip, *Major, Sask.*

Raising Log Building

A man on a farm who wishes to build a log structure and is short of help, can get along mighty well alone by taking log poles, and notching them every

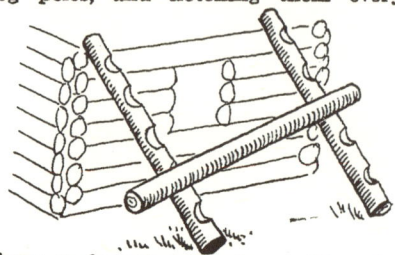

three or four feet, as shown. Then he sets one up at each corner of the building and slides the logs up them, one end at a time. This is a still greater help when nearing the top of the wall. The skids can be made 30 or more feet long.

Making Concrete Walk

The appearance of a concrete walk can be improved at very little expense by making a flagstone effect. This is done by using a form which is so divided that it will make the different sizes of blocks needed. These are then set into a kind of pattern, such as is shown below. The blocks are set in sand just thick enough upon the clay to help get the blocks all to the same level. A pleasing effect is gained if there is distance enough between the blocks for the grass to grow.

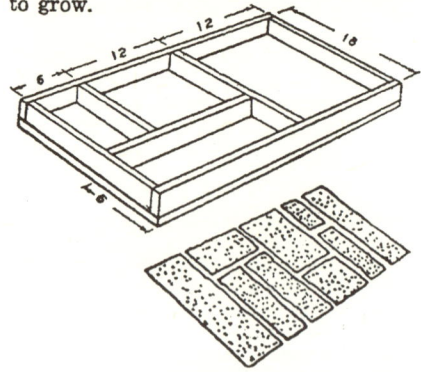

Concrete Bird Bath

The pedestal of this bird-bath is made on the square. It is made in two halves which are then bolted together, the heads and nuts of the bolts being countersunk and the holes covered with concrete. The column is five inches across at the top and nine inches at the bottom. The same form is used in making the two halves. Note that reinforcing rods are used. The bowl is made upside

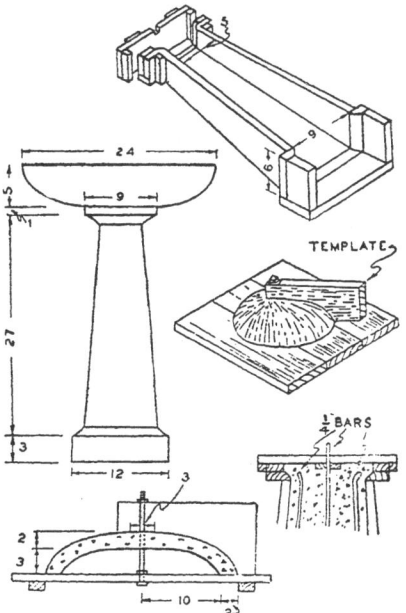

down using templates. The form for the inside of the bowl is made of clay, shaped as shown in the centre right diagram. On top of this the concrete is worked, as is shown in the lower left. Where there are children it is a good precaution to make the base longer and sink it into the ground so that they cannot pull the bird bath over on themselves.

Coloring Cement

Coloring concrete, for example the blocks in a patterned walk, is best done by mixing the colors with the dry cement. Different shades may be secured by using different amounts of coloring matter. The method followed is to mix small trial batches until the shade desired is obtained. For blues, use ultramarine blue; for buffs, yellow ochre or oxide; for pinks and reds, small quantities of red oxide of iron. Greens can be obtained by using a mixture of yellow oxide and ultramarine blue.

Outdoor Pit and Dumb Waiter

IN any house where it is impossible to install an ordinary dumb waiter like the one described in the October issue of The Country Guide, this outdoor pit and dumb waiter combined will be found quite satisfactory.

No dimensions are given, as size and proportions will vary.

Where there is no danger of seepage the pit could be dug deeper—of course the pit is cribbed—the top could be slanted in order to shed the rain—the ladder could be replaced by a curved stairway starting above the chute and ending in the opposite corner.

Shelves are placed along the sides, hooks near the top are handy to hang meat on—in winter it remains frozen, in summer it stays cool. "D" shows the door at the lower end of the chute open for cleaning or for loading the carrier from the pit. This door has a small inset of screening for ventilation, there is also an inset near the top of the north wall for the same purpose. The carrier "A" is shown in two positions, ready to start on its journey and at its journey's end. The small drawing "B" shows how the rope is attached at the front of the carrier. At the top of the chute the loop of rope is passed over the hook on the door "E." This door is on hinges and is held level by two straps fastened to the inside upper extremity of the chute. When the door is closed it is folded inside. The door is secured in place when closed by cupboard catches. The wheels at the back of the carrier are old toy wagon wheels. "C" shows the construction of the back of the carrier. The position of the axle depends on the size of the wheels. The front of the waiter runs on castors. No special track is necessary but reasonable care must be taken when loading the carrier to see that the load is balanced properly. If a more elaborate installation is planned a passage could be constructed beneath the floor of the chute, and weights and pulleys attached as in an ordinary dumb waiter.

Care should be taken in constructing the chute to make it waterproof. At "F" the soil should be banked up so that the rain is shed away.

This outdoor pit is very convenient in summer as it remains cool, being on the north side of the house. To the busy woman who has to prepare three meals a day and a few lunches for good measure this step saver will be a boon. Trips down cellar for butter, milk, cream, bread, cake, etc., will be elimin-

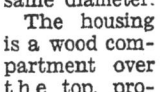

ated, and countless unnecessary steps saved. During the winter the pit will be found an ideal place for that quarter of beef, those turkeys or other meats.—E. C. Lundy, Scapa, Alta.

Editor's Note: Now will some Guide reader design and try out this chute idea in connection with an ice well? We have seen and described in these columns, shallow ice wells that were giving satisfaction, but they were not accessible from inside the house, as this outdoor dry pit is, by means of the chute. If any one tries the combination, and it works, let us know about it.

Dumb Waiter Keeps Food Cool

This dumb waiter which is used in hot weather in place of an ice box, is quite efficient. One small oil drum slides up and down in a hollow cylinder made from larger oil drums. The small drum, which is the dumb waiter proper, has a hinged door and shelves made from the ends of other barrels of the same diameter.

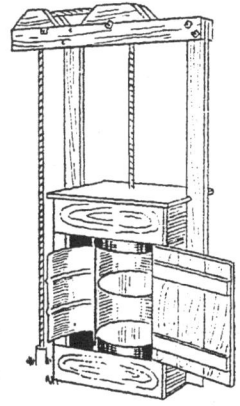

The housing is a wood compartment over the top, provided with a hinged door on one side and is quite dust proof. An extension above and to the left carries two pulleys over which the rope from the dumb waiter to the counter weight pass. The counter weight is a pipe filled with enough lead to make easy work of raising and lowering the dumb waiter. This lead weight slides up and down in a larger pipe whose top lies flush with the top of the ground.—Dale Van Horn.

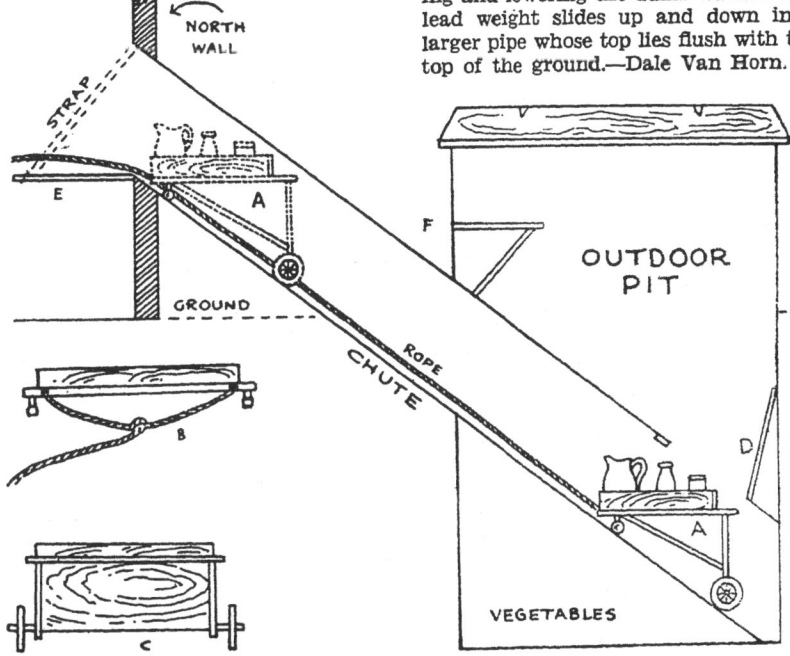

Advertisers' Index
Farm Workshop Guide
1947 Edition

"x" before the number indicates that the advertiser offers sample, trial offer or free literature.

Product	Page
x 1. Automobile Guide	40
x 2. Baby Chicks	45
3. Batteries, Dry Cell	81
4. Batteries, Radio, Flashlight, etc.	54, 55
5. Batteries, Radio	150
6. Batteries, Storage	59
7. Batteries, Storage	83
8. Boots, Men, Boys, etc.	89
9. Cigarette Tobacco	66
x10. Cook Book, Flour	35
11. Cream Separators, Feed Grinders, etc.	143
12. Electric Fencer, Farm Tools	51
x13. Engineering Opportunities	4
14. Farm Improvement Loans	93
15. Farm Improvement Loans	111
x16. Farm Lands	3
17. Farm Machinery	39
18. Farm Machinery	71
19. Farm Machinery	90, 91
x20. Farm Machinery, Engines, Tools, etc.	115
21. Farm Machinery Rubber Tires, etc.	76
22. Feed Grinders	43
23. Feed Grinders, Elevators, Pumps	118
x24. Feed Grinders, Milkers	133
25. Feeds for Livestock, Linseed Oil	10
x26. Feeds for Poultry	23
27. Feeds for Poultry	47
x28. Feeds for Poultry and Livestock	27
29. Feeds for Poultry and Livestock	31
30. Feeds for Poultry and Livestock	57
31. Feeds for Poultry and Livestock	84
x32. Feeds for Poultry and Livestock	92
33. Feeds for Poultry and Livestock	113
x34. Files	7
35. Fuels and Lubricants	63
x36. Fungicide	37
37. Fungicides, Insecticides, Etc.	65
x38. Garden Seeds	50
x39. Garden and Field Seeds	141
x40. Gas Engines	Fr. Cover
x41. Grain Cleaning Equipment	123
42. Grain Handling Service	92
43. Grain Handling Service	149
44. Hack Saw Blades	73
45. Implement Seat Cushions, Overalls	30
x46. Lighting Plants	137
47. Livestock Salt and Salt Blocks	29
48. Lubricants	87
x49. Lubricants	135
50. Lubricants, Batteries, Paints	125
x51. Lubricant and Grease Gun	89
52. Lumber	16
53. Lumber	91
54. Machinery Drive Belts	9
55. Machinery Drive Belts	121
56. Machinery Parts Service	49
57. Mackinaw Jackets	18
58. Magneto Service	67
59. Meat-curing Compound	21
60. Overalls	20
61. Pants, Tweed	19
62. Paints	129
63. Paints, Varnishes, Enamels	131
64. Patent Attorneys	53
65. Piston Rings	105
66. Plastic Wood, Household Cement	56
67. Plumbing Supplies, etc.	44
68. Pumps, Pressure Systems	107
x69. Roofing and Building Material	95
70. Roofing Material	33
71. Rubber Boots	101
72. Saws and Files	116
73. Shirts, Wool	17
74. Steel Sheets, Metal Fasteners	41
75. Tools	103
76. Tools, Paints, etc.	117
x77. Tractor Accessories	97
x78. Tractor Repair Parts	110
79. Trucks	25
80. Trucks	85
81. Trucks, Parts, etc.	99
82. Wallboard	119
x83. Well Curbing	60
x84. Wood Preservatives	109

THE FARM WORKSHOP GUIDE, 1947
Winnipeg, Canada.

From the items numbered I have selected the following in which I am interested in the literature, etc., offered.

Name ....

P.O. ....

Province ....

Numbers ....

Please print plainly.

Dumb Waiter in Ice Well

In The Guide for August you ask your readers if they have tried out the idea of a dumb waiter arrangement in connection with an ice well. I have had one here for a number of years, and it is really fine. We have used it for cream and every other thing that requires to be kept cool in summer. It is accessible from our back kitchen, from which a door opens to the ice well. The well is about 12 feet deep and 3x4 feet in size. It is cribbed with lumber in such a way that

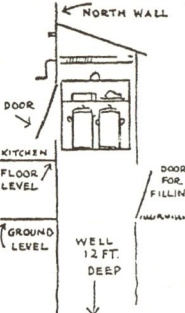

the dumb waiter can slide up and down on the inside of the crib. I have 1x2's fastened at the corners to guide the dumb waiter. The cribbing comes up to about six feet above the kitchen floor. There is also a door at the ground level on the outside for filling the well in winter. I put in some snow, then a few pails of water, and let it freeze and continue until the well is full to the ground level. The well is then filled with a solid block of ice. The dumb waiter sits on the ice all the time. It is a box, made of half-inch lumber, just large enough to hold two cream cans on the bottom with shelves above for butter and other household foods. For raising up and down I use a windlass with a bolt to hold the box at the floor level. This is one of the best arrangements I know of for keeping cream and other things cool in the summer and the cost is merely a bit of lumber and the labor of building it. I am sure it pays for itself twice over every year, especially where cream has to be kept.—F. Clear, Polworth, Sask.

Quilting Frame

Can your wife or mother use to good advantage a quilting frame that stretches and really holds a quilt, one that can be knocked down and stored in a small space in a closet or other room between quilting jobs? If so, why not get busy right away and construct

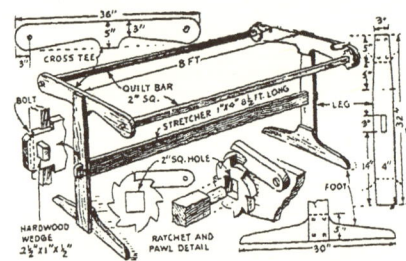

the improved frame shown below. Start by cutting the feet and legs. Spruce or white pine, or in fact almost any kind of wood is suitable. After the legs are assembled, bore holes to take the quilt bars. Fit the stretcher and round the ends of the quilting bars. A wood chisel and pocket or drawing knife can be used on this job. Next, fit ratchets on the bars, noting that they oppose each other. Place the bars in the frames and fit the pawls, using round-head wood screws to serve as pivots. Sand all the parts carefully and give them a coat of shellac.

Rest in Comfort

This wooden rocker is not at all hard to make. Here are some of the measurements: The rockers are 30 inches long and the arms 25 inches. The cross pieces that form the seat are 23x3½ inches, and the two centre pieces of the back 36x3½ inches. Assemble the rockers and the seat first, using 1½-inch

flathead screws and 2½-inch nails. Drill the holes for the screws to avoid splitting the wood. Nail the back together and fasten it in place and then put in the arm supports and the arms. If you happen to have a little white or green paint on the place give the finished product a couple of coats to protect it from the weather.

A Lawn Settee

This lawn settee is constructed entirely of inch lumber. The curved edges are cut with a key-hole saw. Here are some of the major measurements, from which the rest can be judged. The front standards are 23 inches high and 46½ inches from outside to outside. The front support is 46½ inches long and six inches wide and is let into the front standards its own thickness. The arms are 27 inches long and

4½ inches wide, and curved on the inside. The runners are three in number, one to support the seat in the middle, and are curved to make the hollow in the seat. The two end ones are bolted to the front standards while behind they rest on the ground. The highest pieces of the back are 33½ inches long and 5 inches wide and the others somewhat shorter as shown. Note that there is a cross-piece to hold the back together near the top. It is 2½ inches wide and 43½ inches long. The arms are supported at the rear by another cross-piece 2½ inches wide and 51 inches long.

INDEX to CONTENTS

Item	Page
Accelerator Accessory	114
Animal Catcher	120
Anvil Support	95
Anvil	28
Anvil Bench, Putting On	26
Auto Chains, Putting On	92
Auto Chains, Storing	94
Auto Horn, Changing Tone	97
Awl from Valve Stem	103
Babbitt, Banding	94
Babbitt, Melting	96
Bait Spreaders	49
Band Cutter for Pitch Fork	96
Band Saw, Power	2
Barge Header	51
Barge, Header for	52
Barn, How to Frame	142
Barn Door, Inside Rolling	108
Barn (or Garage) Door, Non-Binding	108
Barn (or Garage) Door, Protecting Track	106
Barn (or Garage) Door Catch (Four)	106, 108, 109
Barn Door Closer	109
Barn, Screen Door for	106
Barn Ventilator	94
Barrel Dog House	103
Basket Repair	93
Battery Cap Lifter	111
Battery Care (Two)	90
Battery Charger	88
Battery Connector	88
Battery Connections	90
Bearings, Rebabbitting	91
Belt Dressing	112
Belt Friction Drives	11
Belt Punch from Hinge	98
Belt Tightener	6
Bench Drill (Five)	9, 26, 27
Bench Grinders (Four)	26
Bench Grinder, Treadle for	26
Bench Saw, Power	1
Binder Canvas Repair	48
Bit, Depth Marker for	103
Bits, Sharpening Auger	32
Board, Dividing into Equal Spaces	33
Bolt Fastening Kink	95
Bolt, Straightening	94
Boulder Remover	48
Bottle Opener	96
Brads, Starting Small	98
Briar Cutter	108
Bridle Hooks	115
Brush Axe from Rolling Coulter	109
Brush Hook from Shovel	104
Buckle Tongue from Split Key	101
Building with Logs	139
Building with Stonebord	144
Building Materials, Estimating	139
Bullets for Rats	97
Cabinet from Oil Cans	29
Calf Weaner	124
Cams and Followers	14
Cart, Single Wheel	112
Cattle, Anti-Jumper for	122
Cattle Breeding Rack	120
Cattle, De-horning and Branding Shute	120
Cattle, Hoof-trimming Rack for	121
Cattle Loading Shute	122
Cattle Poke	124
Cattle, Self-feeder for	124
Cattle Stanchions	125
Chain Repair (Two)	101, 102
Chart for Calculating Pulley Sizes and Belt Speeds	35
Chopping Block, Handle for	98
Chimney Care (Three)	132, 133, 135
Chimney Extension	135
Chimney, Heat Saving	136
Chisel for Cutting Drums	32
Cistern Filter	132
Clamp, Carpenter's	28
Clothes Hook	117
Clothesline Pole	86
Clothes Pin Holder	100
Clutches, Simple Types of	12
Cold Chisel	33
Concrete Bird Bath	144
Concrete, Coloring	145
Concrete Floor Scraper	114
Concrete, Mixing Aid	102
Concrete Rocker Mixer	102
Concrete Walk	144
Corn Cutter	48
Cotter Pin, Using	102
Cranks and Eccentrics	13
Cream Can Holder	115
Cultivator, Field	46
Dehorning, Pencil Holder	121
Devices for Adjusting	15
Devices for Designing Machines	10
Disc, One Way	45
Discs, Sharpening Tiller	32

Item	Page
Dog, Stop Car Chasing	120
Door, Folding	115
Door Handles, Assorted	114
Door Hook	110
Door, Sagless	91
Drain Pipe Opener	136
Drill, Depth Gauge for	119
Drill, Pressure Springs, use for	92
Drills, Sharpening Twist	31
Drill Wheels, Repairing	103
Drum, Cart for	113
Drum, Plug Wrench for	112
Drum Tipper	102
Dumb-waiter in Ice Well	146
Dumb-waiter from Oil Drum	145
Dumb-waiter for Out-Door Pit	145
Dusting Machine (Three)	104
Electric Element Repair	88
Electric Persuader	88
Electric Wiring Tips	90
Engine, Deisel	37
Engine, Farm	36
Engine, Frozen Block	42
Engine, Fuel Cost per Horse-power Hour	38
Engine Repair	59
Engine Timing	90
Engine Timing, Correcting Faulty	87
Fence, Electrifying Farm	87
Fence, Guard for Electric	83
Fence, Loop Hole in	83
Fence, Mending Break in (two)	76, 83
Fence Post Puller (five)	81, 84, 86, 115
Fence Wire Tightener (four)	74, 76, 84
Fertilizer Distributor	42
File for Loose Papers	97
File Handle	117
Firewood Holder	91
Flooring Aid	140
Flue Hole Cover	136
Fodder Shock Binder	114
Foot Scraper (two)	97
Forge	31
Fuel Strainer	115
Furnace, Pipeless	132
Furnace Pipes, Insulating	133
Fur Stretchers	93
Garage, Lengthening	140
Garden Cultivator	108
Garden Cultivator, Power	59
Garden Harrow	108
Garden Hoe, Improvised	108
Garden Rake from Fork	109
Garden Weeder	119
Gas Drum Loader	98
Gas Line Tubing, Repairing	102
Gasoline Pail	110
Gasoline Screen	118
Gate Anchor	84
Gate, Cantilever	75
Gate, Cattle Preferred	81
Gate, Chain	75
Gate, Child Proof	81
Gate, Dogs Preferred	75
Gate, Easily Climbed	83
Gate, Emergency	75
Gate from Two by Fours	75
Gate Hanger	83
Gate, Improved Wire	75
Gate Latch (eight)	81, 83, 84, 86
Gate, People Preferred	82
Gate Pole	75
Gate, Safety Device	121
Gate, Sagless	75
Gate, Self-Closing	81
Gate, Squeeze	82
Gate, Texas	81
Gate, Wide	75
Gear Shift Repair	98
Generator Repair	114
Grain Elevator (two)	52, 57
Granary Door	114
Grate for Burning Wood	135
Grates, Mending	132
Grease Cup, Pressure	114
Grinder from Generator	90
Grindstone, Re-shaping	29
Hack Saw Frame	32
Hack Saw Handle	112
Hack Saw Power	5
Hammer, Lead	33
Hand Protector	96
Hangers for Workshop Shaft	26
Harness Repairs (two)	105, 106
Harness Scrubbing Board	100
Harrow, Disc	46
Harrow, One-way Disc	46
Harrow, Rotary	46
Harrow, Spike Toothed	46
Harrow, Spring Toothed	46

Item	Page
Hay Rope Repairs	107
Hay Unloader	72
Heater Cap	102
Heater from Cream Can	132
Heater from Oil Drum	132
Hedge Trimmer	100
Hinge for Small Box	111
Hitch—Drill after Tiller	128
Hitch—Eight-horse Tandem	129
Hitch—Five-horse Abreast Evener	128
Hitch—Five-horse Tandem Evener	128
Hitch—Four-horse for 14 or 16 Sulky Plow	128
Hitch—Frame Tractor	127
Hitch—28 and 20-Run Drills	127
Hitch—One-way Behind Binder	131
Hitch—Safe Binder	131
Hitch—Seven-horse for 14 Triple Gang	127
Hitch—Six-horse for Drill	126
Hitch—Six-horse Tandem	128
Hith—Six-horse Tandem for 14 Gang	127
Hitch—Tandem Disc	126
Hitch—Tandem Plow	129
Hitch—Ten-horse for One-Way	127
Hitch—Three Implement	126
Hitch—Three-horse Eveners (four)	126
Hitch—Three-team Tandem	128
Hitch—Tractor for Two Binders (two)	129, 130
Hitch—Tractor for Two Mowers	127
Hitch—Two Drill	127
Hitch—Two Implement for Hilly Land	130
Hitch—V-Iron Tractor	126
Hitch—Weeders for Tractor	126
Hoe from Fork	119
Hog Catcher	121
Hogs, Field Self-Feeder for	124
Hogs, Inside Self-Feeder for	123
Hog Scalder	122
Hoist, Butchering	67
Hoist, Sampson	67
Hot Water Bottle Holder	111
House, How to Finish	140
House, How to Frame	143
Insecticide Duster	98
Kindling Wood Holder	104
Knife Rack	28
Knives from Old Files	113
Lace Leather Cutter	118
Ladder, Rubber Guard for	92
Ladder Rungs, Fitting	100
Ladder—Wheeled Step	98
Lathe, Power (two)	6, 8
Lawn Rocker	146
Lawn Settee	146
Leather Knife from Saw	103
Level, Handy Farm	101
Level, Improvised	103
Litter Carrier for Straw	58
Lock and Doorstop Combination	96
Log Holder	117
Log Holder for Saw-horse	96
Machine Jack	101
Magnet as Pin Cushion	104
Magpie Trap	122
Mail Box Signal	97
Manure Spreader, Preserving	117
Match Striker	96
Merry-go-Round	94
Mitre Box	26
Moisture Conservation under Different Fallow Treatments	47
Monkey Wrench, Using	118
Mouse Trap	118
Needle for Sewing Bags (two)	100
Nut, Loosening Rusty	115
Nut, Loosening Tight	118
Oilstone, Box for	96
Packer, After One-way	49
Pail Handle Repair	111
Paint Brushes, Protecting	94
Paint Can Holder	112
Parts Rack	112
Pencil Sharpener	117
Peony Support	113
Picture Frame Fastener	91
Pipe Connections	135
Pipe Reducer	97
Pipe Repair (two)	133, 136
Pipes, Thawing Frozen (three)	133, 134, 136

Item	Page
Pipe Wrench, Improvised (two)	104, 117
Pitch Fork Repair	115
Planes and Chisels, Sharpening	32
Playground Device	94
Plough, Moldboard	43
Post Hole Auger (two)	109, 110
Poultry Grit Feeder	120
Poultry Mash Feeder	120
Poultry Waterer, Cover for	102
Poultry Waterer, Protector	121
Poultry Waterer, Support	122
Power Required per Acre for Various Tillage Machines	48
Pump, Drinking Fountain for	134
Punch from Tine	92
Pump Jack	136
Pump Jack Pin	114
Pump Operation	134
Pump Pipe, Draining	134
Pump, Warming	135
Push Hoe (three)	109, 110
Quilting Frame	146
Raising Logs for Building	144
Razor Blade, Using	112
Road Tar Spots, Removing	112
Roof, How to Frame	138
Sack Holder	111
Salt Block Stand	114
Sanding Block	118
Saw for Big Logs, Power	73
Saw Frame	73
Saw Frame, Tilting	74
Saw Gauge	74
Saw Horse Accessory	112
Saw Horse Extension	97
Saw Horse, Non-pinching	103
Saw Horse, One-Man	104
Saw Horse, Open Top	29
Saw, Ice	74
Saw, Power Drag	73
Saw, Protecting Circular	94
Saw Set, Circular	26
Saw, Sharpening Circular	34
Saw, Sharpening Circular Rip	34
Saw, Sharpening Hand	33
Saw, Sharpening Hand Rip	33
Saw, Swing	74
Scissors, Sharpening	110
Scoop Shovel, Protecting	118
Screen Door Opener	98
Screw Driver from Bit Shank	32
Screw Nail, Starting	117
Seed and Feed Building	136
Seed De-Awner and Dresser	53
Seed Grader, Wind Blast	58
Sewage System	131
Sewing Horse (three)	28
Sheep-Salting Trough	120
Sheet Metal Bender	27
Sheet Metal Cutter	94
Shelf, Handy Milk	96
Shelf for Small Parts	29
Shelves, Hanging	114
Shingling Scaffold	140
Shock Absorber for Towing	98
Shovel from Ford Fender	110
Silage Wagon, Long	64
Silage Wagon, Short	64
Snow Plow	57
Soil Drifting Control	48
Soldering	16
Soldering a Seam	21
Soldering Fluxes	21
Soldering Enameled Ware	22
Soldering—Heating the Copper	16
Soldering Iron	90
Soldering Iron, Heating	100
Soldering—Sweated Joints	21
Solder into Ribbons	104
Spray Pump	100
Sprockets, Removing	92
Stacker, Cable	72
Stacker, Beaver Slide	70
Stacker, Braced Boom	66
Stacker, Overhead	68
Stacker, Roll-in	68
Stacker, Roll-up	72
Stacker, Single Pole	71
Stacker, Swinging Boom	68
Stacker, Two-Pole	72
Stack, Holding Top on	117
Staple Puller	103
Stiles (eight)	82
Stone Boat	57
Stove Pipe Holder	131
Stove Pipe Lock	134
Stump Burner	98
Swath Lifter	59
Sweep, Power	56
Switch, Alarm Clock (three)	86, 89
Switch for Pump Engine (Mercury)	87
Switch for Van	89
Syringe Needle Marker	100

INDEX to CONTENTS---Continued

Item	Page
Tap, Repairing Noisy	134
Tell-Tale Tank Float	102
Tender for Tractor	62
Tillage and Western Agriculture	47
Tillage Equipment—Use and Adjustment	43
Tire Spreader	117
Tool Box	32
Tool from Fork Tines	103
Tool Rack	31
Tools, Fitting Handles for	94
Towel Drying Rack	98
Trace, Drop Preventor	96
Tractor—Daily Service	38
Tractor, Gasoline Gauge for	60
Tractor Lighting System	88
Tractor Light, Mounting Rear	88
Tractor, Maintenance and Overhaul	38
Tractor—Major Overhaul	40
Tractor—Minor Overhaul	40
Tractor—Periodic "Tune-up"	39
Tractor, Pick-up Box for	57
Tractor, Radiator Front for	42
Tractor, Rear Light for	90
Tractor, Starting Cold	115
Tractor—Weekly Service	39
Trailer, Auto	61
Trailer, Four Wheel	64
Trailer, Livestock	62
Trailer or Buggy	62
Trailer, Steering Device for	64
Trailer, Tractor	61
Trailer for Moving Brooder	62
Trough Control Valve	125
Trough Cover	124
Trough, Insulating Concrete	122
Trough, Water Heater (three)	120, 122, 125
Tube Repair Rack	112
Turnstile (two)	82
Two by Fours, Straightening	91
Ventilator for Ice-Cellar	144
Vise for Auto Jack	28
Vise for Thin Wall Pipes	28
Vise, Saw	28
Vise, Smooth Jaw	28
Vise, Soldering	28
Vise, Workbench	32
Wagon from Auto Chassis	64
Wagon from Auto Parts	63
Wall, Moving	95
Water Pipe, Insulating	135
Water System for Kitchen (two)	133, 134
Weeder, Blade	46
Weeder, Rod	49
Weeder, Rod and Wire	46
Wedge, Strengthening Wooden	92
Welding	22
Welding, Electric Arc	22
Welding, Forge	22
Welding, Oxyacetylene	22
Well Bucket Tipper	136
Well Pipe Holder	136
Well, Protection from Frost	134
Well, Windlass	134
Wheelbarrow	102
Wheel Hoe	109
Wind Charger, Tower for	89
Wind Indicator	102
Windmill Cutout (two)	100
Window Crack Repair	96
Windows, Imitation Frosting for	92
Wire Snipper from Scissors	110
Wiring Model T Engine	86
Workbench and Vise	26
Workbench Catch-All	31
Workbench Drawer, Swinging	29
Workbench, Portable	29
Workshop and Garage Plan	27
Woodbox for Kitchen	138
Wood Carrying Sling	104
Wood Joints	140
Wood Splitter	56

INDEX to SECTIONS

Section	Page
1—Work Bench and Machine Shop	1
2—Engines, Implements, Heavy Equipment	36
3—Trailers, Wagons, Sleighs	61
4—Derricks, Hoists, Presses	66
5—Power Saws	73
6—Gates, Fences, Clotheslines	74
7—Electricity and Electric Wiring	86
8—Constructing Simple Gadgets	91
9—Concerning Livestock	120
10—Hitches—Tractor and Horse	126
11—Water, Drainage and Heating Systems	131
12—Building, Carpentry, Concrete	136

Published by THE COUNTRY GUIDE LTD., 290 Vaughan Street, Winnipeg. Printed by THE PUBLIC PRESS LIMITED, Winnipeg.

Co-operative for OVER 40 YEARS

UNITED GRAIN GROWERS LIMITED

The history of United Grain Growers Limited—the farmer-owned company—is that of an organization born and bred in the highest traditions of co-operation.

No narrow concept, nor sectional interest has ever been served by this farmers' company. Its directors, in carrying out their trust and responsibility, have consistently sought to serve the national interest of agriculture as well as that of the Company's shareholders and customers.

Thus in 40 years the U.G.G. has initiated and advocated many of the policies which are today accepted as wise and prudent and in the general interest.

The influence of the unified voice of Canadian agriculture in the councils of world agriculture is in the U.G.G. tradition of achieving strength and making progress through co-operation.

The co-operative movement in Canada has made great strides: its record of achievement is, in the main, solid and strong. In the framing of co-operative policy, the long experience of United Grain Growers Limited in handling the farmers' business through its elevators and terminals, has been and will continue to be a leading factor in helping to maintain the stability and progress of the organized farmer in business.

The company could not, however, have succeeded and progressed without the loyal support and patronage of thousands of shareholders and customers. To these loyal friends and supporters, the President and Board of Directors of United Grain Growers Limited wish to pay their tribute of warm and sincere appreciation.

United Grain Growers Ltd.

Publications by Algrove Publishing Limited

The following is a list of titles from our popular *"Classic Reprint Series"* as well as other publications by Algrove Publishing Limited.

ARCHITECTURE, BUILDING, AND DESIGN

Item #	Title
49L8096	A GLOSSARY OF TERMS USED IN ENGLISH ARCHITECTURE
49L8137	AUDELS CARPENTERS AND BUILDERS GUIDE - *VOLS. 1-4*
49L8016	BARN PLANS & OUTBUILDINGS
49L8046	BEAUTIFYING THE HOME GROUNDS
49L8112	BUILDING WITH LOGS AND LOG CABIN CONSTRUCTION
49L8092	DETAIL, COTTAGE AND CONSTRUCTIVE ARCHITECTURE
49L8015	FENCES, GATES & BRIDGES
49L8706	FROM LOG TO LOG HOUSE
49L0720	HOMES & INTERIORS OF THE 1920'S
49L8111	LOW-COST WOOD HOMES
49L8030	SHELTERS, SHACKS & SHANTIES
49L8139	THE STAIR BUILDERS GUIDE
49L8050	STRONG'S BOOK OF DESIGNS
49L8064	THE ARCHITECTURE OF COUNTRY HOUSES
49L8023	THE OPEN TIMBER ROOFS OF THE MIDDLE AGES

CLASSIC CATALOGS

Item #	Title
49L8004	BOULTON & PAUL, LTD. 1898 CATALOGUE
49L8098	CATALOG OF MISSION FURNITURE 1913 – *COME-PACKT FURNITURE*
49L8097	MASSEY-HARRIS CIRCA 1914 CATALOG
49L8089	OVERSHOT WATER WHEELS FOR SMALL STREAMS
49L8079	WILLIAM BULLOCK & CO. – *HARDWARE CATALOG, CIRCA 1850*

GARDENING

Item #	Title
49L8082	CANADIAN WILD FLOWERS (C. P. TRAILL)
49L8113	COLLECTING SEEDS OF WILD PLANTS AND SHIPPING LIVE PLANT MATERIAL
49L8029	FARM WEEDS OF CANADA
49L8056	FLORA'S LEXICON
49L8705	REFLECTIONS ON THE FUNGALOIDS
49L8057	THE WILDFLOWERS OF CANADA

HUMOR AND PUZZLES

Item #	Title
49L8074	ARE YOU A GENIUS? WHAT IS YOUR I.Q.?
49L8106	CLASSIC COWBOY CARTOONS, VOL. 1 (J.R. WILLIAMS)
49L8109	CLASSIC COWBOY CARTOONS, VOL. 2 (J.R. WILLIAMS)
49L8118	CLASSIC COWBOY CARTOONS, VOL. 3 (J.R. WILLIAMS)
49L8119	CLASSIC COWBOY CARTOONS, VOL. 4 (J.R. WILLIAMS)
49L8072	CLASSIC PUZZLES AND HOW TO SOLVE THEM
49L8103	GRANDMOTHER'S PUZZLE BOOK 1
49L8142	GRANDMOTHER S PUZZLE BOOK 2
49L8127	JOIN THE DOTS PUZZLE BOOKS
49L8081	MR. PUNCH WITH ROD AND GUN – *THE HUMOUR OF FISHING AND SHOOTING*
49L8073	NAME IT! THE PICTORIAL QUIZ BOOK
49L8126	OUR BOARDING HOUSE WITH MAJOR HOOPLE – *1927*
49L8125	OUT OUR WAY– SAMPLER 20s, 30s & 40s (J.R. WILLIAMS)
49L8044	SAM LOYD'S PICTURE PUZZLES
49L8084	THE ART OF ARTHUR WATTS
49L8071	THE BULL OF THE WOODS, VOL. 1 (J.R. WILLIAMS)
49L8080	THE BULL OF THE WOODS, VOL. 2 (J.R. WILLIAMS)
49L8104	THE BULL OF THE WOODS, VOL. 3 (J.R. WILLIAMS)
49L8114	THE BULL OF THE WOODS, VOL. 4 (J.R. WILLIAMS)
49L8115	THE BULL OF THE WOODS, VOL. 5 (J.R. WILLIAMS)
49L8116	THE BULL OF THE WOODS, VOL. 6 (J.R. WILLIAMS)
49l8128	THE NIGHT BEFORE CHRISTMAS WITH PUZZLE PICTURES
49L8107	U.S. CAVALRY CARTOONS (J.R. WILLIAMS)

NAVAL AND MARINE

Item #	Title
49L8090	BOAT-BUILDING AND BOATING
49L8707	BUILDING THE NORWEGIAN SAILING PRAM *(MANUAL AND PLANS)*
49L8708	BUILDING THE SEA URCHIN *(MANUAL AND PLANS)*
49L8138	HOW SAILS ARE MADE AND HANDLED
49L8078	MANUAL OF SEAMANSHIP FOR BOYS AND SEAMEN OF THE ROYAL NAVY, 1904
49L8129	OLD SHIP FIGURE-HEADS & STERNS
49L8095	SAILING SHIPS AT A GLANCE
49L8134	SAILING VESSEL SILHOUETTES
49L8099	THE SAILOR'S WORD-BOOK
49L8605	THE SAILOR S POCKET BOOK OF KNOTS
49L8058	THE YANKEE WHALER
49L8025	THE YOUNG SEA OFFICER'S SHEET ANCHOR
49L8061	TRADITIONS OF THE NAVY

REFERENCE

Item #	Title
49L8024	1800 MECHANICAL MOVEMENTS AND DEVICES
49L8093	507 MECHANICAL MOVEMENTS
49L8055	970 MECHANICAL APPLIANCES AND NOVELTIES OF CONSTRUCTION
49L8602	ALL THE KNOTS YOU NEED
49L8083	AMERICAN MECHANICAL DICTIONARY – KNIGHT VOL. I, VOL. II, VOL. III
49L8077	CAMP COOKERY
49L8001	LEE'S PRICELESS RECIPES
49L8135	MUSSON S IMPROVED LUMBER AND LOG POCKET BOOK
49L8018	THE BOY'S BOOK OF MECHANICAL MODELS
49L8019	WINDMILLS AND WIND MOTORS

TRADES

Item #	Title
49L8014	BOOK OF TRADES
49L8086	FARM BLACKSMITHING
49L8031	FARM MECHANICS
49L8141	FARM WORKSHOP GUIDE
49L8087	FORGING
49L8027	HANDY FARM DEVICES AND HOW TO MAKE THEM
49L8002	HOW TO PAINT SIGNS & SHO' CARDS
49L8054	HOW TO USE THE STEEL SQUARE
49L8094	THE YOUNG MILL-WRIGHT AND MILLER'S GUIDE

WOODWORKING AND CRAFTS

Item #	Title
49L8130	50 POPULAR WOODWORKING PROJECTS
49L8012	BOY CRAFT
49L8110	CHAIN SAW AND CROSSCUT SAW TRAINING COURSE
49L8048	CLAY MODELLING AND PLASTER CASTING
49L8005	COLONIAL FURNITURE
49L8065	COPING SAW WORK
49L8032	DECORATIVE CARVING, PYROGRAPHY AND FLEMISH CARVING
49L8091	FURNITURE DESIGNING AND DRAUGHTING
49L8049	HANDBOOK OF TURNING
49L8020	MISSION FURNITURE, HOW TO MAKE IT
49L8710	QUEEN ANNE FURNITURE - *HISTORY, DESIGN & CONSTRUCTION*
49L8033	ORNAMENTAL AND DECORATIVE WOOD CARVINGS
49L8003	RUSTIC CARPENTRY
49L8085	SKELETON LEAVES AND PHANTOM FLOWERS
49L8068	SPECIALIZED JOINERY
49L8052	STANLEY COMBINATION PLANES – *THE 45, THE 50 & THE 55*
49L8034	THE ART OF WHITTLING
49L8131	TIN-CAN PROJECTS AND ART-METAL WORK
49L8042	TURNING FOR AMATEURS
49L8067	WOOD HANDBOOK – *WOOD AS AN ENGINEERING MATERIAL*
49L8060	WOODEN PLANES AND HOW TO MAKE THEM
49L8013	YOU CAN MAKE IT
49L8035	YOU CAN MAKE IT FOR CAMP & COTTAGE
49L8036	YOU CAN MAKE IT FOR PROFIT

Algrove Publishing Limited, 36 Mill Street, P.O. Box 1238, Almonte, Ontario, Canada K0A 1A0
Telephone: (613) 256-0350   Fax: (613) 256-0360   Email: sales@algrove.com